AF389098

New Horizons in Time-Domain Diffuse Optical Spectroscopy and Imaging

New Horizons in Time-Domain Diffuse Optical Spectroscopy and Imaging

Special Issue Editor

Yoko Hoshi

MDPI • Basel • Beijing • Wuhan • Barcelona • Belgrade • Manchester • Tokyo • Cluj • Tianjin

Special Issue Editor
Yoko Hoshi
Hamamatsu University School
of Medicine
Japan

Editorial Office
MDPI
St. Alban-Anlage 66
4052 Basel, Switzerland

This is a reprint of articles from the Special Issue published online in the open access journal *Applied Sciences* (ISSN 2076-3417) (available at: https://www.mdpi.com/journal/applsci/special_issues/Diffuse_Optical_Spectroscopy).

For citation purposes, cite each article independently as indicated on the article page online and as indicated below:

LastName, A.A.; LastName, B.B.; LastName, C.C. Article Title. *Journal Name* **Year**, *Article Number*, Page Range.

ISBN 978-3-03936-100-7 (Pbk)
ISBN 978-3-03936-101-4 (PDF)

Contents

About the Special Issue Editor

Yoko Hoshi graduated from the Akita University School of Medicine (MD) in 1981 and completed her PhD at Hokkaido University in 1990. She is a pediatrician (child neurologist), and she has also been participating in the development of NIRS and researching cognitive neuroscience. She has been a professor of the Department of Biomedical Optics at the Hamamatsu University School of Medicine since April 2015. Her recent research interest is the development of diffuse optical tomography.

Editorial

Special Issue on New Horizons in Time Domain Diffuse Optical Spectroscopy and Imaging

Yoko Hoshi

Department of Biomedical Optics, Institute for Medical Photonics Research, Preeminent Medical Photonics Education & Research Center, Hamamatsu University School of Medicine, Hamamatsu, Shizuoka 431-3192, Japan; yhoshi@hama-med.ac.jp; Tel.: +81-53-345-2329

Received: 17 March 2020; Accepted: 13 April 2020; Published: 16 April 2020

1. Time Domain Measurements

In 1977, Jöbsis first described the in vivo application of near-infrared spectroscopy (NIRS) [1], also called diffuse optical spectroscopy (DOS). Originally, NIRS was designed for clinical monitoring of tissue oxygenation, and today it has also become a useful tool for neuroimaging studies (functional near-infrared spectroscopy (fNIRS)) [2–4]. However, difficulties in the selective and quantitative measurements of tissue hemoglobin (Hb), which have been central issues in the NIRS field for over 40 years, are yet to be solved. To overcome these problems, time domain (TD) [5,6] and frequency domain (FD) [7,8] measurements have been tried. Presently, a wide range of NIRS instruments are available, including commercial commonly available instruments for continuous wave (CW) measurements based on the modified Beer–Lambert law (steady-state domain measurements). Among these measurements, the TD measurement is the most promising approach, although, compared with CW and FD measurements, TD measurements are less common due to the need for large and expensive instruments with poor temporal resolution and limited dynamic range. However, thanks to technological developments, TD measurements are increasingly being used in research and also in various clinical settings [9,10].

2. Light Propagation in Biological Tissue and Time Domain Diffuse Optical Spectroscopy

In TD DOS, also termed time-resolved spectroscopy (TRS), tissue is irradiated by ultrashort (picosecond order) laser pulses, and the intensity of the emerging light at the tissue surface is recorded over time to show a temporal point spread function (TPSF) with picosecond resolution. The mean total length of the light path is determined by multiplying the light speed in the media by the mean transit time of the scattered photons, which is calculated with the TPSF [5]. The TPSF reflects the propagation of light in biological tissue, which is characterized by the optical properties of absorption, scattering, scattering anisotropy, and refractive indexes. It is widely accepted that the radiative transfer equation (RTE) correctly describes the light propagation in biological tissue [11,12]. Since, however, the computational cost is extremely high in numerically solving the RTE, the photon diffusion equation (PDE), a diffusion approximation to the RTE, is often used. Based on the PDE, it is possible to estimate the absorption (μ_a) and reduced scattering (μ_s') coefficients with the TPSF, and to calculate concentrations of biological chromophores, including Hb. The TPSF also carries information about depth-dependent attenuation based on the correlation of the detection time with the penetration depth of photons. Accurate numerical modelling of light propagation is critical for the quantification of TD measurements and the image reconstruction of diffuse optical tomography (DOT), as will be described below.

3. Time Domain Diffuse Optical Tomography

Diffuse optical tomography, one of the most sophisticated near-infrared optical imaging techniques for observations through biological tissue, allows 3-D quantitative imaging of optical properties which include functional and anatomical information [13]. The DOT image reconstruction can be approximately divided into two kinds—one is a linearization approach; the other is a non-linear iterative approach. With DOT, especially the non-linear iterative DOT, it is expected that it will become possible to overcome the limitations of conventional NIRS as well as it offers the potential for diagnostic optical imaging. The DOT algorithm essentially consists of two parts—one is a forward model to calculate the light propagation and the resultant outward re-emissions at the boundary of the tissue, typically based on the PDE or the RTE. The other is an inverse model to search for the distribution of optical properties. The implementation of DOT is possible with CW, TD, and FD measurements, where the TD measurements provide more of the information required for image reconstruction.

4. Cutting Edge Time Domain Diffuse Optical Spectroscopy and Imaging

This Special Issue highlights the issues at the cutting edge of TD DOS and DOT. It covers all aspects related to TD measurements described above, including advances in hardware, methodology, theory of light propagation, and clinical applications. The Special Issue has two reviews and 10 original research papers. One review paper by Yamada, Suzuki and Yamashita provides a comprehensive review of the past and current status of TD DOS and TD DOT, with chronological summaries of the major events in instrument and theoretical method developments [14]. This paper will help readers who are new to NIRS and also experts to obtain an overview of TD measurements and broaden their knowledge and understanding. The second review is by Lange and Tachtsidis and focuses on clinical applications in brain monitoring, where they also describe recent developments in instrumentation and methodologies that have the potential to affect and broaden the clinical use of TD measurements [15].

Five of 10 original papers deal with clinical applications that utilize the strengths of TD measurements. Ueda and Saeki applied TD DOS to three studies on breast cancer, reporting the detection rate of breast cancer, tumor hemodynamic responses to neoadjuvant chemotherapy, and antiangiogenic therapy [16]. Kinoshita et al., who measured skeletal muscle oxygenation in the lower extremities during 3 h of continuous sitting, report that compression stockings suppress increases in extracellular water in the lower extremities, leading to reduced blood volume and oxygenation levels in skeletal muscles [17]. The study by Morimoto et al. measured cerebral blood volume and optical properties in five term neonates from 2–3 min to 15 min after birth, and demonstrate that TD DOS can stably measure the cerebral hemodynamics of neonates in the labor room [18]. Sakatani et al. examined the effects of aging, cognitive dysfunction, and brain atrophy on Hb concentrations and optical pathlengths at rest in the prefrontal cortex in 202 elderly subjects, concluding that TD DOS enables an evaluation of the relation between prefrontal oxygenation at rest and cognitive function [19]. Kuroiwa et al. employed TD DOS to test their hypothesis that total Hb concentration in abdominal subcutaneous adipose tissue correlates negatively with risk factors for developing metabolic diseases and were able to verify the hypothesis [20].

Ohmae et al. conducted basic research into the application of TD measurements to breast cancer, and report that the validity of TD DOS measurements of the lipid and water contents of the breast is confirmed by a comparison of the TD DOS values to the values measured by dual-energy computed tomography [21]. The paper by Di Sieno et al. deals with hardware, a TD fast gated NIRS system, and presents results showing that the gating approach can improve the contrast and contrast–noise ratio for the detection of absorption changes, irrespective of the source–detector separation distance [22].

Three papers present theoretical studies. To recover the optical properties from boundary measurements, iterative inversion schemes, where a theoretical TPSF is derived from the analytical solution to the PDE and fitted to the measured temporal profile of detected light intensity, are often used [23]. However, in these schemes, the initial guesses need to be close to the true values. Jiang et al. propose a scheme combining Markov chain Monte Carlo and iterative methods to overcome this weakness in iterative schemes [24]. In TD DOT, regarding the datatypes obtained from the TPSF, such as temporal windows and Fourier transformations, determining which datatypes are used for image reconstruction is crucial for computational efficiency as well as for image quality. Orive-Miguel et al. propose a new process for the efficient computation of long sets of temporal windows in the FD and demonstrate that the absorption quantification of the inclusions in a rectangular medium is improved at all depths in numerical experiments by the proposed method [25]. The M-th order delta-Eddington equation (dEM) is used as one effective approach to reduce the computational cost of a numerical solution to the RTE. The final paper in the issue by Fujii et al. examined photon transport in 3D, homogeneous, highly forward-scattering media with different optical properties by using time-dependent RTE, dEM, and PDE and estimated the length and time scales in which the dEM is valid [26].

5. Future Prospects and Challenges

The ultimate goal of developing TD measurements is to establish an optical-based diagnosis. Further studies on the numerical modeling of light propagation in biological tissue, developing accurate and efficient inverse solutions and high-quality instruments are required for reaching this goal. Although these tasks are challenging, recent advances in computer and optical technologies will advance the efforts to solve these bottlenecks.

Funding: This research received no external funding.

Acknowledgments: I wish to express my very great appreciation to all of the authors and peer reviewers for their valuable contributions to this Special Issue. I also wish to thank the editorial team of *Applied Sciences* for their assistance. I am particularly grateful for the assistance given by Marin Ma, a section managing editor.

Conflicts of Interest: The authors declare no conflict of interest.

References

1. Jobsis, F. Noninvasive, infrared monitoring of cerebral and myocardial oxygen sufficiency and circulatory parameters. *Science* **1977**, *198*, 1264–1267. [CrossRef]
2. Hoshi, Y.; Tamura, M. Detection of dynamic changes in cerebral oxygenation coupled to neuronal function during mental work in man. *Neurosci. Lett.* **1993**, *150*, 5–8. [CrossRef]
3. Kato, T.; Kamei, A.; Takashima, S.; Ozaki, T. Human Visual Cortical Function during Photic Stimulation Monitoring by Means of near-Infrared Spectroscopy. *Br. J. Pharmacol.* **1993**, *13*, 516–520. [CrossRef]
4. Villringer, A.; Planck, J.; Hock, C.; Schleinkofer, L.; Dirnagl, U. Near infrared spectroscopy (NIRS): A new tool to study hemodynamic changes during activation of brain function in human adults. *Neurosci. Lett.* **1993**, *154*, 101–104. [CrossRef]
5. Delpy, D.T.; Cope, M.; Van Der Zee, P.; Arridge, S.R.; Wray, S.; Wyatt, J. Estimation of optical pathlength through tissue from direct time of flight measurement. *Phys. Med. Boil.* **1988**, *33*, 1433–1442. [CrossRef]
6. Chance, B.; Leigh, J.S.; Miyake, H.; Smith, D.S.; Nioka, S.; Greenfeld, R.; Finander, M.; Kaufmann, K.; Levy, W.; Young, M. Comparison of time-resolved and -unresolved measurements of deoxyhemoglobin in brain. *Proc. Natl. Acad. Sci. USA* **1988**, *85*, 4971–4975. [CrossRef]
7. Lakowicz, J.R.; Berndt, K. Frequency-domain measurements of photon migration in tissues. *Chem. Phys. Lett.* **1990**, *166*, 246–252. [CrossRef]
8. Duncan, A.; Whitlock, T.L.; Cope, M.; Delpy, D.T. A multiwavelength, wideband, intensity modulated optical spectrometer for near infrared spectroscopy and imaging. *Proc. SPIE* **1993**, *1888*, 248–257.
9. Wabnitz, H.; Taubert, D.R.; Mazurenka, M.; Steinkellner, O.; Jelzow, A.; Macdonald, R.; Milej, D.; Sawosz, P.; Kacprzak, M.; Liebert, A.; et al. Performance assessment of time-domain optical brain imagers, part 1: Basic instrumental performance protocol. *J. Biomed. Opt.* **2014**, *19*, 086010. [CrossRef] [PubMed]
10. Abdalmalak, A.; Milej, D.; Diop, M.; Shokouhi, M.; Naci, L.; Owen, A.M.; Lawrence, K.S. Can time-resolved NIRS provide the sensitivity to detect brain activity during motor imagery consistently? *Biomed. Opt. Express* **2017**, *8*, 2162–2172. [CrossRef] [PubMed]
11. Duderstadt, J.J.; Martin, W.R.; Aronson, R. Transport Theory. *Phys. Today* **1982**, *35*, 65. [CrossRef]

12. Ishimaru, A. *Wave Propagation and Scattering in Random Media*; Institute of Electrical and Electronics Engineers (IEEE): New York, NY, USA, 1999.

13. Hoshi, Y.; Yamada, Y. Overview of diffuse optical tomography and its clinical applications. *J. Biomed. Opt.* **2016**, *21*, 91312. [CrossRef]

14. Yamada, Y.; Suzuki, H.; Yamashita, Y. Time-Domain Near-Infrared Spectroscopy and Imaging: A Review. *Appl. Sci.* **2019**, *9*, 1127. [CrossRef]

15. Lange, F.; Tachtsidis, I. Clinical Brain Monitoring with Time Domain NIRS: A Review and Future Perspectives. *Appl. Sci.* **2019**, *9*, 1612. [CrossRef]

16. Ueda, S.; Saeki, T. Early Therapeutic Prediction Based on Tumor Hemodynamic Response Imaging: Clinical Studies in Breast Cancer with Time-Resolved Diffuse Optical Spectroscopy. *Appl. Sci.* **2018**, *9*, 3. [CrossRef]

17. Kinoshita, M.; Kurosawa, Y.; Fuse, S.; Tanaka, R.; Tano, N.; Kobayashi, R.; Kime, R.; Hamaoka, T. Compression Stockings Suppressed Reduced Muscle Blood Volume and Oxygenation Levels Induced by Persistent Sitting. *Appl. Sci.* **2019**, *9*, 1800. [CrossRef]

18. Morimoto, A.; Nakamura, S.; Sugino, M.; Koyano, K.; Htun, Y.; Arioka, M.; Fuke, N.; Mizuo, A.; Yokota, T.; Kato, I.; et al. Measurement of the Absolute Value of Cerebral Blood Volume and Optical Properties in Term Neonates Immediately after Birth Using Near-Infrared Time-Resolved Spectroscopy: A Preliminary Observation Study. *Appl. Sci.* **2019**, *9*, 2172. [CrossRef]

19. Sakatani, K.; Hu, L.; Oyama, K.; Yamada, Y. Effects of Aging, Cognitive Dysfunction, Brain Atrophy on Hemoglobin Concentrations and Optical Pathlength at Rest in the Prefrontal Cortex: A Time-Resolved Spectroscopy Study. *Appl. Sci.* **2019**, *9*, 2209. [CrossRef]

20. Kuroiwa, M.; Fuse, S.; Amagasa, S.; Kime, R.; Endo, T.; Kurosawa, Y.; Hamaoka, T. Relationship of Total Hemoglobin in Subcutaneous Adipose Tissue with Whole-Body and Visceral Adiposity in Humans. *Appl. Sci.* **2019**, *9*, 2442. [CrossRef]

21. Ohmae, E.; Yoshizawa, N.; Yoshimoto, K.; Hayashi, M.; Wada, H.; Mimura, T.; Asano, Y.; Ogura, H.; Yamashita, Y.; Sakahara, H.; et al. Comparison of Lipid and Water Contents by Time-domain Diffuse Optical Spectroscopy and Dual-energy Computed Tomography in Breast Cancer Patients. *Appl. Sci.* **2019**, *9*, 1482. [CrossRef]

22. Di Sieno, L.; Mora, A.D.; Torricelli, A.; Spinelli, L.; Re, R.; Pifferi, A.; Contini, D. A Versatile Setup for Time-Resolved Functional Near Infrared Spectroscopy Based on Fast-Gated Single-Photon Avalanche Diode and on Four-Wave Mixing Laser. *Appl. Sci.* **2019**, *9*, 2366. [CrossRef]

23. Patterson, M.S.; Chance, B.; Wilson, B.C. Time resolved reflectance and transmittance for the non-invasive measurement of tissue optical properties. *Appl. Opt.* **1989**, *28*, 2331–2336. [CrossRef]

24. Jiang, Y.; Hoshi, Y.; Machida, M.; Nakamura, G. A hybrid inversion scheme combining Marcov chain Monte Carlo and iterative methods for determining optical properties of random media. *Appl. Sci.* **2019**, *9*, 3500. [CrossRef]

25. Orive-Miguel, D.; Herve, L.; Condat, L.; Mars, J. Improving Localization of Deep Inclusions in Time-Resolved Diffuse Optical Tomography. *Appl. Sci.* **2019**, *9*, 5468. [CrossRef]

26. Fujii, H.; Ueno, M.; Kobayashi, K.; Watanabe, M. Characteristic Length and Time Scales of the Highly Forward Scattering of Photons in Random Media. *Appl. Sci.* **2019**, *10*, 93. [CrossRef]

Review

Time-Domain Near-Infrared Spectroscopy and Imaging: A Review

Yukio Yamada [1,*]**, Hiroaki Suzuki** [2] **and Yutaka Yamashita** [2]

1   Center for Neuroscience and Biomedical Engineering, The University of Electro-Communications, 1-5-1 Chofuga-oka, Chofu, Tokyo 182-8585, Japan
2   Central Research Laboratory, Hamamatsu Photonics K.K., Hamamatsu, Shizuoka 434-8601, Japan; hiro-su@crl.hpk.co.jp (H.S.); yutaka@crl.hpk.co.jp (Y.Y.)
*   Correspondence: yukioyamada@uec.ac.jp; Tel.: +81-42-443-5220

Received: 27 December 2018; Accepted: 5 March 2019; Published: 17 March 2019

Abstract: This article reviews the past and current statuses of time-domain near-infrared spectroscopy (TD-NIRS) and imaging. Although time-domain technology is not yet widely employed due to its drawbacks of being cumbersome, bulky, and very expensive compared to commercial continuous wave (CW) and frequency-domain (FD) fNIRS systems, TD-NIRS has great advantages over CW and FD systems because time-resolved data measured by TD systems contain the richest information about optical properties inside measured objects. This article focuses on reviewing the theoretical background, advanced theories and methods, instruments, and studies on clinical applications for TD-NIRS including some clinical studies which used TD-NIRS systems. Major events in the development of TD-NIRS and imaging are identified and summarized in chronological tables and figures. Finally, prospects for TD-NIRS in the near future are briefly described.

Keywords: time-domain spectroscopy; near-infrared spectroscopy; radiative transfer equation; diffusion equation; biological tissue; time-domain instruments; light propagation in tissue; optical properties of tissue; diffuse optical tomography; fluorescence diffuse optical tomography

1. Introduction

The time-domain (TD) technique in near-infrared spectroscopy (NIRS) and imaging technology has the greatest potential among the continuous wave (CW), frequency-domain (FD), and time-domain modalities owing to having the richest information contained in the measured TD data. Although the advantages of the TD technique are widely recognized, only two TD-NIRS systems were commercially available for brain and/or muscle NIRS oximeters, while more than ten CW-NIRS systems were commercially available by 2011, according Ferrari and Quaresima [1]. These limited usages of the TD-NIRS technique are due to its drawbacks, to which some articles have referred, as follows.

In 2014, Torricelli et al. [2] described the situation of the TD-NIRS technique as follows (with modification), "TD-NIRS and TD techniques in general had the reputations of being cumbersome, bulky, and very expensive when compared with commercially available continuous wave (CW) functional-NIRS systems. These disadvantages cannot be ignored, and a gap between CW and TD-NIRS technology still exists."

In 2018, Papadimitriou et al. [3] also described the drawbacks of TD-NIRS systems more in details (with modification), "TD-NIRS instruments used so far are bulky and expensive, and typically employ sensitive optoelectronics which are susceptible to vibrations. When solid state lasers are used, switching from one wavelength to another is slow about 10 s. When pulsed laser diodes are used, long *warm-up* time (about 60 min) is required to achieve a stability of pulse timing in the picosecond range. These factors limit the usages of TD-NIRS not only in hospitals but also in laboratories."

The drawbacks of TD-NIRS systems described above still exist to some extent and have not been resolved completely, but improvements and developments of the TD-NIRS technique have been progressing steadily. Torricelli et al. [2] continued the above description with expectations, "Recent advances in photonic technologies might allow to bridge the gap between TD-NIRS and CW-NIRS and potentially to overtake CW-NIRS." Also, Papadimitriou et al. [3] expressed expectations for advanced TD-NIRS systems, describing that smaller and more robust instruments might lead to wider applications, for example in emergency medicine.

In this article for TD-NIRS and imaging, we first review the theoretical background of TD-NIRS and provide an overview of TD-NIRS instruments. Then, we proceed to the advanced theories and methods of the TD-NIRS technique and review studies on the clinical applications of the TD-NIRS technique including TD diffuse optical tomography (TD-DOT) and TD fluorescence tomography (TD-FT). Before summarizing this article, some results of the clinical applications of commercially available TD-NIRS systems by Japanese researchers are reviewed. Please note that many studies of CW and FD approaches were intentionally excluded in this article, although they developed the basis of the TD approach. In the summary, the major/key developments in TD-NIRS and imaging technology are listed in chronological tables and figures to help the reader view the whole TD-NIRS and imaging picture. Finally, we summarize this review article with the expectations of wider usages for the TD-NIRS technique in the near future. There must be many important TD studies which were not referred to in this article, and there may be descriptions with misunderstandings of the references. The authors are grateful for any comments provided by readers.

2. Theoretical Background of TD-NIRS

The principles, concepts, and theoretical background of TD-NIRS are described in this section by showing the fundamental equation of light propagation, the radiative transfer equation (RTE), followed by its approximate equations with analytical solutions, numerical solving methods, and quantities featuring TD-NIRS. With an understanding of the theoretical background and the featuring quantities, it will be easier to reasonably interpret the results of calculations of light propagation in in vitro and in vivo experiments and the clinical applications of TD-NIRS. The theoretical background, particularly for TD-DOT, is comprehensively reviewed by Arridge [4].

2.1. Radiative Transfer Equation (RTE)

We start from the radiance, or the specific intensity, $I(\mathbf{r}, \hat{s}, t)$, defined as the average radiant power flowing at position $\mathbf{r}$ and at time t through the unit area oriented in the direction of the unit vector $\hat{s}$ and through the unit solid angle along $\hat{s}$ in a medium as shown in Figure 1. The most fundamental equation describing light propagation in biological tissue, which is accepted in this field, is the radiative transfer equation in time-domain (TD-RTE) (or the Boltzmann transport equation) for radiance [5],

$$\frac{\partial}{c\partial t} I(\mathbf{r}, \hat{s}, t) + \hat{s} \cdot \nabla I(\mathbf{r}, \hat{s}, t) + [\mu_a(\mathbf{r}) + \mu_s(\mathbf{r})] I(\mathbf{r}, \hat{s}, t) = \mu_s(\mathbf{r}) \int_{4\pi} p(\hat{s}, \hat{s}\prime) I(r, \hat{s}\prime, t) d\Omega\prime + q(\mathbf{r}, \hat{s}, t) \quad (1)$$

where c is the velocity of light in the medium, ∇ is the spatial gradient operator, $\bullet$ is the scalar product operator, $\mu_a(\mathbf{r})$ and $\mu_s(\mathbf{r})$ are the absorption and scattering coefficients, respectively, $p(\hat{s}, \hat{s}\prime)$ is the scattering phase (angular) function describing the probability of scattering from direction $\hat{s}'$ into direction $\hat{s}$, $d\Omega'$ is the solid angle for integration, and $q(\mathbf{r}, \hat{s}, t)$ is the light source. Here we assume that the radiance is for a specific wavelength and that the velocity of light is constant throughout the medium. The RTE is an energy conservation equation, and each term has physical meanings; the total temporal change in the radiance, the energy inflow due to the gradient of the radiance (or the diffusion of the radiance), the energy gain by absorption and scattering, the energy inflow to direction $\hat{s}$ by scattering from direction $\hat{s}'$ over the entire solid angle, and the energy gain by light sources. Here we note: if the radiance, $I(\mathbf{r}, \hat{s}, t)$, having a dimension of W/(m^{2}sr) is divided by the velocity of light, c, it has a dimension of $J/(m^3 sr)$ and is often called as the photon energy density, $u(\mathbf{r}, \hat{s}, t) = I(\mathbf{r}, \hat{s}, t)/c$. Equation (1) is sometimes expressed as $u(\mathbf{r}, \hat{s}, t)$ instead of $I(\mathbf{r}, \hat{s}, t)$.

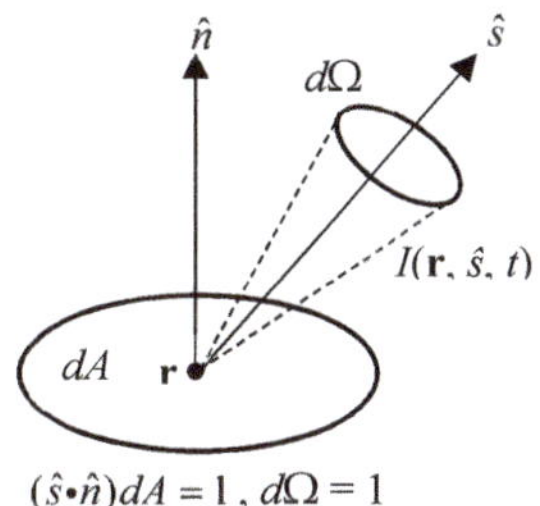

Figure 1. Definition of the radiance.

The RTE is an integro-differential equation and it is not easy to calculate even with the use of modern computers due to the integral term on the right-hand side. Many studies have been carried out on how to solve the RTE by numerical and analytical methods, and by equivalently statistical methods. Analytical and statistical methods are briefly reviewed in the following, while numerical methods are briefly reviewed in a later section.

2.2. Expansion of the RTE by Spherical Harmonics and the P_N Approximations

One way to simplify the RTE is to expand $I(\mathbf{r}, \hat{s}, t)$, $q(\mathbf{r}, \hat{s}, t)$, and $p(\hat{s}, \hat{s}\prime)$ into a series of spherical harmonics, $Y_l^m(\hat{s})$ ($l = 0, 1, 2, \ldots, m = -l, -l + 1, \ldots, l - 1, l$), to separate the angular dependences of $I(\mathbf{r}, \hat{s}, t)$ and $q(\mathbf{r}, \hat{s}, t)$ on $\hat{s}$ from the dependences on $\mathbf{r}$ and t [5,6] [7] (pp. 282–288). After some mathematical manipulations, the RTE is rewritten in terms of a series of spherical harmonics for the expansion coefficients $i_{lm}(\mathbf{r}, t)$ of $I(\mathbf{r}, \hat{s}, t)$,

$$\sum_{l=0}^{\infty} \sum_{m=-l}^{l} \left\{ \left[\frac{1}{c}\frac{\partial}{\partial t} + \hat{s}\nabla + [\mu_a(r) + \mu_s(r)(1 - p_l)] \right] i_{lm}(r, t) - q_{lm}(r, t) \right\} Y_l^m(\hat{s}) = 0 \tag{2}$$

where $q_{lm}(\mathbf{r}, t)$ and p_l are the expansion coefficients which are known. For a particular combination of (l, m) = (L, M), Equation (2) is transformed to an infinite number of coupled partial-differential equations for $i_{LM}(\mathbf{r}, t)$ with L ranging from 0 to ∞ and M ranging from $-L$ to $+L$ [7] (pp. 282–288).

By retaining the equations for L from 0 to N with $M = -L, -L + 1, \ldots, L - 1, L$, the number of the retained coupled partial-differential equations is [summation of ($2L + 1$) from $L = 0$ to N] = $(N + 1)^2$, the same as the number of unknown functions, $i_{LM}(\mathbf{r}, t)$. The system of these $(N + 1)^2$ equations for $(N + 1)^2$ unknowns is the P_N approximation. For example, for the P_1 approximation, there are four unknowns of $i_{LM}(\mathbf{r}, t)$, and for the P_3 approximation, there are 16 unknowns of $i_{LM}(\mathbf{r}, t)$. Even-order P_N approximations are not useful and only odd-order P_N approximations are considered.

2.3. The P_1 Approximation

In the P_1 approximation, four unknowns, $i_{00}(\mathbf{r}, t)$, $i_{1-1}(\mathbf{r}, t)$, $i_{10}(\mathbf{r}, t)$, and $i_{11}(\mathbf{r}, t)$ are related to the fluence rate, $\phi(\mathbf{r}, t)$, and to the flux vector of the fluence rate, $\mathbf{F}(\mathbf{r}, t)$, expressed as,

$$\phi(\mathbf{r}, t) = \int_{4\pi} I(r, \hat{s}, t)d\Omega, \ \mathbf{F}(\mathbf{r}, t) = \int_{4\pi} I(r, \hat{s}, t)\hat{s}d\Omega, \tag{3}$$

and the P_1 approximation is expressed by two coupled equations for $\phi(\mathbf{r}, t)$ and $\mathbf{F}(\mathbf{r}, t)$, including the reduced scattering coefficient, $\mu_s'(\mathbf{r}) = [1 - g]\mu_s(\mathbf{r})$, and the anisotropy parameter, $g(\mathbf{r})$, defined as the average cosine of the phase function as,

$$g = \frac{\int_{4\pi} (\hat{s}\hat{s}\prime)p(\hat{s}, \hat{s}\prime)d\Omega}{\int_{4\pi} p(\hat{s}, \hat{s}\prime)d\Omega} = 2\pi \int_0^\pi p(\theta) \cos\theta d\theta \tag{4}$$

where θ is the angle between the directions $\hat{s}$ and $\hat{s}\prime$, and the phase function $p(\hat{s}, \hat{s}\prime)$ is assumed symmetric to the azimuthal angle, ϕ, and dependent on the polar angle, θ, only. Here, the phase function is normalized as $(1/4\pi) \int_{4\pi} p(\hat{s}, \hat{s}') d\Omega = 1$.

From the two coupled equations, the equation for $\phi(\mathbf{r}, t)$ is derived as,

$$\frac{3D(\mathbf{r})}{c^2} \frac{\partial^2}{\partial t^2} \phi(\mathbf{r}, t) + [1 + 3D(\mathbf{r})\mu_a(\mathbf{r})] \frac{1}{c} \frac{\partial}{\partial t} \phi(\mathbf{r}, t) + \mu_a(\mathbf{r})\phi(\mathbf{r}, t) - \nabla \cdot [D(\mathbf{r})\nabla\phi(\mathbf{r}, t)]$$
$$= \frac{3D(\mathbf{r})}{c} \frac{\partial}{\partial t} q_0(\mathbf{r}, t) + q_0(\mathbf{r}, t) - 3\nabla \cdot [D(\mathbf{r})\mathbf{q}_1(\mathbf{r}, t)] \tag{5}$$

where $D(\mathbf{r}) = 1/[3(\mu_s'(\mathbf{r}) + \mu_a(\mathbf{r}))]$ is the diffusion coefficient, $q_0(\mathbf{r}, t)$ and $\mathbf{q}_1(\mathbf{r}, t)$ are related to $q_{lm}(\mathbf{r}, t)$ in Equation (2) describing the isotropic and anisotropic components of the light source, $q(\mathbf{r}, \hat{s}, t)$. Equation (5) is the equation for $\phi(\mathbf{r}, t)$ in the P_1 approximation having a form of the telegraph equation (TE) which is an elliptic type of partial differential equation including the second derivatives with respect to both time and space indicating a phenomenon of wave propagation in the medium.

2.4. Diffusion Approximation and Diffusion Equation (DE)

The TE of the P_1 approximation is further simplified to the time-domain diffusion equation (TD-DE) by adding conditions of (i) strong scattering meaning $\mu_s' \gg \mu_a$ or $D\mu_a \ll 1$, (ii) slow temporal changes in the fluence rate and the light source leading to,

$$\frac{3D(\mathbf{r})}{c} \left| \frac{\partial^2}{\partial t^2} \phi(\mathbf{r}, t) \right| \ll \left| \frac{\partial}{\partial t} \phi(\mathbf{r}, t) \right|, \quad \frac{3D(\mathbf{r})}{c} \left| \frac{\partial}{\partial t} q_0(\mathbf{r}, t) \right| \ll |q_0(\mathbf{r}, t)|, \tag{6}$$

and (iii) the source emission is isotropic meaning $\mathbf{q}_1(\mathbf{r}, t) = 0$. Then, Equation (5) reduces to the TD-DE,

$$\frac{1}{c} \frac{\partial}{\partial t} \phi(\mathbf{r}, t) + \mu_a(\mathbf{r})\phi(\mathbf{r}, t) - \nabla \cdot [D(\mathbf{r})\nabla\phi(\mathbf{r}, t)] = q_0(\mathbf{r}, t), \tag{7}$$

Equation (7) is a parabolic type of partial differential equation describing diffusion phenomena, and the net flux of the fluence rate is given by Fick's law for diffusion phenomena, $\mathbf{F}(\mathbf{r}, t) = -D(\mathbf{r})\nabla\phi(\mathbf{r}, t)$.

The regime map of Figure 2 shows the valid region of the DE where the demarcating curve is drawn as $t = 10/(\mu_s' c) \approx 10(3D/c)$ under the condition of $\mu_s' \gg \mu_a$. Here, $3D/c$ is the characteristic time of interaction. For a case of $\mu_s' = 1.0$ mm^{-1} typical for biological tissue, the DE fails for light propagation within times shorter than 0.05 ns (50 ps), and the RTE is required in this period of times.

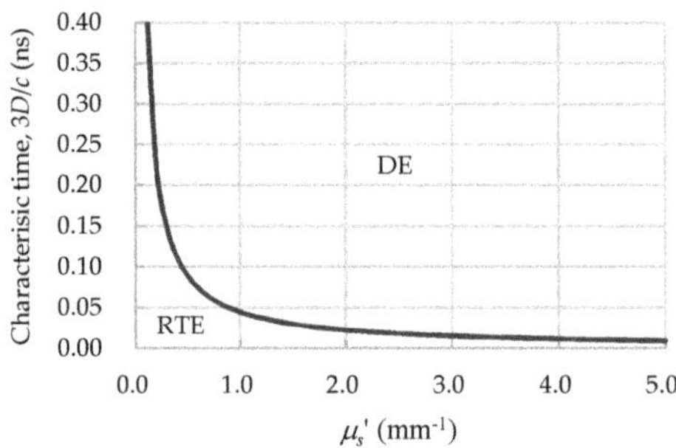

Figure 2. Regime map for the DE and the RTE.

The net fluxes of the fluence rate, $\mathbf{F}(\mathbf{r}, t)$, can be measured by time-resolved (TR) detectors and the measured fluxes such as TR reflectance and TR transmittance are often expressed as the time-of-flight distribution (TOF-distribution) in this article.

2.5. Diffusion Coefficient Independent of the Absorption Coefficient in TD-DE

The diffusion coefficient is given as $D(\mathbf{r}) = 1/[3(\mu_s'(\mathbf{r}) + \mu_a(\mathbf{r}))]$ in the process of deriving the DE in the framework of the P_1 approximation stated above. However, there was a long controversy about the expression of the diffusion coefficient and whether it depends on μ_a or not. Furutsu and Yamada [8] first discussed that in time-domain, $D(\mathbf{r})$ is independent of $\mu_a(\mathbf{r})$, i.e., $D(\mathbf{r}) = 1/[3\mu_s'(\mathbf{r})]$, while in the CW-domain $D(\mathbf{r})$ may depend on $\mu_a(\mathbf{r})$. For optically homogeneous media, the radiance in the RTE for an impulse source can be written as,

$$I(\mathbf{r}, \hat{s}, t) = I_0(\mathbf{r}, \hat{s}, t) \exp(-\mu_a ct), \tag{8}$$

where $I_0(\mathbf{r}, \hat{s}, t)$ is the radiance for non-absorbing medium without absorption, Equation (9),

$$\frac{\partial}{c\partial t} I_0(\mathbf{r}, \hat{s}, t) + \hat{s}\cdot\nabla I_0(\mathbf{r}, \hat{s}, t) + \mu_s I_0(\mathbf{r}, \hat{s}, t) = \mu_s \int_{4\pi} p(\hat{s}, \hat{s}\prime) I_0(r, \hat{s}\prime, t) d\Omega\prime. \tag{9}$$

This is easily understood by substituting Equation (8) into Equation (1) neglecting the impulse source term, and Equation (8) is sometimes referred to as expressing the microscopic Beer–Lambert law [9]. The process of the P_1 approximation described above can be applied to Equation (9), then the diffusion coefficient is clearly given as $D = 1/(3\mu_s')$ independent of μ_a. For inhomogeneous media, the derivation of $D(\mathbf{r}) = 1/[3\mu_s'(\mathbf{r})]$ is not straightforward like for homogenous media, but Furutsu and Yamada [8] proved it mathematically.

Then, the controversy about the diffusion coefficient started regarding whether it should be included μ_a or not [10–12]. But experimental and numerical studies supported $D(\mathbf{r}) = 1/[3\mu_s'(\mathbf{r})]$ [13–15], and mathematical studies using processes different from the P_1 approximation derived $D(\mathbf{r})$ independent of μ_a for the TD-DE and $D(\mathbf{r})$ dependent on μ_a for the CW-DE [16,17]. Finally, the following expressions are likely to be accepted in this field,

$$D(\mathbf{r}) = \frac{1}{3\mu_s\prime(\mathbf{r})} \text{ for the TD} - \text{DE}, \tag{10}$$

$$D(\mathbf{r}) = \frac{1}{3[\mu_s\prime(\mathbf{r}) + \alpha\mu_a(\mathbf{r})]} \quad \alpha = 0.2\text{\textasciitilde}0.8 \text{ for the CW} - \text{DE}. \tag{11}$$

The absorption coefficient, μ_a, of biological tissue is much smaller than μ_s' in the NIR range (typically μ_a ~0.01 mm^{-1} and μ_s' ~1.0 mm^{-1}), and the change in the magnitude of D upon including μ_a is very small. Therefore, $D(\mathbf{r}) = 1/[3\mu_s'(\mathbf{r})]$ is a good choice for all cases.

2.6. Boundary Condition for DE

The DE requires initial and boundary conditions to be solved. A boundary condition at the interface between two different media with mismatched refractive indexes is expressed by the following,

$$\phi(\mathbf{r}_b, t) + 2AD(\mathbf{r}_b)\frac{\partial\phi(\mathbf{r}_b, t)}{\partial n} = 0, \tag{12}$$

where $\mathbf{r}_b$ is a position on the interface, n indicates the direction outward normal to the interface, and A is a reflection parameter given by $A = (1 + R_{in})/(1 - R_{in})$, with R_{in} denoting the internal diffusive reflectivity estimated by the Fresnel reflection coefficient or other empirical models. In the process of obtaining analytical solutions of the DE, extrapolated boundary conditions are often employed with the mirror image method to satisfy Equation (12). For simplicity, the zero-boundary condition where the fluence rate at the interface is given as zero is sometimes used, and this is the case for $A = 0$ in Equation (12).

2.7. Analytical Solutions for TD-DE

Analytical solutions for TD-DE for a point and impulse source are very useful because they are Green's functions, and they have been obtained for simple geometries such as infinite media, semi-infinite media, slabs, cylinders, and spheres with homogeneous optical properties. The most fundamental analytical solution for TD-DE is for infinite homogeneous media given by Equation (13),

$$\phi(r,t) = \frac{c}{(4\pi Dct)^{3/2}} \exp(-\frac{r^2}{4Dct} - \mu_a ct), \tag{13}$$

where r is the distance from a point impulse source located at the origin.

Patterson et al. [18] first derived analytical solutions for TD-DE for semi-infinite and slab media by use of the mirror image source method under a zero-boundary condition. In the extrapolated boundary condition, the fluence rate is zero at an extrapolated surface with a distance of $2AD$ from the physical boundary, as shown in Figure 3, and the fluence rate for semi-infinite homogeneous media is given by Equation (14),

$$\phi(r,t) = \frac{c}{(4\pi Dct)^{3/2}} \exp(-\mu_a ct) \left[\exp(-\frac{r_1^2}{4Dct}) - \exp(-\frac{r_2^2}{4Dct}) \right], \tag{14}$$

where $r_1 = [(z - z_0)^2 + \rho^2]^{1/2}$ and $r_2 = [(z + z_0 + 2z_b)^2 + \rho^2]^{1/2}$ are the distances from the position of interest, $P(\rho, z)$, to the positive and negative point impulse sources, respectively, The reflectance at the surface, $P(\rho, 0)$ with the source-detector (SD) distance of ρ, is given by Fick's law as Equation (15),

$$R(\rho,t) = \frac{1}{16(\pi Dc)^{3/2} t^{5/2}} \exp(-\mu_a ct) \left[z_0 \exp(-\frac{r_{10}^2}{4Dct}) + (z_0 + 2z_b) \exp(-\frac{r_{20}^2}{4Dct}) \right], \tag{15}$$

where $r_{10} = (z_0^2 + \rho^2)^{1/2}$ and $r_{20} = [(z_0^2 + 2z_b)^2 + \rho^2]^{1/2}$. Equation (15) is the Green's function for semi-infinite homogeneous media and is often used in the derivation of analytical solutions using the perturbation theory.

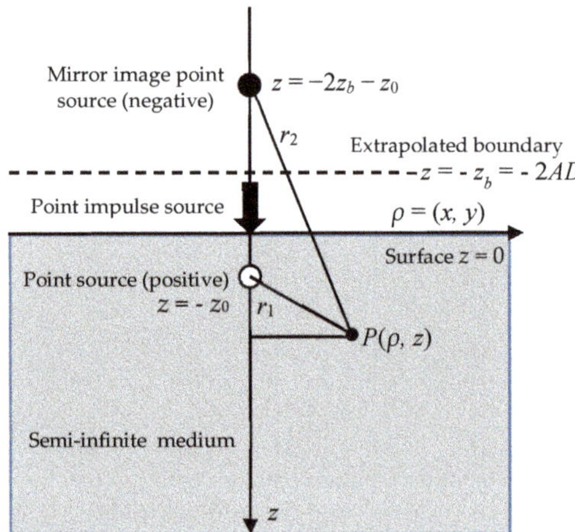

Figure 3. Concept of the extrapolated boundary and mirror image source for a semi-infinite medium.

Analytical solutions, or the Green's functions, for TD-DE for other simple geometries are comprehensively summarized in Reference [19].

When the medium scatters light non-isotropically but anisotropically, the diffusion coefficient is dependent on the direction and is expressed by a diffusion tensor. The analytical solution for TD-DE for the anisotropic diffusion in a slab medium is given by Kienle et al. [20] using a diffusion tensor for the diffusion coefficient.

2.8. Monte Carlo Simulations

Instead of directly solving the RTE or the DE with analytical and numerical methods, stochastic methods have been used to obtain statistically equivalent solutions of the RTE or the DE. One of them is random walk theory, which describes light propagation as a sequence of steps on a regular cubic lattice [21]. But this method oversimplifies the phenomenon, and the most common stochastic methods are Monte Carlo (MC) simulations. Wilson and Adam [22] first introduced an MC simulation into this field, and Wang et al. [23] provided free software for an MC simulation to calculate light propagation in multi-layered tissue with their published paper explaining the details. Now free software codes for MC simulations are available from the website of the Oregon Laser Center in Portland [24], and these codes are used by many researchers in this field.

The concept of MC simulations is to track the trajectories of energy packets, which are often called simply photons for brevity, while the trajectories of the photons are determined to statistically satisfy the optical properties of the medium, as shown in the following equations for Figure 4. The injected photon with an energy of E_i is scattered with a scattering path length of L to the direction of the polar and azimuthal angles of θ and φ, respectively, and the energy is attenuated to $E_{i+1} = E_i[\mu_s/(\mu_s + \mu_a)]$. L, θ, and φ are determined using uniformly distributed random numbers in the range of (0, 1), R_1, R_2, and R_3 by the following equations,

$$L = -\frac{\ln(R_1)}{\mu_t} = -\frac{\ln(R_1)}{\mu_s + \mu_a}, \ \theta = f_2^{-1}(R_2), \ f_2(\theta) = \frac{1}{2}\int_0^\theta p(\theta\prime)\sin\theta\prime d\theta\prime, \ \varphi = 2\pi R_3, \tag{16}$$

where the phase function $p(\theta)$ is assumed to be independent of φ, $f_2(\theta)$ is the accumulated phase function of $p(\theta)$, f_2^{-1} means the inverse of $f_2(\theta)$. When $p(\theta)$ is constant, meaning that $g = 0$ and the diffusion approximation is applicable, $f_2(\theta)$ becomes as simple as $f_2(\theta) = (1 - \cos\theta)/2$, and resultantly θ is simply determined by $\theta = \cos^{-1}(1 - R_2)$.

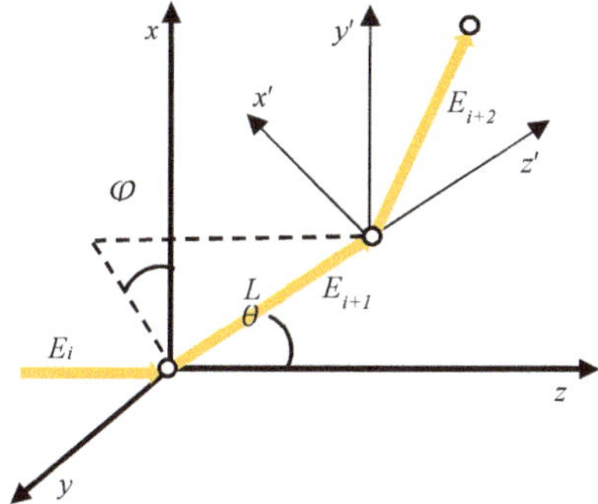

Figure 4. Schematic of a Monte Carlo simulation.

Monte Carlo simulations can equivalently provide solutions to the RTE even under the conditions for which the DE breaks down, and due to its flexibility to geometry, Boas et al. [25] succeeded in developing an MC code for predicting 3D and TD light propagation in the adult human head, which is a complex heterogeneous medium including a non-scattering and non-absorbing cerebro-spinal fluid (CSF) layer. However, for obtaining accurate solutions, long computation times are needed because the statistical noises are inversely proportional to the square root of the number of injected photons. Some techniques have been studied to overcome this drawback of MC simulations and will be reviewed in a later section.

2.9. Time-Domain Sensitivity Functions of Optical Signals

Assume that the TR fluence rate, $\phi_0(\mathbf{r}, t)$, is a solution of the TD-DE in a homogeneous medium with the absorption and diffusion coefficients, μ_{a0} and D_0, for the impulse point source at the origin and time zero. If μ_{a0} and D_0 in a small volume of the medium, V_p, are perturbed to $\mu_a = \mu_{a0} + \delta\mu_a$ and $D = D_0 + \delta D$, the perturbations in the fluence rate due to the perturbations of $\delta\mu_a$ and δD, $\delta\phi_a(\mathbf{r}, t)$, and $\delta\phi_D(\mathbf{r}, t)$, respectively, are derived by the first order perturbation theory (explained in detail in a later section), and the TD sensitivities (weights or Jacobians) of the fluence rate to the changes in μ_a and D, defined as the limits of $\delta\mu_a \to 0$ and $\delta D \to 0$ of the absolute values of $\delta\phi_a(\mathbf{r}, t)/\delta\mu_a$ and $\delta\phi_D(\mathbf{r}, t)/\delta D$, respectively, are given as,

$$W_a^{(\phi)}(\mathbf{r}, t) = \int_{V_p} d\mathbf{r}\prime \int_0^t dt\prime \{\phi_0(\mathbf{r}\prime, t\prime) G(\mathbf{r}, \mathbf{r}\prime, t - t\prime)\}, \tag{17}$$

$$W_D^{(\phi)}(\mathbf{r}, t) = \int_{V_p} d\mathbf{r}\prime \int_0^t dt\prime \{[\nabla_{\mathbf{r}\prime}\phi_0(\mathbf{r}\prime, t\prime)] \cdot [\nabla_{\mathbf{r}\prime} G(\mathbf{r}, \mathbf{r}\prime, t - t\prime)]\} \tag{18}$$

where $G(\mathbf{r}, \mathbf{r}', t - t')$ is the Green's function which is the solution of $\phi(\mathbf{r}, t)$ in the TD-DE for the impulse point source at position $\mathbf{r}'$ and time t'. The right-hand sides of Equations (17) and (18) are calculated analytically using the analytical solutions of TD-DE for simple geometries or numerically for complex geometries. Equations (17) and (18) are the TD sensitivities of the fluence rate to the changes in μ_a and D, respectively, but the TD sensitivities of any optical signals such as the reflectance and transmittance can be defined.

For example, when the reflectance, $R(\mathbf{r}_d, \mathbf{r}_s, t)$ ($\mathbf{r}_d$ and $\mathbf{r}_s$ are the detector and source positions, respectively) is considered, the TD sensitivity of the reflectance to the change in μ_a is expressed approximately as,

$$W_a^{(R)}(\mathbf{r}_d, \mathbf{r}_s, t) = \int_{V_p} d\mathbf{r}\prime \int_0^t dt\prime \left\{ G^{(R)}(\mathbf{r}_d, \mathbf{r}\prime, t\prime) G(\mathbf{r}\prime, \mathbf{r}_s, t - t\prime) \right\}, \tag{19}$$

where $G^{(R)}(\mathbf{r}_d, \mathbf{r}', t)$ is the Green's function of the reflectance measured at $\mathbf{r}_d$ for the impulse point source at position $\mathbf{r}'$ and time t', and $G(\mathbf{r}', \mathbf{r}_s, t - t')$ is the Green's function of the fluence rate measured at position $\mathbf{r}'$ and time $t - t'$ for the impulse point source at position $\mathbf{r}_s$ and time t'.

Now the physical meaning of Equation (19) should be considered. By integrating the right-hand side of Equation (19) for infinitesimal volume of $V_p = \delta(\mathbf{r}' - \mathbf{r})$ around $\mathbf{r}$, and after some mathematical manipulations, a three-point function of $\mathbf{r}_d$, $\mathbf{r}_s$, and $\mathbf{r}$ can be defined as Equation (20),

$$H(\mathbf{r}_d, \mathbf{r}_s, \mathbf{r}, t) = \alpha \int_0^t \{G(r, r_s, t\prime) G(r_d, r, t - t\prime)\} dt\prime = \alpha \int_0^t \{G(r, r_s, t\prime) G(r, r_d, t - t\prime)\} dt\prime, \tag{20}$$

where α is a proportionality constant and the second equality is derived by the reciprocity between the position of the detector and the position of interest, or the adjoint formulation. Here, $G(\mathbf{r}, \mathbf{r}_s, t')$ means the fluence rate at position $\mathbf{r}$ and at time t' generated by an impulse point source at position $\mathbf{r}_s$ and time 0, and $G(\mathbf{r}_d, \mathbf{r}, t - t')$ in the first equality means the fluence rate at position $\mathbf{r}_d$ at time $t - t'$ generated by an impulse point source at position $\mathbf{r}$ and time t' as indicated in Figure 5a. The function $H(\mathbf{r}_d, \mathbf{r}_s, \mathbf{r}, t)$ indicates the probability of photons that exist at position $\mathbf{r}$ after being injected from the source at position $\mathbf{r}_s$ at time 0 and finally being detected by the detector at position $\mathbf{r}_d$ at time t, and when illustrated in 2D or 3D images at varying times t, they show the temporal evolution of probabilistic light paths in the medium between the source and detector. In case of reflectance from semi-infinite media, the probabilistic light paths exhibit so-called banana shapes, and the banana shapes grow as time evolves. In case of transmittance through slabs, similar formulations are derived as in the case of reflectance, and the probabilistic light paths exhibit so-called spindle-like shapes.

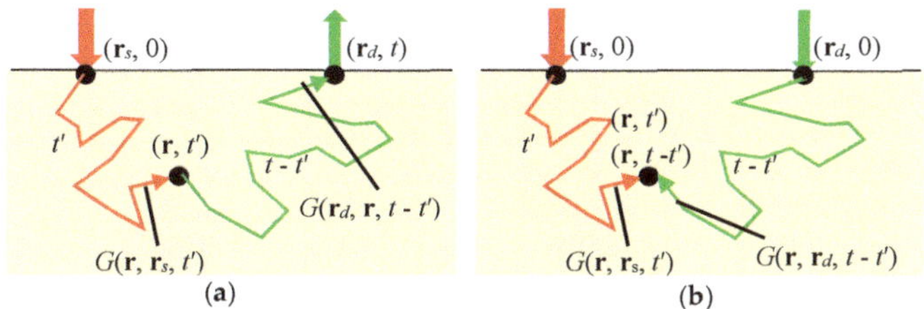

Figure 5. (**a**) Light propagation from the source position, $\mathbf{r}_s$, to the detector position, $\mathbf{r}_d$, through an arbitrary position, $\mathbf{r}$. (**b**) Light propagation from the source position, $\mathbf{r}_s$, to the position, $\mathbf{r}$, and from the detector position, $\mathbf{r}_d$, to the position, $\mathbf{r}$.

The function $H(\mathbf{r}_d, \mathbf{r}_s, \mathbf{r}, t)$ is called various terminology such as the sensitivity to absorption, photon probability distribution [26], photon hitting density [27], photon measurement density function [28], relative photon visit probability [29], photon visiting probability [30], photon sampling function, photon weight function, Jacobian with respect to absorption, etc., with variation in the proportionality constant, α. The sensitivity to scattering changes can also be derived from Equation (18), but its physical meaning is not as clear as $H(\mathbf{r}_d, \mathbf{r}_s, \mathbf{r}, t)$ The sensitivities play important roles not only in understanding the propagating path of scattered light but also in the inversion process of reconstructing absorption and scattering images in DOT as explained later.

Sawosz et al. [30] experimentally estimated $H(\mathbf{r}_d, \mathbf{r}_s, \mathbf{r}, t)$ by time-gated measurements of the fluence rates through a homogeneous semi-infinite medium. The second equality in Equation (20) means that $H(\mathbf{r}_d, \mathbf{r}_s, \mathbf{r}, t)$ is given by the convolution between the Green's function for the input position at the source, $\mathbf{r}_s$, and that for the input position at the detector, $\mathbf{r}_d$, as shown in Figure 5b. Sawosz et al. [30] used this characteristic of $H(\mathbf{r}_d, \mathbf{r}_s, \mathbf{r}, t)$ to acquire the images in Figure 6 showing the banana shapes growing bigger as time evolved. Continuous wave sensitivity is obtained by extending the upper time limit in Equation (20) to infinity.

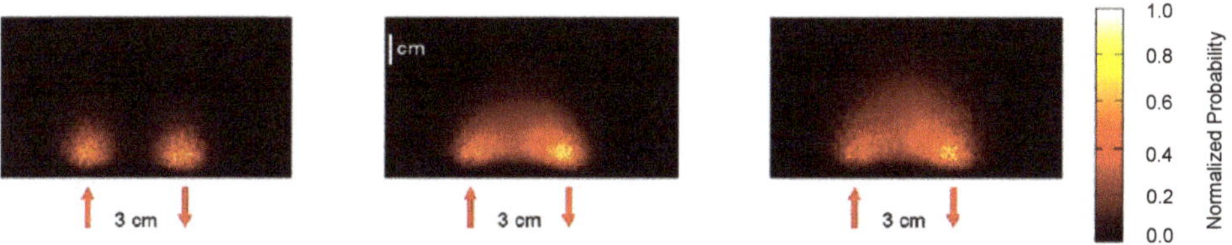

Figure 6. Distributions of the time-resolved sensitivity of the reflectance from a homogeneous semi-infinite medium with the source-detector (SD) distance, ρ, of 30 mm, $\mu_a = 0.01$ mm^{-1}, and $\mu_s' = 0.5$ mm^{-1} for different delays of the time windows of 0.86 ns, 1.66 ns, and 2.46 ns from left to right. Reproduced from Reference [30].

2.10. Time-Resolved (TR) Mean Depth of Light Propagation

From sensitivity to absorption, $H(\mathbf{r}_d, \mathbf{r}_s, \mathbf{r}, t)$, it is possible to calculate the TR mean depth of light propagation, $<z>(\mathbf{r}_d, \mathbf{r}_s, t)$, in a semi-infinite homogeneous medium by,

$$<z>(\mathbf{r}_d, \mathbf{r}_s, t) = \frac{\int_V z H(\mathbf{r}_d, \mathbf{r}_s, \mathbf{r}, t)dr}{\int_V H(\mathbf{r}_d, \mathbf{r}_s, \mathbf{r}, t)dr} = \frac{\int_V z dr \int_0^t dt'\{G(\mathbf{r}, \mathbf{r}_s, t')G(\mathbf{r}, \mathbf{r}_d, t - t')\}}{\int_V dr \int_0^t dt'\{G(\mathbf{r}, \mathbf{r}_s, t')G(\mathbf{r}, \mathbf{r}_d, t - t')\}}, \tag{21}$$

where the integral with respect to position is over the volume, V, with finite values of $H(\mathbf{r}_d, \mathbf{r}_s, \mathbf{r}, t)$.

As a note for the CW domain, the mean depth (or penetration depth) in semi-infinite homogeneous media with the source-detector (SD) distance of ρ is approximately expressed using the effective attenuation coefficient, $\mu_{eff} = (3\mu_s'\mu_a)^{1/2}$, as $<z>(\rho) = (1/2)(\rho/\mu_{eff})^{1/2}$ [31].

2.11. Time-Resolved (TR) Pathlength

Time-Resolved pathlength, expressed as $l(t, \rho)$, is defined as the pathlength of the trajectory that light takes from the impulse source position at time zero to the position where light exists at time t on the way to a detector with the SD distance of ρ, and is equal to the TR mean TOF, $<t>(t, \rho)$, multiplied by the speed of light, c. The TR mean TOF, $<t>(t, \rho)$, is calculated for the measured TOF distribution, expressed by $F(t, \rho)$, as the following,

$$l(t,\rho) = c<t>(t,\rho) = \frac{c\int_0^t t\prime F(t\prime,\rho)dt\prime}{\int_0^t F(t\prime,\rho)dt\prime}. \tag{22}$$

If the medium is multi-layered, the TR partial pathlength of the i-th layer is defined as,

$$l_i(t,\rho) = c<t_i>(t,\rho) = \frac{c\int_0^t t_i\prime F(t_i\prime,\rho)dt_i\prime}{\int_0^t F(t_i\prime,\rho)dt_i\prime}, \tag{23}$$

where t_i is time when the light is propagating inside the i-th layer. For CW light, the (partial) pathlengths are obtained by evolving the upper limit of time t to infinity, ∞.

In the i-th layer of a multi-layered medium or in a homogeneous medium, the optical properties are assumed to be constant. Then $F(t, \rho)$ can be generally expressed by $f(t_i, \rho)\exp(-\mu_{ai}ct)$ as seen from the TR reflectance of Equation (15), and Equation (23) is modified as follows,

$$l_i(t,\rho) = c<t_i>(t,\rho) = \frac{c\int_0^t t_i\prime f(t_i\prime,\rho)\exp(-\mu_{ai}ct_i)dt_i\prime}{\int_0^t f(t_i\prime,\rho)\exp(-\mu_{ai}ct_i)dt_i\prime} = -\frac{1}{F(t,\rho)}\frac{\partial F(t,\rho)}{\partial \mu_{ai}} = -\frac{\partial \ln[F(t,\rho)]}{\partial \mu_{ai}} \tag{24}$$

This equation holds not only for the DE but also for the RTE [32] (p.37), and is very useful for estimating total and partial pathlengths in various media.

Differential pathlength factor (DPF) is often referred to as, which is defined as the ratio of the CW pathlength, $l = c<t>$, to the SD distance, ρ, i.e., $DPF = l/\rho$, and is approximately expressed as $DPF(\rho) = [\mu_{eff}/(2\mu_a)][1+1/(\rho\mu_{eff})]^{-1}$ for your reference [31].

2.12. Physiological Information and Optical Properties

Once the optical properties of biological tissues are estimated or determined by TD-NIRS physiological information, particularly, the hemodynamic information can be estimated.

The absorption coefficient of tissue is the sum of the absorption coefficients of the chromophores such as hemoglobin, myoglobin, melanin, water, lipid, cytochrome, etc., and expressed as,

$$\mu_a(\lambda) = \sum_{i=1}^{N_C} \varepsilon_i(\lambda)C_i, \tag{25}$$

where ε_i is the extinction coefficient or the molar absorption coefficient of the component i, C_i is the molar concentration or the molarity of the component i, and N_C is the number of components. Note that ε_i is often based on common logarithm while μ_a is based on natural logarithm, necessitating correction between the two logarithms. If $\varepsilon_i(\lambda)$ are known for all the components, measurements of $\mu_a(\lambda)$ at N_C (or more than N_C) wavelengths will determine C_i of all the components. Oxygenated hemoglobin (oxy-Hb) and deoxygenated hemoglobin (deoxy-Hb) are two dominant chromophores in NIRS with known spectra of $\varepsilon_{oxy\text{-}Hb}(\lambda)$ and $\varepsilon_{deoxy\text{-}Hb}(\lambda)$. Then $C_{oxy\text{-}Hb}$ and $C_{deoxy\text{-}Hb}$ are determined and total hemoglobin (total-Hb) concentration, $C_{total\text{-}Hb} = C_{oxy\text{-}Hb} + C_{deoxy\text{-}Hb}$, and oxygen saturation, $S_{O2} = C_{oxy\text{-}Hb}/C_{total\text{-}Hb}$, can be calculated to show hemodynamic statuses in brain, muscle, breast, skin, etc.

In contrast to μ_a, the scattering properties, μ_s and μ_s', indicate the statuses of microstructures and sub-cellular components, because scattering is caused by gradients and discontinuities of refractive indexes in tissue.

3. Instruments for TD-NIRS

3.1. Overview of TD-NIRS Instruments

In the mid 1980s, NIRS succeeded in in vivo and non-invasive monitoring of the changes in cerebral hemoglobin concentrations in human heads [33]. However, the monitoring method based on the modified Beer–Lambert law is subject to the problem of quantitativeness due to the unknown optical pathlength which is necessary for applying the modified Beer–Lambert law.

To cope with this problem in the late 1980s, studies were conducted to estimate the optical pathlengths in biological tissues by measuring TOF distributions of reemitted light after being multiply scattered in tissues using picosecond-pulsed lasers and detected by detectors having picosecond temporal resolutions such as streak cameras. Delpy et al. and van der Zee et al. used an ultrafast dye laser excited by a krypton-laser for a light source (a pulse width less than 6 ps and repetition rate of 76 MHz at wavelengths of 761 nm and 783 nm) and a streak camera for a detector to estimate the mean optical pathlengths from the measured TOF distributions [34,35], and found that the absorbance changes of the measured transmittance were related to the absorption changes in the media through the estimated mean optical pathlengths. Chance et al. reported that the logarithmic slopes of TOF distributions in the decaying period were proportional to hemoglobin concentrations in biological tissues by TD experiments using two dye lasers excited by the second harmonics of an Nd:YAG laser for pulsed light sources and time-correlated single-photon counting (TCSPC) technique for a detector which was widely employed for measurement of fluorescence life time [36,37]. Furthermore, Nomura et al. measured TOF distributions through rat heads using a Ti–Sapphire laser for a pulsed light source and a streak camera for a detector and concluded that the Beer–Lambert law held every time in the TOF distributions [38].

These pioneering studies promoted wide use of time-domain near-infrared spectroscopy (TD-NIRS) systems for understanding light propagation and quantitative measurements of hemoglobin concentration in biological tissue. However, the TD-NIRS systems used in these pioneering studies had the disadvantages of being very expensive, large, non-portable, and highly sophisticated, and resultantly, their systems were unacceptable for clinical applications.

To mitigate the disadvantages stated above, Cubeddu et al. developed a compact dual-wavelength multichannel tissue oximeter using two pulsed-laser diodes (pulse width less than 100 ps, repetition rate of 80 MHz, wavelengths of 672 nm and 818 nm) for light sources and multi-anode photomultiplier tubes (PMTs) for detectors [39]. Oda et al. developed a compact three-wavelength one-channel tissue oxygenation monitor using three pulsed-laser diodes (pulse width less than 100 ps, repetition rate of 5 MHz, wavelengths of 759 nm, 797 nm and 833 nm) for light sources and PMTs for detectors based on the TCSPC technique [40]. This tissue oxygenation monitor was marketed but only in Japan. Grosenick et al. also developed a single-wavelength single-channel TD optical mammograph using a pulsed laser diode (pulse width of about 400 ps, repetition rate of 80 MHz, wavelength of 785 nm) for light source and a PMT-TCSPC for detector, and measured TR-transmittance through healthy breasts as well as breasts carrying tumor [41].

Currently, TD-NIRS systems with a few channels employing the TCSPC technique are actively developed for the purpose of applications in clinical usage, and also multi-channel TD-NIRS systems have been developed for TD-DOT.

3.2. Time-Correlated Single Photon Counting (TCSPC) Technique

3.2.1. Principle, Components, Characteristics, and Operation of the TCSPC Technique

The TCSPC technique is a time-resolved spectroscopic measurement method based on a single photon counting, which is a way of light detection by counting photons one by one, combined with a pulsed light source with a narrow pulse width in the order of picoseconds and a stabilized repetition of pulse generation in the order of a mega-hertz, and has the capability of measuring temporal changes in very weak light intensity with a high-temporal resolution in the order of picoseconds. At time, t, after the emission of a single ultra-short light pulse, photons reaching a detector with a probability of less than one photon per one light pulse are detected. This detection of single photon at time t is repeated many times in the order of a mega-hertz, and a histogram of the numbers of detected single photons as a function of time, t, is produced, i.e., a TOF distribution is obtained [37].

As shown in Figure 7, a standard TCSPC instrument consists of a pulsed light source, a single-photon counting detector, such as a PMT, a time pick-off circuit, such as a constant fraction discriminator (CFD), a time-to-amplitude converter (TAC), an A/D converter, a histogram memory, etc. A pulsed light source must emit ultrashort light pulses with a pulse width much less than the characteristic response time of the optical phenomenon in a sample and with a stabilized and fast repetitive pulse generation. The TAC, one of the key components, receives output pulses (start signals) from the detector and reference pulses (stop signals) synchronized to the ultrashort light pulses from the light source, and outputs pulses with the intensities proportional to the time differences between the start and stop signals. The output pulses from the TAC are processed by the A/D converter and the histogram memory.

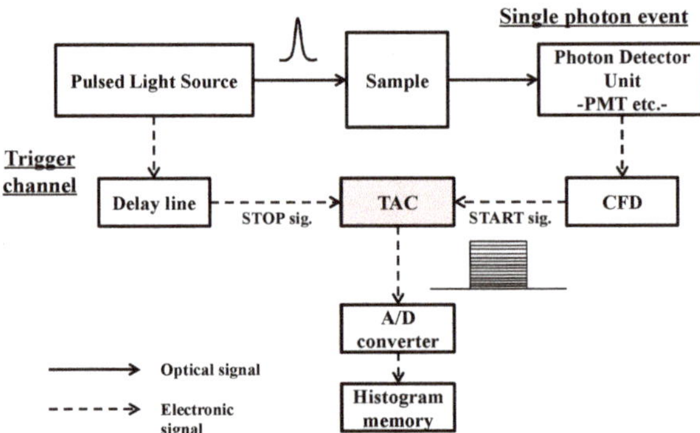

Figure 7. Block diagram of a standard time-correlated single-photon counting (TCSPC) instrument.

The important factor in the measurement of the TCSPC technique is that the light intensity detected by the detector is adjusted so that the detection probability in a time-gate for one input pulse is constant at any time-gate. If more than one photon is detected in the time-gate for one input pulse, the TAC starts working at the arrival of the start signal generated by the first detected photon and neglects the start signals generated by the second, third, ... detected photons until the next stop signal arrives. Then, measurement pile-up takes place causing a distortion in the TOF distribution. Resultantly, earlier times are enhanced because the detection probability in the time-gate is no longer constant, and a correct TOF distribution cannot be obtained due to a serious distortion. To solve this problem of pile-up distortion, the detection rate of the detector must be sufficiently smaller than the repetition rate of the input light pulses.

When m photons are detected in an average of one input pulse, the probability of detecting n photons for one input pulse is given by a Poisson distribution,

$$P_n = \frac{m^n e^{-m}}{n!} \tag{26}$$

As an example, when only one photon for 100 input pulses is detected in an average ($m = 0.01$), the probability that more than one photon for one input pulse are detected becomes very small as,

$$\frac{1 - P_0 - P_1}{1 - P_0} = \frac{1 - e^{-0.01} - 0.01e^{-0.01}}{1 - e^{-0.01}} = 0.0050 \ (0.5\%). \tag{27}$$

It is obvious that the appropriate value of m depends on the acceptable distortion of the measured TOF distributions, and it is a common understanding that m less than 5% (0.05) provides negligible distortions of the measured TOF distributions. Actually, m is often kept at small values of about 1% (0.01) or less than 2% (0.02). In principle, the errors will decrease with the decrease in m, but at the same time the measuring time will be prolonged to assure good S/N ratios. Therefore, it is important to choose a light source capable of emitting light pulses stably with a narrow pulse width and a high repetition rate. To shorten measuring times, Ida and Iwata [42] devised a method correcting the pile-up distortion mathematically for the measured data with m larger than 0.1 (10%).

In the TCSPC technique, there also exists a dead time when the detector system is unable to count photons due to processing signals after receiving photons. If the dead time is longer than the time period between two successive input pulses, counting losses take place. In standard TCSPC systems, the dead times are longer than the time period between two successive input pulses. Then the CW intensity, I, given by integration of a TOF distribution with respect to time must be corrected for the counting losses by $I_{true} = I_{meas}/(1 - \alpha I_{meas})$ where I_{true} and I_{meas} [s^{-1}] are the corrected and integrated intensities, respectively, and α is the dead time [s] when the measuring systems are non-paralyzed models.

3.2.2. Single- and Dual-Channel TD-NIRS Systems Based on the TCSPC Technique

In 1999, Oda et al. developed a single-channel TD-NIRS system based on the TCSPC technique (TRS-10, Hamamatsu Photonics K.K., Japan), and in 2009 it was extended to a dual-channel system (TRS-20, Hamamatsu Photonics K.K., Japan) [43], both for research use. Figure 8 shows a photograph and block diagram of the TRS-20 consisting of three pulsed laser diodes with wavelengths of 760 nm, 800 nm, and 830 nm, a TCSPC unit with a CFD/TAC, A/D converter, histogram memory, etc.

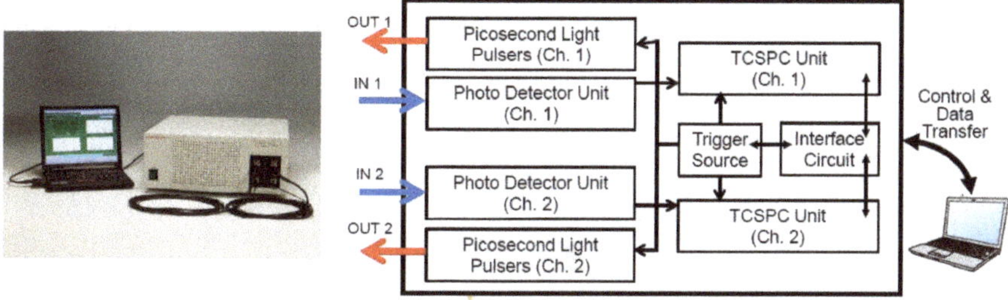

Figure 8. Photograph and block diagram of the TRS-20 system. The picosecond light pulser emit lights with wavelengths of 760, 800, and 830 nm; the photo detector unit consists of a variable optical attenuator, the photomultiplier tube (PMT), and a fast amplifier; the TCSPC unit consists of a CFD/TAC, an A/D converter, a histogram memory, etc.

From the TR reflectances measured by the TRS-10 and TRS-20, μ_a and μ_s' of the media are estimated under the assumption that the media are homogeneous and semi-infinite. The TR reflectance from a homogeneous semi-infinite medium is given by the analytical solution of the TD-DE as Equation (15). For the zero-boundary condition, Equation (15) is slightly simplified as the following,

$$R(\rho, t) == \frac{z_0}{(4\pi Dc)^{3/2}t^{5/2}} \exp(-\mu_a ct) \exp(-\frac{\rho^2 + z_0^2}{4Dct}), \tag{28}$$

here, the wavelength dependences are not explicitly written. Equation (28), convoluted by the instrumental response function (IRF), is fitted to the measured TD reflectance to estimate μ_a and μ_s' of the medium using a non-linear least-squared technique based on the Levenberg–Marquardt method. Applying Equation (25) to the case of the two components of oxy-Hb and deoxy-Hb with background absorption gives the following equation with the wavelength dependence written explicitly,

$$\mu_a(\lambda) = \varepsilon_{oxy\text{-}Hb}(\lambda)C_{oxy\text{-}Hb} + \varepsilon_{deoxy\text{-}Hb}(\lambda)C_{deoxy\text{-}Hb} + \mu_{a,BG}(\lambda) \tag{29}$$

where $\mu_{a,BG}$ is the absorption coefficient of the background medium. Solving the simultaneous equations of Equation (29) for the three wavelengths used in the instrument provides $C_{oxy\text{-}Hb}$ and $C_{deoxy\text{-}Hb}$, and $C_{total\text{-}Hb}$, and $S_{O2} = C_{oxy\text{-}Hb}/C_{total\text{-}Hb}$ is calculated.

For verification of the quantities obtained by the TRS-10, experiments were performed using a blood phantom consisting of an Intralipid suspension and blood with the controlled S_{O2} from 0% to 100% [44]. The values of $C_{oxy\text{-}Hb}$, $C_{deoxy\text{-}Hb}$, and $C_{total\text{-}Hb}$ measured by the TRS-10 agreed well with the values calculated from the measurements of pH, P_{O2}, and P_{CO2} in the phantom by conventional methods.

3.2.3. Multi-Channel TD-NIRS Systems Based on TCSPC for DOT

Multi-channel TD-NIRS systems have been developed for the purpose of TD-DOT. In 1999, a Japanese group developed a 64-channel TD-NIRS system and studied the image reconstruction algorithm as well as the performance of the system [45]. Figure 9 shows the photograph and block diagram of the system which had 64 independent detector units, while light sources of three pulsed laser diodes with three wavelengths operated in turn using an optical switch for selecting a wavelength and a mechanical switch for selecting a source position. Results using this system are reviewed in a later section [46–48].

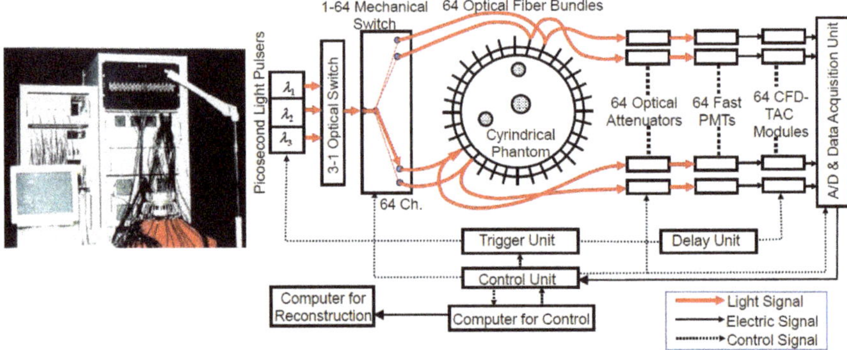

Figure 9. Photo and block diagram of the 64-channel TD-NIRS system developed by the group in Japan. Modified from Reference [45].

A 32-channel TD-NIRS system named MONSTIR (Figure 10) was developed by a group from the University of London for reconstructing TD-DOT images of brain functions in premature infants at bedside [49]. The DOT images obtained using these systems are reviewed later in Section 5.3.4.

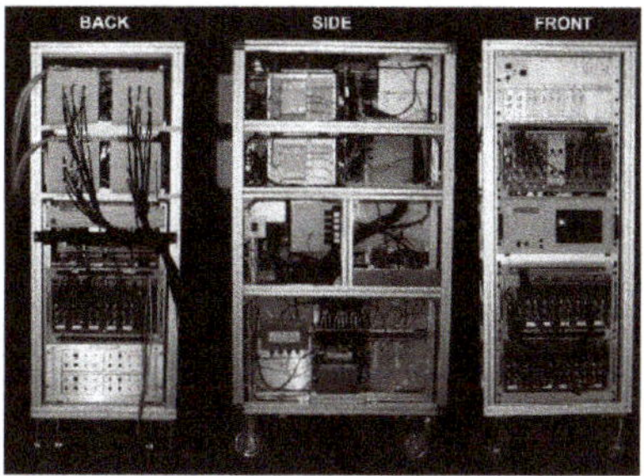

Figure 10. Photo of the 32-channel TD-NIRS system, "MOSTIR", developed by a group in the UK [49]. Reproduced from Reference [49].

Ueda et al. [50] developed a 16-channel TD-NIRS system based on the TCSPC technique to reconstruct TD-DOT images from measured TR reflectances at an adult human forehead. Results of TD-DOT using this system are reviewed later in Section 5.3.4.

3.3. Other New TD-NIRS Systems

Various research institutes have updated and improved TD-NIRS systems by incorporating recently developed and useful apparatuses for the components of TD-NIRS systems. In the following, newly developed TD-NIRS systems and new measurement methods for TD-NIRS systems including ones which do not use the TCSPC technique are briefly reviewed.

3.3.1. MONSTIR II

In 2014, a group from the University of London developed an updated version of MONSTIR named MONSTIR II employing a supercontinuum (SC) laser for the light source with an acoustic-optic tunable filter (AOTF), which enables to select four wavelengths arbitrarily in a wavelength range from 650 nm to 950 nm [51].

3.3.2. TD-DOT and TD-NIRS Systems for Optical Mammography

In 2011, Hamamatsu Photonics developed a 48-channel three-wavelength optical mammography as shown in Figure 11a [52]. Breasts were immersed in a hemispherical gantry filled with a matching fluid, and DOT images were reconstructed by an algorithm based on the microscopic Beer–Lambert law [9,50]. Furthermore, a 12-channel TD-NIRS system for optical mammography with a hand-held probe shown in Figure 11b was developed to be used at a bed side [53]. Another TD-NIRS system for optical mammography, as shown in Figure 11c, was developed by a group from Italy to obtain TR transmittance images of compressed breasts [54]. The results of these systems are briefly described later in Sections 5.2.1 and 5.3.5.

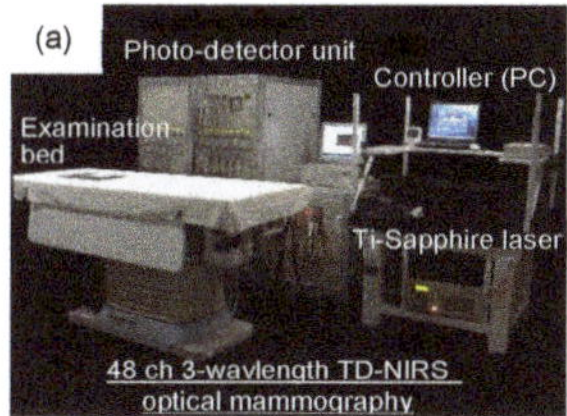

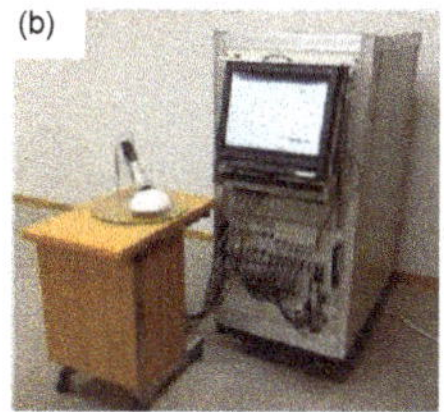

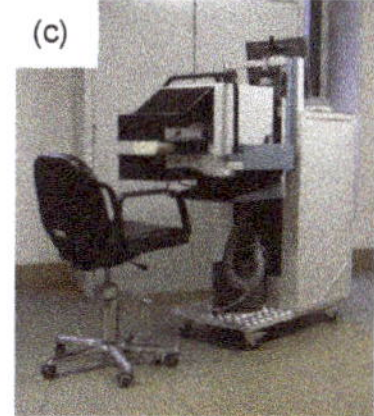

Figure 11. Photos of (**a**) the 48-channel TD-DOT for optical mammography [52]; (**b**) the 12-channel TD-NIRS system for optical mammography with a hand-held probe [53], and (**c**) the transmittance TD-NIRS imaging system for optical mammography with breast compression [54]. (**b**) Reproduced from Reference [53]. (**c**) Reproduced from Reference [54].

3.3.3. Compact TD-NIRS Systems Using MPPCs (or SiPMs)

A multi-pixel photon counter (MPPC), or a silicon photomultiplier (SiPM) for another name, is a new type of photon counting semiconductor device consisting of a high-density matrix of avalanche photodiodes, enabling single-photon detection and offering high-photon detection efficiency, excellent timing resolution, low bias voltage operation, ruggedness, resistance to excess light, and immunity to magnetic fields. Using MPPCs (or SiPMs), compact TD-NIRS systems have been developed and commercialized recently.

A dual-channel TD-NIRS system for non-invasive monitoring of oxygenation in the brain (tNIRS-1, Hamamatsu Photonics K.K., Japan) was manufactured and marketed as a medical instrument in 2014 [55]. Figure 12 shows its photograph. It employs semi-conductor devices of cooled multi-pixel photon counters (MPPCs) for light detection and time-to-digital converters (TDCs) for TR measurement with a design concept of compactness and usability, resulting in a size of 292 mm (width) × 291 mm (height) × 207 mm (depth) and a weight of about 7.5 kg.

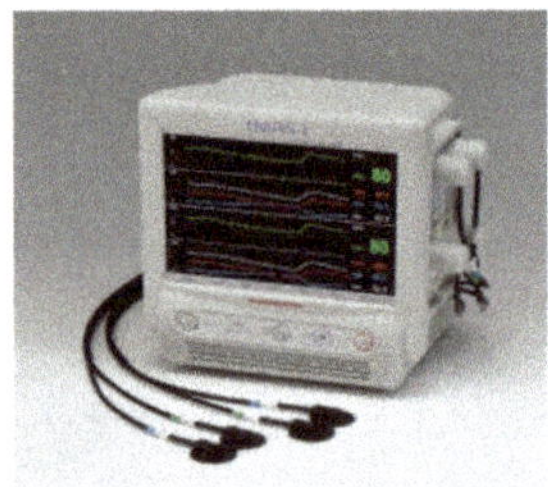

Figure 12. Photo of tNIRS-1.

In 2016, the group from Politecnico di Milano, Italy, developed a compact (size of 160 mm (width) × 50 mm (height) × 200 mm (depth)), low power consumption, dual-wavelength TD-NIRS system as shown in Figure 13 [56]. It employs two pulsed laser diodes with wavelengths of 670 nm and 830 nm for the light sources, a silicon photomultiplier (SiPM) with an active area of 1 mm^2 for the detector, and a home-made TDC with a temporal resolution of 10 ps for TR measurements. For TD-NIRS, the same group also developed a compact detector probe integrating a fast SiPM and its electronics, which can be directly put in contact with the skin [57]. Many compact detector probes can be installed into a head-cap without the need for optical fibers for collecting light.

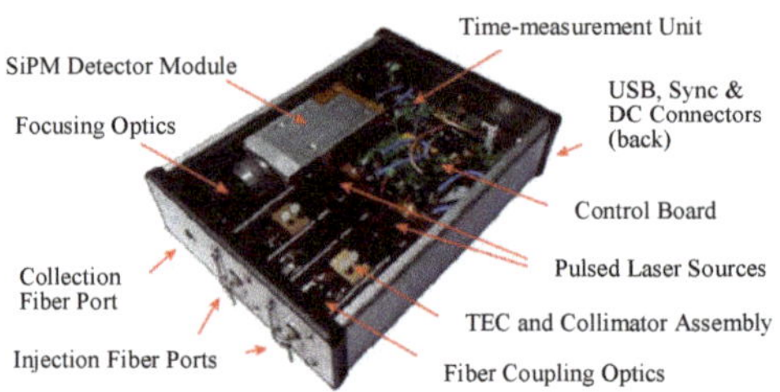

Figure 13. Photo of the compact dual-wavelength TD-NIRS system developed by the group in Italy. Modified from Reference [56].

3.3.4. TD-NIRS Systems Using SPADs

Puszka et al. [58] proposed and developed a new TD-NIRS system incorporating single-photon avalanche diodes (SPADs) operating in a fast-gating mode into a TCSPC unit for use in TD-DOT. Through phantom experiments, they compared the results with and without time-gating and concluded that the proposed method not only extended the dynamic range of the detector but also improved the depth sensitivity due to the higher sensitivity to late photons. Dalla Mora et al. [59] also developed a system using fast-pulsed vertical cavity surface emitting lasers and fast-gated SPADs, both embedded into probes with SD distances of 5 mm and 30 mm, indicating the feasibility of application to wearable imaging instruments which should be compact, inexpensive, and quantitative. Furthermore, Di Sieno et al. [60] modified the above system using an SC laser with an AOTF for the light source and fast-gated SPADs for the detector, aiming the application at non-contact measurements. They performed phantom experiments to evaluate the performances of the system and discussed the possibility of TD-DOT imaging with non-contact measurements.

Sinha et al. [61] proposed an early photon approach to improve the spatial resolution of TD-DOT images using a developed TD-NIRS system based on the TCSPC technique incorporating SPADs. To enhance the detection efficiency of early photons with less scattering events, measurements were made with a detection rate much higher than that leading to counting losses due to the dead time.

Kalyanov et al. [62] reported a TD-DOT system using a state-of-the-art SPAD camera chip for detection, which is an array of 32×32 SPADs including a time-to-digital converter, and a super-continuum laser. The source light can be guided to 24 channels at most, and a total of 1024 TOF distributions at the maximum can be obtained without bulky instruments using conventional TD-NIRS systems.

3.3.5. TD-NIRS Systems Using Pseudo-Random Bit Sequences

Chen and Zhu [63] developed a new TD-DOT system based on the spread spectrum approach using a laser diode modulated with pseudo-random bit sequences, which replaced picosecond or femtosecond ultrashort pulsed lasers usually employed for light sources in the TCSPC technique. Phantom experiments verified the improvements of the spatial resolution and S/N ratio of the 2D scanning images through phantoms with a thickness of 5.5 cm. Mo and Chen [64] developed a fast TD-DOT system using the pseudo-random bit sequences. They reported that the IRF of the system was about 800 ps and that 2D maps of the optical properties could be obtained within a few seconds.

3.3.6. TD-NIRS Systems Using ICCD

Time-domain near-infrared spectroscopy systems using a time-gated intensified charge coupled device (ICCD) instead of using the TCSPC technique have been developed. For the purpose of measuring TOF distributions in several different time-gates, Selb et al. [65] developed a system employing optical fibers with different lengths leading to time differences for the time-gates. Zhao et al. [66] developed a TD-NIRS system employing a time-gated ICCD enabling non-contact measurement with a wide dynamic range of detected signals. They reported the possibility of imaging absorbers deep in media with liquid phantom experiments.

Lange et al. [67] developed a broadband multichannel TD-NIRS system using a supercontinuum laser for a light source and an ICCD camera coupled with an imaging spectrometer for a detector. The spectral range was from 600 nm to 900 nm and the IRF was about 660 ps. The performances of the system were demonstrated by in vivo experiments to monitor the hemodynamic responses in adult arms during a cuff occlusion and in adult brains during a cognitive task.

3.3.7. Compact TD-NIRS System Incorporating Devices Used in Telecommunications

Very recently, Konstantinos et al. [3] developed a single-channel TD-NIRS system using a spread spectrum technique for TOF measurements. As a light source, the system incorporates a low-cost commercially available optical transceiver module, widely used in telecommunications applications. The system can generate an IRF under 600 ps, exhibits good accuracy and low-noise properties, requires very short warm-up times in contrast to conventional pulsed laser diodes, and is very compact compared to traditional TD-NIRS systems.

3.3.8. TD-NIRS Systems for Measurement of Water, Lipid, and Collagen Contents

Most TD-NIRS systems employ light sources with a wavelength range from about 700 nm to about 900 nm by focusing attention on hemodynamics. By extending the wavelength range up to 1100 nm, it becomes possible to measure the contents of other tissue components. Taroni et al. [54] developed a TR reflectance imaging system for optical mammography, as shown in Figure 11c, to obtain images of the contents of water, lipid, and collagen in addition to oxy- and deoxy-hemoglobin using seven wavelengths from 635 nm to 1060 nm by employing seven pulsed laser diodes. Ohmae et al. [68] developed a compact TD-NIRS system using six wavelengths in the wavelength range from 760 nm to 980 nm for measuring water and lipid content in addition to hemoglobin oxygenation. The results from using these instruments are briefly reviewed later in Section 5.2.1.

3.4. Future Trend of TD-NIRS Instruments

As stated above in Section 3.3, many new techniques have been developed to improve the performances of TD-NIRS instruments toward faster data acquisition within shorter measurement times, with contactless measurements, more compact devices, lower costs, faster image reconstruction algorithms with higher image quality, and more physiological information by extending the wavelength range, etc. These improvements will expand clinical applications of TD-NIRS systems and will lead to the development of 3D imaging with TD-DOT.

4. Advanced Theories and Methods for TD-NIRS

In this section, advanced theories and methods based on Section 2 are briefly described.

4.1. Solving the TD-RTE and TD-DE Numerically

The TD-DE is a parabolic type of partial differential equation which can be easily solved by commercially available software codes even for geometrically complicated media such as the human head and small animal models using the finite element method (FEM). Schweiger et al. [69] generally described the numerical method using FEM to solve the TD-DE as a forward calculation, as well as an inverse calculation for time-domain diffuse optical tomography (TD-DOT).

Solving the TD-RTE is not an easy task even by numerical computation due to the term of the scattering integral in Equation (1). The discrete ordinates method (DOM) is a standard technique to solve the RTE and some good monographs have been published for the method [70]. A review of solving the TD-RTE numerically is beyond the scope of this article, and only a few articles, which were arbitrarily chosen, are reviewed below.

Das et al. [71,72] solved the TD-RTE using the DOM and the calculated TOF distributions were compared with measured TOF distributions by TD experiments using tissue phantoms, rat tissue samples in vitro, and anesthetized rats in vivo.

Fujii et al. [73] solved the TD-RTE using the DOM to understand light propagation in the human neck including the thyroid and trachea for diagnosis of thyroid cancer by DOT. Light propagation inside the trachea cannot be expressed by the TD-DE because the trachea is a perfect void region of air, and reflection and refraction at the tissue–trachea interface should be considered. The results showed an interesting pattern of light propagation inside the trachea.

Some methods have been proposed to calculate the TD-RTE with smaller computation loads. Only two studies are reviewed briefly for your information. One is to employ a cell-vertex finite volume method for discretization of the space for faster computing of the 2D TD-RTE [74]. The other is to employ a new renormalization approach to calculate the scattering integral for accurate and efficient computation of the 3D TD-RTE [75].

4.2. Analytical Solutions for the TD-RTE

Paasschens [76] derived an almost exact analytical solution for the TD-RTE for the impulse point source in an infinite medium using the Fourier and its inverse transforms. He succeeded in correctly expressing the ballistic component which is never expressed by solutions for the TD-DE.

Martelli et al. developed an approximate Green's function for the TD-RTE for TD reflectance from a semi-infinite medium by heuristically employing the mirror image method to the analytical solution of the TD-RTE by Paasschens [76,77]. The closed-form Green's function is useful for measurement of optical properties or DOT based on the TD-RTE using inversion analyses.

Liemert and Kienle derived a Green's function of the TD-RTE for the radiance and fluence rate in an infinite medium by expanding the radiance into Legendre polynomials similarly to the P_N approximation and by expressing the expansion coefficients analytically [78]. They also derived a Green's function of the TD-RTE for the radiance in a 3D anisotropically-scattering medium with an impulse point unidirectional source [79]. The solution involves a spherical Hankel transform necessitating numerical calculation, but the dependences on all the variables were found analytically.

Simon et al. extended the analytical solution derived by Liemert and Kienle to a semi-infinite medium by employing the mirror image method as Martelli et al. did [77,78,80]. The mirror image method correctly describes the boundary condition for the DE, but only approximately for the RTE. Nevertheless, the solutions agree well with the results of MC simulations and those of Martelli et al. and are applicable for anisotropically-scattering media.

4.3. The TD-RTE with Spatially Varying Refractive Index

The TD-RTE of Equation (1) assumes a constant refractive index throughout the medium, but the refractive index may spatially vary inside the medium. Ferwerda formulated the RTE for media with a spatially varying refractive index [81]. Khan and Jiang then derived the DE with a spatially varying refractive index from the TD-RTE formulated by Ferwerda [81,82].

Tualle and Tinet derived the TD-RTE with varying refractive indexes, which satisfy energy conservation while the RTE by Ferwerda does not satisfy energy conservation [81,83]. They also derived the TD-DE with varying refractive index and verified the equation by comparison with the results of MC simulations.

4.4. Solutions of the Telegraph Equation (TE)

Kumar et al. [84] numerically solved a 1D case of the TE, Equation (5), using the method of characteristics and finite difference. The solutions of the TE were compared with those of the TD-DE, and the ballistic components were clearly observed in the solutions of the TE while the solutions of the TD-DE violated the causality in the early times.

Analytical solutions of the TE were obtained by Durian and Rudnick, although their TE was a little bit different from that derived by the P_1 approximation, Equation (5) [85]. They obtained analytical solutions for infinite slabs and semi-infinite media as the forms of Laplace transform, and the solutions obtained by inverse Laplace transform cannot be expressed explicitly. Their analytical solutions agreed well with the results of MC simulations, although small differences were observed in very early times.

4.5. Perturbation Theory

Perturbation approaches have widely been employed for imaging inclusions with the optical properties different from those of the background homogeneous medium. The perturbation approach is a general concept applicable to any equations describing physical phenomena, and it is natural to apply the perturbation approach to the phenomenon of light propagation, or, photon migration, described by the TD-RTE, the TE, and the TD-DE. Actually, Polonsky and Box [86] reported general formulations of the perturbation approach applied to the RTE. However, the perturbation approach using the TD-RTE has difficulties in obtaining analytical solutions such as the Green's functions, and most studies of the perturbation approach are based on the TD-DE for which the Green's functions are available for simple geometries such as the infinite, semi-infinite, infinite slab, cylindrical, and spherical media [19].

4.5.1. Formulation of the TD-Perturbation Based on the TD-DE

Formulation of the TD-perturbation starts from the TD-DE (Equation (7)) for the unperturbed fluence rate, $\phi_0(\mathbf{r}, t)$, under the unperturbed $\mu_{a0}(\mathbf{r})$ and $D_0(\mathbf{r}) = 1/3\mu_{s0}'(\mathbf{r})$, and the TD-DE for the perturbed fluence rate, $\phi_p(\mathbf{r}, t) = \phi_0(\mathbf{r}, t) + \delta\phi(\mathbf{r}, t)$, under the perturbed coefficients, $\mu_{ap}(\mathbf{r}) = \mu_{a0}(\mathbf{r}) + \delta\mu_a(\mathbf{r})$ and $D_p(\mathbf{r}) = D_0(\mathbf{r}) + \delta D(\mathbf{r})$. By subtracting the unperturbed equation from the perturbed equation, the TD-DE for $\delta\phi(\mathbf{r}, t)$ is obtained as,

$$\left\{ \frac{1}{c}\frac{\partial}{\partial t} - \nabla\cdot[D_0(\mathbf{r})\nabla] + \mu_{a0}(\mathbf{r}) \right\}\delta\phi(\mathbf{r}, t) = \{-\delta\mu_a(\mathbf{r}) + \nabla\cdot[\delta D(\mathbf{r})\nabla]\}\phi_p(\mathbf{r}, t). \tag{30}$$

If the right-hand side of Equation (30) is considered as a virtual source, $Q_v(\mathbf{r},t) = \{-\delta\mu_a(\mathbf{r}) + \nabla\bullet[\delta D(\mathbf{r})\nabla]\}\phi_p(\mathbf{r},t)$, and Equation (30) has the same form as the unperturbed equation. Then, if the Green's function of the unperturbed equation is known as $G(\mathbf{r}, \mathbf{r}', t - t')$, the solution of Equation (33) for the virtual source of $Q_v(\mathbf{r}, t)$ is given by the Green's function method as,

$$\delta\phi(\mathbf{r}, t) = \int_{V_p} d\mathbf{r}\prime \int_0^t dt\prime Q_v(\mathbf{r}\prime, t\prime)G(\mathbf{r}, \mathbf{r}\prime, t - t\prime), \tag{31}$$

where V_p indicates the volume occupied by the perturbations. Substitution of $Q_v(\mathbf{r},t)$ into Equation (31) and some mathematical operations leads to the following equation,

$$\phi_p(\mathbf{r}, t) = \phi_0(\mathbf{r}, t) + \delta\phi(\mathbf{r}, t) = \phi_0(\mathbf{r}, t) + \delta\phi_a(\mathbf{r}, t) + \delta\phi_D(\mathbf{r}, t), \tag{32}$$

where

$$\delta\phi_a(\mathbf{r}, t) = -\int_{V_p} d\mathbf{r}\prime \int_0^t dt\prime\{\delta\mu_a(\mathbf{r}\prime)\phi_p(\mathbf{r}\prime, t\prime)G(\mathbf{r}, \mathbf{r}\prime, t - t\prime)\}, \tag{33}$$

$$\delta\phi_D(\mathbf{r}, t) = -\int_{V_p} d\mathbf{r}\prime \int_0^t dt\prime\{\delta D(\mathbf{r}\prime)[\nabla_{\mathbf{r}\prime}\phi_p(\mathbf{r}\prime, t\prime)]\cdot[\nabla_{\mathbf{r}\prime}G(\mathbf{r}, \mathbf{r}\prime, t - t\prime)]\}. \tag{34}$$

Now, the key subject is how to calculate the integrals of Equations (33) and (34) including the unknown $\phi_p(\mathbf{r}, t)$ (note: the integral equation composed of Equations (32), (33), and (34) is called the Fredholm equation of the second kind), and various assumptions and approximations have been employed. Usually, the perturbations are assumed constant inside the inclusions, i.e., $\delta\mu_a(\mathbf{r}) = \delta\mu_a =$ const and $\delta D(\mathbf{r}) = \delta D =$ const, and the simplest approximation is to substitute $\phi_0(\mathbf{r}, t)$ for $\phi_p(\mathbf{r}, t)$ by assuming $\phi_p(\mathbf{r}, t) \approx \phi_0(\mathbf{r}, t)$ resulting in the followings,

$$\delta\phi_a^{(1)}(\mathbf{r}, t) = -\delta\mu_a \int_{V_p} d\mathbf{r}\prime \int_0^t dt\prime\{\phi_0(\mathbf{r}\prime, t\prime)G(\mathbf{r}, \mathbf{r}\prime, t - t\prime)\}, \tag{35}$$

$$\delta\phi_D^{(1)}(\mathbf{r}, t) = -\delta D \int_{V_p} d\mathbf{r}\prime \int_0^t dt\prime\{[\nabla_{\mathbf{r}\prime}\phi_0(\mathbf{r}\prime, t\prime)]\cdot[\nabla_{\mathbf{r}\prime}G(\mathbf{r}, \mathbf{r}\prime, t - t\prime)]\}. \tag{36}$$

These solutions are called the first order perturbation or the Born approximation, and many TD studies have reported analytical formulations in the first order perturbation using the TD-Green's functions for homogeneous infinite media, semi-infinite media, infinite slabs, multi-layered media [32] (pp. 131–157) [28,87–91].

Solutions of the second order perturbation are derived by utilizing the first order perturbation in the manner of $\phi_p(\mathbf{r}, t) \approx \phi_0(\mathbf{r}, t) + \delta\phi_a^{(1)} + \delta\phi_D^{(1)}$ resulting in the followings,

$$\delta\phi_a^{(2)}(\mathbf{r}, t) = \delta\phi_a^{(1)}(\mathbf{r}, t) - \delta\mu_a \int_{V_p} d\mathbf{r}\prime \int_0^t dt\prime\left\{ [\delta\phi_a^{(1)} + \delta\phi_D^{(1)}](\mathbf{r}\prime, t\prime)G(\mathbf{r}, \mathbf{r}\prime, t - t\prime) \right\}, \tag{37}$$

$$\delta\phi_D^{(2)}(\mathbf{r}, t) = \delta\phi_D^{(1)}(\mathbf{r}, t) - \delta D \int_{V_p} d\mathbf{r}\prime \int_0^t dt\prime\left\{ [\nabla_{\mathbf{r}\prime}(\delta\phi_a^{(1)}(\mathbf{r}\prime, t\prime) + \delta\phi_D^{(1)}(\mathbf{r}\prime, t\prime))]\cdot[\nabla_{\mathbf{r}\prime}G(\mathbf{r}, \mathbf{r}\prime, t - t\prime)] \right\}. \tag{38}$$

The third, fourth, and higher order perturbation can be formulated in a similar manner, and the cross talks between $\delta\mu_a$ and δD begin to play a role.

The first order perturbation, i.e., the Born approximation, is believed to be valid for inclusions with very small sizes and very small changes in μ_a and D from those of the background. For large inclusions with large changes in μ_a and D such as in the case of breast cancers, the second or third order perturbation may be necessary. For this purpose, many studies of higher order perturbation theory have been reported [92–95].

4.5.2. First Order TD-Perturbation Using the TD-DE

Arridge [28] reported the analytical formulations of the first order TD perturbation approach based on the TD-DE for homogeneous infinite media, semi-infinite media and infinite slabs. The paper describes the analytical formula of the kernels of the integrals in Equation (33) and (34), so called the absorption and diffusion kernels, respectively. The kernels in the inversion problems are essentially the Jacobians which have a physical meaning of probability density function, or sensitivity of a measurement by the perturbation, and the images of the TD Jacobians for various geometries are shown. Hebden and Arridge reported experimental results using a breast-like slab phantom to show reconstructed DOT images of inclusions in the phantom by use of the perturbation method in the preceding paper [28,87]. The spatial resolution of scattering inclusions was found to be better than that of absorbing inclusions.

Morin et al. experimentally demonstrated the performance of a TD perturbation approach for detection of cylindrical inclusions embedded in a 20-mm thick scattering slab simulating compressed breasts measuring the TR transmittance in a collinear geometry of the source, inclusion, and detector [88]. They reported that the changes in the transmittance by absorbing perturbations were two-orders of magnitudes stronger than those by scattering perturbations.

Carraresi et al. [89] studied the accuracy of a TD perturbation model to predict the effects of absorbing and scattering inclusions on light propagation. They compared the results of the perturbation model with those of Monte Carlo simulations and phantom experiments for the TR-transmittance through a 40-mm thick scattering slab simulating breasts and verified that the changes in the TR transmittance caused by absorbing perturbations were independent of those by scattering perturbations.

Spinelli et al. [90] employed Padé approximants to extend the application of the first order TD perturbation to large-volume inclusions using a non-linear contrast of the measured TR transmittance as a function of absorbing and scattering perturbations. Their perturbation model and experimental setup were similar to those of Carraresi [89], and the employment of Padé approximants improved the accuracy of the reconstructed images for the large-volume inclusions.

Martelli et al. [91] developed formulations of the TD perturbation model for two- and three-layered media with absorbing perturbations supposing human heads. The eigen-function solutions of the TD-DE for two- and three-layered media were used [96,97], and results of their model agreed well with those of MC simulations, and TD-sensitivity maps were also shown.

4.5.3. Higher Order TD Perturbation Using the TD-DE

Sassaroli et al. proposed a higher order TD perturbation model for absorbing perturbation by introducing a probability distribution function and defining TD mixed-pathlength moments (mixed-moments) [94,95]. The fundamental equation is valid for both the TD-RTE and TD-DE, but analytical solutions for the TD-RTE are not available while those of the TD-DE are available. Also, calculations of the mixed-moments are not easy, but because the contributions of the mixed-moments are negligibly small compared with those of the self-moments, calculations using the self-moments only is found to provide good results. The fourth order perturbation using the Green's functions for the TD-DE and the Padé approximants have shown excellent results for large-volume inclusions with large absorption contrasts to the background.

Wassermann [92] derived general formulations of a higher order TD perturbation model for absorbing perturbations, and the author discussed that higher order perturbation is effective when the characteristic parameter, Λ, is smaller than 3, i.e., $\Lambda = |\delta\mu_a| R_{inc}^2/D_0 < 3$ where R_{inc} is the radius of a spherical inclusion. The paper provided calculation results of the second and third perturbations of the TR transmittance from a 60-mm thick slab having an absorbing inclusion with a diameter of 10 mm and a contrast of $\delta\mu_a/\mu_{a0}$ about unity, and it summarized that the second perturbation is sufficient for accurate predictions.

Grosenick et al. reported explicit formula of higher order perturbations similar to those derived by Wassermann [92,93], not only for absorbing perturbations but also for scattering perturbations up to the third order corrections although calculation results for scattering perturbations were not shown. The authors also refer to the characteristic parameter, Λ, as Wassermann for the criterion for convergence of the Born series for absorbing perturbations [92].

Most of the above reports for higher order perturbations are applicable to absorbing perturbations only, and the development of higher order perturbation models for scattering perturbations is desirable.

4.5.4. TD-Perturbation Using the TD-RTE

For more general treatment of the perturbation theory, Polonsky and Box started from the TD-RTE for the Stokes vector of the radiance, $\mathbf{I} = \{I, Q, U, V\}$, where I, Q, U, and V are the Stokes parameters describing the polarization state [86]. For simplicity of discussions, they first defined the operator of the RTE, $\mathbf{L}$, for $\mathbf{I}(t, \mathbf{r}, \mathbf{n})$ at point $\mathbf{r}$ in the direction $\mathbf{n}$ at time t, and its adjoint operator, $\mathbf{L}^*$, for the Stokes vector of the adjoint radiance, $\mathbf{I}^*(t, \mathbf{r}, \mathbf{n})$, and then expressed an optical measurement of E (such as reflectance or transmittance) as the scalar product of a response function (equivalent to the Green's function) of $\mathbf{R}$ and the radiance of $\mathbf{I}$, $E = <\mathbf{R}^+, \mathbf{I}>$ (superscript + denotes the transpose of the vector). Finally, they derived two basic equations for the perturbation approach as,

$$\delta E \approx - <\mathbf{I}_b^{*+}, \delta\mathbf{L}^*\mathbf{I}_b > \text{ and } \frac{dE}{dp} = - <\mathbf{I}_b^{*+}, \frac{d\mathbf{L}}{dp}\mathbf{I}_b > \tag{39}$$

where subscript b denotes the background medium and p is a parameter such as μ_a and μ_s'. It is very difficult to explicitly give analytical solutions of the TD-RTE for the response function of $\mathbf{R}$, because Equation (39) is expressed implicitly, but the perturbations of Equation (39) are given by the direct and adjoint solutions of the RTE, $\mathbf{I}_b$ and $\mathbf{I}^*_b$, and for the background medium.

4.6. Multi-Layered Media

Most biological organs can be categorized in multi-layered media from an optical point of view. Two typical examples are the human head and the tissues above the skeletal muscle, the former consisting of five layers of scalp (skin), skull, cerebrospinal fluid (CSF), gray matter, and white matter layers, the latter consisting of three layers of skin, fat, and muscle layers. The analytical solutions of the equations describing light propagation in multi-layered media will be very useful to recover the optical properties of each layer from measurements of TR reflectance. No analytical solution for the TD-RTE for multi-layered media has been reported so far, but the analytical solutions for the TD-DE have been obtained as follows.

Kienle et al. [98] reported the analytical solution of TR reflectance for two-layered semi-infinite media based on the TD-DE to recover μ_a and μ_s' of the layers. But the analytical solutions derived using the Fourier-transform approach were not explicitly given in TD. Tualle et al. [99] reported another analytical solution of TR reflectance for two-layered semi-infinite media based on the TD-DE by applying the mirror image method for semi-infinite homogeneous media to two-layered media using distributed sources. Further, Tualle et al. extended Kienle et al. study to multi-layered media and obtained explicitly the asymptotic solution of TR reflectance in the very late time when the asymptotic solution is dependent on μ_a and μ_s' of the deepest layer [98,100]. Liemert and Kienle extended the method for two-layered media to general N-layered media although explicit analytical solutions were not given [98,101]. The TD solutions were obtained using a fast Fourier transform from the FD solutions for N-layered media at many (512 for example) frequencies.

The analytical solutions of TR reflectance using the eigenfunction method were reported in series by Martelli et al. [96,97,102,103]. The geometries of the objects were a two-layered semi-infinite medium [102], two-layered parallelepiped [96], two-layered finite cylinder [97], and three-layered finite cylinder [103]. Using the separation of variables for space ($\mathbf{r}$) and time (t) based on the microscopic Beer–Lambert law, the analytical solution of the TD-DE for a three-layered finite cylinder shown in Figure 14 are obtained as [103],

$$\Phi(\mathbf{r},t) = \sum_{l,n=1}^{\infty} cJ_0(K_l\rho)\sin(K_{lni}z + \gamma_{lni})\sin^*(K_{lni}z_0 + \gamma_{ln1})\exp[-(K_i^2 D_i + \mu_{ai})ct]/N_{ln}^2 \tag{40}$$

where $i = 1, 2, 3$ denote the first, second, and third layer, respectively, J_0 is the 0th order Bessel function, K_l are the roots of $J_0(K_lL) = 0$, and for K_{lni}, γ_{lni}, and N_{ln} refer to the reference [103]. The TR reflectance, $R(\rho, t)$, is obtained from Equation (40) by use of Fick's law at $z = 0$, and the TR partial mean pathlength inside each layer, $<l_i(\rho, t)>$, is given analytically by the following equation,

$$< l_i(\rho,t) >= -\frac{1}{R(\rho,t)}\frac{\partial R(\rho,t)}{\partial \mu_{ai}}. \tag{41}$$

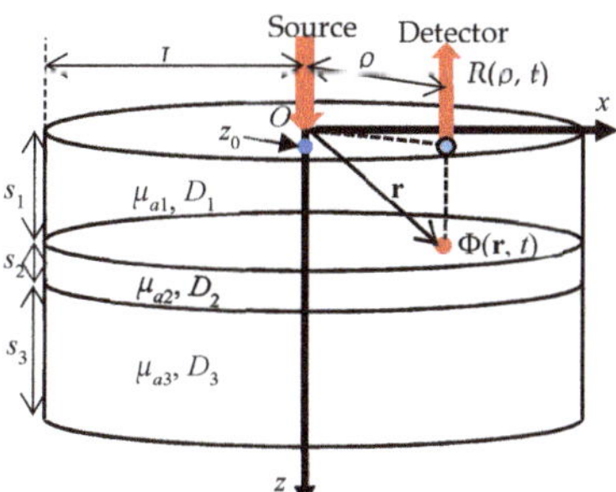

Figure 14. Geometry of a three-layered finite cylinder simulating the human head. Modified from Reference [103].

The results are shown for the human head model having three layers consisting of the combined scalp and skull layer, the CSF layer, and the brain. The mean partial pathlength, given by temporal integration of Equation (41), of the third (brain) layer is also shown to vary from 0.3 mm to about 50 mm with ρ varying from 10 mm to 50 mm, respectively. Software for the two-layered finite cylinders is given in the Reference [32] (pp. 109–125).

Another analytical solution of the TD-DE for finite cylindrical stratified media was reported by Barnett based on the separation of variables [104]. This method was verified by comparing the results for two-layered media with those of Tualle et al. [99], and reportedly the computation was fast enough for application to DOT.

4.7. Advanced Monte Carlo Simulations

Kienle and Patterson proposed a method to produce results of MC simulations for many combinations of μ_a and μ_s by correcting the result of one MC simulation for the reference combination of $\mu_a = 0$ and $\mu_s = \mu_{s0}$ [105]. When the TR reflectance from a semi-infinite homogeneous medium with $\mu_a = 0$ and $\mu_s = \mu_{s0}$ is $R_0(\rho, t)$, the TR reflectance, $R(\rho, t)$, for a medium with μ_a and μ_s is given by,

$$R(\rho,t) = \left(\frac{\mu_s}{\mu_{s0}}\right)^3 R_0(\rho\frac{\mu_s}{\mu_{s0}}, t\frac{\mu_s}{\mu_{s0}})\exp(-\mu_a ct). \tag{42}$$

Many results of $R(\rho, t)$ produced from a single MC simulation were used for estimating μ_a and μ_s' from the measured TR reflectance. A similar MC method was employed by Alterstam et al. [106] where they called the method white Monte Carlo.

Sassaroli et al. [107] proposed another correction method to produce many results from a single Monte Carlo simulation, called perturbation Monte Carlo (pMC) method. At the first step, a standard MC simulation is performed for homogeneous and non-absorbing medium with $\mu_a = 0$ and $\mu_s = \mu_{s0}$, and all the coordinates of scattering positions only for detected photons are stored. The weights of the detected photons are unity at this step. At the second step, another medium with an inhomogeneity is considered with μ_{ai} and μ_{si} for the inhomogeneity and with μ_{a0} and μ_{s0} for the surroundings. Then the corrections of the weights of the detected photons are given by the following two scaling relationships, w_a for absorption and w_s for scattering,

$$w_a = \exp(-\mu_{ai}l_i - \mu_{a0}l_0)w_s = \left(\frac{\mu_{si}}{\mu_{s0}}\right)^{K_i} \exp[-(\mu_{si} - \mu_{s0})l_i] \tag{4-14}$$

where l_i and l_0 are the pathlength of the photons inside and outside of the inhomogeneity, respectively, and K_i is the number of scattering events inside the inhomogeneity. The two scaling relationships were combined into one equation by Hayakawa et al. [108]. By this method the computation time is significantly reduced. Later, Sassaroli proposed a faster perturbation MC method by storing the seed values of random number series only for the detected photons [109]. This can speed up computation by 1000 times under some conditions.

Standard MC methods are inappropriate for inverse processes due to long computation times, but the pMC method can be used in retrieving the optical properties from measured TR reflectance [110], in functional imaging in small animals using time-gated technique [111], and in computing the Jacobians for DOT [112]. However, still faster MC methods are desirable for TD-DOT, and an MC method using a graphics processing unit (GPU) and two parallel random number generators was developed to report 300 times faster calculations than using conventional CPUs [113]. GPU-based MC methods are developed to simulate light propagation in teeth where the optical properties vary in a wide range due to different tissues such as enamel and tubules in dentin [114].

The equivalence between MC simulations and solving the RTE is further studied by Voit et al. [115] to compare the results of MC simulations with those of the Maxwell equation in light propagation with polarization.

4.8. Hybrid RTE and DE Models

Hybrid RTE and DE models have been developed to overcome the limitation of the DE which is not applicable to early time regimes and regions close to the boundary, source, and with low-scattering, and to save the computation loads of calculating the RTE for the whole space and time. In the fields of DOT and fluorescence diffuse optical tomography (FT) for small animals and small organs like human fingers and teeth, the early-time regime plays an important role, and simple application of the DE may be inappropriate. Several hybrid models have been proposed for CW [116], FD [117,118], and TD [119–121]. In the FD version of the coupled RTE-DE model [117], the whole computation domain is divided into two subdomains of near-field regions using the RTE and far-field regions using the DE, and the continuity of the intensity and its derivatives at the interface between the two subdomains are given for connecting the solutions of the RTE and DE. In the FD version of the overlapped or buffered RTE-DE model [118], the subdomains overlap in an artificially defined buffer zone where smooth transition from the RTE to the DE is achieved. One example of the overlapped RTE-DE model for TD was proposed by Fujii et al. [118]. In their model, the RTE and DE regions are separated spatially and temporally by using a crossover length for spatial separation and a crossover time for temporal separation, succeeding in describing light propagation accurately and reducing computational load by a quarter when compared with full computation of the RTE. Further development of hybrid RTE-DE models for TD is expected in the future.

4.9. Anisotropic Light Propagation in TD

Anisotropic light propagation in biological tissues was experimentally observed such as in human skin by Nickell et al. [122]. Heino et al. [123] investigated anisotropic light propagation in the TD-RTE with the scattering phase function dependent on both the polar and azimuthal angles and derived the anisotropic TD-DE with the diffusion tensor instead of the diffusion coefficient. Kienle et al. [20] also studied the anisotropic light propagation caused by structural anisotropic media using MC simulations and the TD-DE with the diffusion tensor, and the analytical expressions of reflectance and transmission from a slab were provided.

5. Studies toward Clinical Applications of TD-NIRS

In this section, fundamental studies toward clinical applications of TD-NIRS including features of TD-light propagation, measurements of the optical properties, TD-DOT, and fluorescence-DOT are reviewed.

5.1. Features of TD-Light Propagation Including Penetration Depth, Optical Pathlength, etc.

Features of TD-light propagation in tissue-like media have extensively been studied, and some of the studies are reviewed below.

5.1.1. Light Propagation Based on the Microscopic Beer–Lambert Law

In 1991, Hasegawa et al. [124] numerically studied the characteristics of TR transmission through homogeneous slabs using MC simulations under the microscopic Beer–Lambert law including the time ranges where the diffusion approximation does not hold. Nomura et al. [125] reported that the microscopic Beer–Lambert law was valid not only for tissue-like phantoms but also for living tissues such as rat brain and thigh muscle using TD measurements. Based on the microscopic Beer–Lambert law, Tsuchiya [9] derived simple equations describing light propagation by use of the photon path distribution which expresses the distribution of the pathlengths inside all voxels in the whole volume.

5.1.2. Mean-TOF, Partial Pathlength, and Sensitivity for the Head Model

Okada et al. [126] investigated light propagation in models of the adult head theoretically using MC simulations as well as solving the DE and experimentally using measurements of TR reflectance, $R(t)$, from phantoms. Four different models of adult brain were constructed to see the effects of the SD distance on the mean-TOF, $<t>$, and partial pathlength of the i-th layer, $<l_i>$, and the sensitivity distribution for CW on the SD distance, and to see the effect of the CSF layer on $<t>$ and $<l_i>$. More realistic head models were used to extend the study [127].

Firbank et al. [128] theoretically studied light propagation in a realistic adult head model constructed from MRI data using a hybrid DE/radiosity technique and a MC simulation to obtain $R(t)$, $<t>$, their TR-sensitivities, $H_R(t)$ and $H_{<t>}(t)$, and $<l_i(t)>$. They found that $H_{<t>}(t)$ has larger values than $H_R(t)$ in deeper regions.

Steinbrink et al. [129] studied $R(t)$ from a multi-layered model to estimate the changes in $R(t)$ due to absorption changes in the layers from $< l_i(t) >$ calculated by MC simulations. Their results were verified by experiments using two- and three-layered phantoms and by in vivo experiments.

5.1.3. Use of Null SD Distance

Del Bianco et al. [130] studied $<l_i(t)>$ and penetration depth, $<z(\rho, t)>$, for semi-infinite two-layered media using the analytical solution of the TD-DE. Based on this study, Torricelli et al. [131] proposed the usage of null SD distance for improving contrast and resolution of diffuse optical imaging from TR reflectance measurements. They concluded that the null SD distance for TR reflectance measurement provides four advantages over the conventional SD distances of a few centimeters, i.e., (i) stronger optical signals, (ii) deeper penetration depths, (iii) larger contrast for absorbing inclusions, and (iv) higher localization of the inclusions, but also discussed a problem of too strong signals of early photons. They argued several ways for solving the problem: (i) gating the detector to measure photons arriving at 200 ps and later, (ii) using SPADs which are not damaged by the burst of initial photons, and (iii) the use of non-null but small SD distance, while keeping the advantages of a null SD distance. The method of the null SD distance for TR reflectance measurements was further studied for two-layered media by Spinnelli et al. [132] using MC simulations and the TD-DE. Later, proof-of principle tests of the null SD distance technique were reported using fast-gated SPAD as well as non-contact probes showing good agreements of the depth sensitivity and spatial resolution between the phantom experiments and MC simulations [133]. These results indicated the feasibility and potentiality of the null SD distance technique for applications to non-contact and high-density diffuse TR reflectance measurements.

5.1.4. Measurement of Mean Pathlength

The mean (CW) pathlengths, l, or the DPFs, can be experimentally obtained in vivo by use of TR reflectance measurements. Zhao et al. reported measured DPF maps of the forehead, somatosensory motor, and occipital regions of 11 adults using a TR system with an SD distance of 30 mm at three wavelengths around 800 nm [134]. The measured DPFs varied from about 6 to 9 depending on the regions and wavelength. These DPF data will be useful for quantitative monitoring of the hemodynamic changes occurring in adult heads. Bonnéry et al. studied the changes in the DPF of the foreheads of adult heads with aging and found that the upper layer including the scalp, skull, and CSF was thicker for older subjects than younger ones [135].

5.2. Measuring Optical Properties

The optical properties of tissue can be determined by various methods in CW, FD and TD, but TD data provide the richest information as a matter of course. Many papers have reported the optical properties of not only homogeneous but also layered tissues, determined by use of TD data. Once the optical properties of multi-layered tissue such as human heads are determined, it becomes possible to estimate the pathlengths and sensitivities inside the heads which are very important parameters for investigating brain activities using NIRS. In the following, TD determination of the optical properties of homogeneous semi-infinite media is described and reviewed first, then those of multi-layered media are reviewed.

5.2.1. Homogenous Semi-Infinite Media or Infinite Slab

Determination of μ_a and μ_s' in homogeneous semi-infinite media by TD techniques is essentially based on fitting analytical or numerical solutions of the TD-DE to measured TR reflectances. The TR-reflectance, $R(\rho, t)$, is expressed by Equation (15) for the TD-DE under the extrapolated boundary condition, and for simplicity the zero-boundary condition is sometimes employed as Equation (28). For the zero-boundary condition, the following two equations are derived [18],

$$\lim_{t \to \infty} \frac{d}{dt} \ln R(\rho, T) = -\mu_a c, \ \mu_s' = \frac{1}{3\rho^2}(4\mu_a c^2 t_{\max}^2 + 10 c t_{\max}) - \mu_a, \tag{43}$$

where t_{max} is the time of maximum $R(\rho, t)$. From these equations, it is possible to determine μ_a and μ_s' of semi-infinite homogeneous media from the measured $R(\rho, t)$. Especially, it should be noted that μ_a is determined from the slope of $\ln R(\rho, t)$ at very late time.

This feature of $R(\rho, t)$ is also stated by Jacques [136]. But $R(\rho, t)$ measured at very late times are usually associated with noises, and it is difficult to determine t_{max} due to the IRF. Therefore, using Equation (43) is not appropriate for accurate determination of μ_a and μ_s', and fittings of the IRF convoluted analytical solutions of Equation (15) or Equation (28) to measured $R(\rho, t)$ are often used to recover μ_a and μ_s'. Note that deconvolution of measured $R(\rho, t)$ by the IRF enhances measurement noises and instead the analytical solutions are convoluted by the IRF.

Determination of μ_a and μ_s' in homogeneous infinite slabs by TR transmittance measurements is conducted using the analytical equation for TR transmittance, which is derived similarly to that for TR reflectance using the mirror image source method as Equation (15) [18,19].

Sassaroli et al. [137] made in vivo measurements of μ_a and μ_s' of a piglet brain using a TD system at three wavelengths (759, 794, 824 nm) by the method stated above. To separate the contributions of different head layers, the measured $R(\rho, t)$ were acquired at the surfaces of skin, skull, dura mater and brain, step by step. The values of ρ were chosen to assure the mean penetration depth within each layer. Measured μ_a and μ_s' were compared with the other in vivo results reported in literatures, and the differences were discussed.

Cornelli et al. [138] estimated μ_a and μ_s' of human foreheads in the wavelength range from 700 nm to 1000 nm. They found that the estimated μ_a and μ_s' were close to those of the superficial (scalp and skull) layers by additional MC simulations for four-layered media simulating the structure of the human head.

Using the optical mammography system shown in Figure 11c, which employed seven wavelengths of 635 nm, 685 nm, 785 nm, 905 nm, 930 nm, 975 nm, and 1060 nm, Taroni et al. conducted a pilot study for optical estimates of tissue components to differentiate malignant from benign breasts from the data of patients with 45 malignant and 39 benign lesions [54]. They measured TR reflectances in vivo through compressed breasts and obtained images of the differences in μ_a from the background at the seven wavelengths, $\Delta\mu_a(\lambda)$, using a time-gated perturbation analysis based on the microscopic Beer–Lambert law under the assumption that the compressed breasts were homogeneous slabs [54,139]. The $\Delta\mu_a(\lambda)$ images were converted to the changes in the contents of oxy-Hb, deoxy-Hb, water, lipid, and collagen using Equation (25), and from statistical analyses they concluded that the collagen content was the most important parameter for discriminating malignant and benign lesions. Ohmae et al. evaluated the performance of the six-wavelength TD-NIRS system using phantoms varying the contents of water, lipid, and an absorber [68]. The performance was confirmed with the measurements using a magnetic resonance imaging system.

Guggenheim et al. [140] recovered the spectra of $\mu_a(\lambda)$ and $\mu_s'(\lambda)$ of irregularly-shaped homogeneous media from TD-NIRS measurements. The analytical solutions of the TD-DE are usually not applicable to irregularly shaped media, and the finite element method (FEM) was employed to solve the TD-DE. The FEM solutions for non-absorbing medium were multiplied by $\exp(-\mu_a ct)$ to incorporate the attenuation by absorption.

5.2.2. Multi-Layered Media

Kienle and Glanzman [141] used TR reflectance measurements to determine the optical properties of human forearms being assumed as a two-layered media consisting of skin-fat (top: subscript 1) and muscle (bottom: subscript 2) layers. The analytical solutions of the TD-DE for two-layered semi-infinite media were fitted to $R(\rho, t)$ measured in vivo at two values of ρ with the wavelength of 830 nm to determine $\mu_{a1}, \mu_{a2}, \mu_{s1}'$, and μ_{s2}' with known thicknesses of the top layer (denoted by s_1) [98]. They concluded that μ_{a2} can be determined accurately even if s_1 is known only approximately. Gagnon et al. used the same analytical solutions for two-layered media as Kienle and Glanzman to estimate the intra- and extra-cerebral hemoglobin concentrations [141,142]. The upper layer included the scalp, skull, and CSF, and the lower layer was the brain. They observed noticeable inter-subject variations in the hemoglobin concentrations and constant oxygen saturation of the cerebral hemoglobin.

Martelli et al. used analytical solutions for the TD-DE using the eigenfunction method for estimating the optical properties of two-layered media [102,143,144]. The analytical solutions were fitted to the TR reflectances provided by a MC simulation or phantom and in vivo experiments to estimate $\mu_{a1}, \mu_{a2}, \mu_{s1}', \mu_{s2}'$, and s_1 accurately [143,144]. The errors of the estimated μ_{a1}, μ_{a2}, and μ_{s1}' were small while those of μ_{s2}' were large.

Sato et al. [145] proposed a method of determining μ_{a2} in two-layered semi-infinite media based on the microscopic Beer–Lambert law. Extending this method, a simple algorithm was developed to recover μ_{a1} and μ_{a2} in a two-layered medium from the TR reflectances measured at two values of ρ and was verified by numerical simulations and in vivo experiments using human foreheads [146,147].

Hoshi et al. [148] reported a method to determine the optical properties of a four-layered human head model using TR reflectance measurements and MC simulations. They used a TRS-10 (Hamamatsu Photonics) to measure the TR reflectances from various head portions of eleven healthy humans with a fixed ρ of 30 mm. For MC simulations, head models based on MRI scans were constructed as four-layered slabs consisting of the scalp, skull, cerebrospinal fluid (CSF), and brain. By a trial-and-error method, they changed the values of μ_a and μ_s' of the four layers to fit the simulated TR reflectances to the measured ones and estimated the partial pathlengths and sensitivities of the four layers. As of the results, they found the followings: (i) the total pathlength ranging from 100 mm to 250 mm was closely related to the thickness of the scalp ranging from 2 mm to 10 mm; (ii) the partial pathlength of the brain ranged from 3% to 13% of the total pathlength; (iii) the brain tissue deeper than 25 mm from the head surface were hardly detected by the near-infrared light; (iv) the scalp tissue at the depth of 4 mm had the highest sensitivity inside the heads; and (v) most of the signals attributed to the brain came from the superficial layer with the thickness of 1 to 2 mm from the brain surface. These findings are important for functional NIRS and could not have been obtained without TD-measurements.

Jäger and Kienle [149] employed a neural network (NN) to estimate μ_a of the brain ($\mu_{a,brain}$) from TR reflectances in the case of five-layered media. Training data of the TR reflectances with a single value of ρ were generated by the analytical solutions of the TD-DE for five-layered media as well as MC simulations which can model the CSF layer more accurately than the TD-DE [101]. One-hundred to 500 training data with different noises were input into the NN, and the NN estimated $\mu_{a,brain}$ under the condition that the optical properties of the five layers except $\mu_{a,brain}$ and the thicknesses of the upper four layers are known with some uncertainties. Resultantly, $\mu_{a,brain}$ was estimated with errors less than 5% even if there are uncertainties of 20% in the other optical properties and the thicknesses.

Using the eigen-function analytical solutions of the TD-DE [97], Martelli et al. employed optimal estimation method of a Bayesian approach to incorporate prior information in the process of recovering the optical properties of the two-layered media [150,151]. The method was verified by recovering of $\mu_{a1}, \mu_{a2}, \mu_{s1}'$, and μ_{s2}' from the TR reflectances provided by MC simulations and phantom experiments with better performances than the standard non-linear least-squares methods. The thickness of the top layer, s_1, and the time origin of the TR reflectance, t_0, were also included in the unknowns. The values of μ_{a2} were very accurately estimated because the TR partial pathlength of the bottom layer, $l_2(t)$, becomes much longer than that of the top layer, $l_1(t)$, after 2 ns in the decaying period. Even if the thickness of the top layer, s_1, is unknown the values of μ_{a2} are obtained with errors less than 10%, and this method will be well suited for applications to human heads where the effect of the CSF is mainly to decrease μ_{s2}'.

Zucchelli et al. and Re et al. utilized the analytical formulations of the partial pathlength of Equation (24) with the analytical solutions of the TD-DE for two-layered media to estimate perturbations of μ_{a2} [32] (pp. 109–125) [152,153]. When the spectroscopic TR reflectance, $R(t, \lambda)$, is divided into many time gates, the ratio of the perturbed and unperturbed $R(t, \lambda)$ averaged over the g-th time-gate is derived from Equation (24) as,

$$\ln \frac{R_g(\lambda)}{R_{0g}(\lambda)} = -\sum_{i=1}^{N} \Delta\mu_{a,i}(\lambda) l_{g,i}(\lambda) \tag{44}$$

where $R_{0g}(\lambda)$ and $R_g(\lambda)$ are the unperturbed and perturbed $R(t, \lambda)$ averaged over the g-th time-gate, respectively, $\Delta\mu_{a,i}(\lambda)$ is the perturbation of $\mu_{a,i}$ in the i-th layer, N is the number of layers, and $l_{g,i}(\lambda)$ is the partial pathlength of the i-th layer averaged over the g-th time-gate, which is given by the analytical solution of the TD-DE in case of two-layered media ($N = 2$). Then a system of linear equations of Equation (44) with the number of the time-gates is constructed and $\Delta\mu_{a,i}(\lambda)$ is obtained by inverting the system of the linear equations.

The optical properties of the human heads were estimated by in vivo measurements using multi-wavelength TD spectroscopy system at a group of laboratories [154]. The analytical solution of the TD-DE for two-layered media was fitted to the measured $R(t, \lambda)$ [103], and five parameters, μ_{a1}, μ_{a2}, $\mu_{s1}{}'$, $\mu_{s2}{}'$, and s_1 were estimated. Although the inter-laboratory differences are rather large, a definite trend was found after comparisons with the results using $R(t, \lambda)$ generated by MC simulations.

Liebert et al. estimated the changes in μ_a of the brain and superficial layers after the intravenous administration of indocyanine-green from multi-distance TR reflectance measurements and MC simulations for a two-layered model [155]. In the estimation process, the information of the partial pathlengths of the two layers was important for calculating the sensitivities to absorption changes in the layers.

5.3. Time-Domain Diffuse Optical Tomography (TD-DOT)

Diffuse optical tomography (DOT) has been investigated since the early 1990s in CW, FD, and TD. Many investigations of DOT have been conducted in CW and FD, but the advantages of TD over CW and FD attracted researchers. First TD-DOT was reported by Benaron and Stevenson [156] in 1993. They demonstrated the advantage of TD-DOT over CW-DOT by showing a TD-DOT image of a mouse. The TD-DOT image using the early photons revealed the internal organs of a mouse while the CW-DOT image was unable to distinguish the internal organs at all. In the late 1990s, several groups developed more sophisticated multi-channel TD-DOT systems for studies of brain, breast, and muscle imaging [45,49,157].

In the following, various methods for TD-DOT are reviewed first, and the applications of TD-DOT for in vivo studies of brain, breast, and muscle imaging are reviewed.

5.3.1. General Concept of TD-DOT

Some studies employed the reconstruction algorithm for X-ray CT to reconstruct the absorption images using the temporally extrapolated absorbance method [158,159] or using the time-gated measurements at early times [160–163]. These methods aimed to detect the ballistic-like components of transmitted ultra-short pulse light, but their applications were limited to small-sized objects because transmitted light was unmeasurable with high signal-to-noise ratios for objects thicker than a few centimeters.

Standard algorithms of TD-DOT for thick or large objects are categorized as a model-based iterative image reconstruction (MOBIIR) which employs a forward model of light propagation and a non-linear iterative inversion process. Arridge et al. proposed an algorithm of MOBIIR based on the DE [164]. They derived the sensitivities (Jacobians) for the mean time-of-flight, <t>, as the data type, from the perturbation theory, and formulated the inverse process to reconstruct μ_a and μ_s' images. The software developed in this framework was named as "TOAST" and its revised version "TOAST++" was presented as an open-source software [165]. A general algorithm of MOBIIR is summarized below.

Figure 15 shows a standard process of MOBIIR mainly consisting of measurements and an inverse process. The measured TD data (TOF distributions), $\Gamma(t)$, can be used fully or converted to featured TD data. Several types of featured TD data characterizing the TOF distributions are often used as the following [4],

$$\text{Time} - \text{gated intensity}: \ M_{DC(T_a,T_b)} = \int_{T_a}^{T_b} \Gamma(t)dt, \tag{45}$$

$$n - \text{th temporal moment}: \ M_n = \int_0^\infty t^n \Gamma(t)dt, \tag{46}$$

$$\text{Mellin–Laplace transform}: \ M_{ML}(n, p) = \int_0^\infty t^n e^{-pt}\Gamma(t)dt, \tag{47}$$

Amplitude and phase of frequency-domain data:

$$M_{amp} = |\hat{\Gamma}(\omega)| = \left| \int_0^\infty \Gamma(t)e^{-i\omega t}dt \right|, \ M_\theta = \text{arg}\hat{\Gamma}(\omega) = \text{arg}\left(\int_0^\infty \Gamma(t)e^{-i\omega t}dt \right). \tag{48}$$

The time-gated intensity with $T_a = 0$ and $T_b = \infty$ corresponds to the CW intensity, and multiple time-gated intensities with a small time-step of $T_b - T_a = \Delta T$ dividing the whole TOF distributions are equivalent to the use of full TOF distributions. The n-th temporal moment with $n = 0$ is also identical to the CW intensity, and those with $n = 1, 2, 3$ are the mean, $M_1 = <t>$, variance, M_2, and skew, M_3, of the TOF distribution, respectively. The Mellin–Laplace transform with $n = 0$ or $p = 0$ corresponds to the simple Laplace or Mellin transform, respectively. Equation (48) describes the transformation of the TD data to the FD data at a frequency of ω.

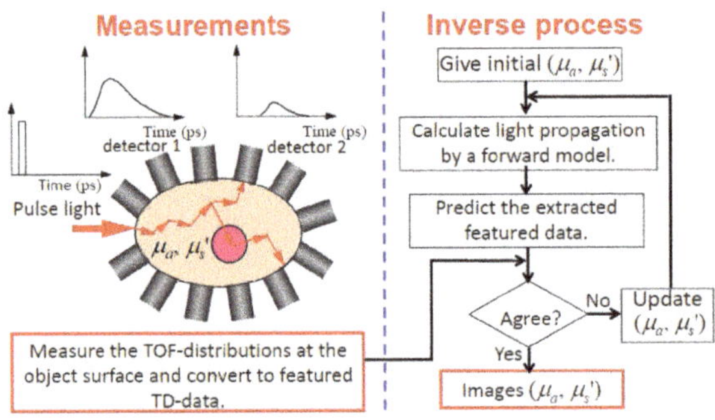

Figure 15. Standard process of MOBIIR for TD-DOT.

Several updating methods are available, and Newton-type non-linear reconstruction is a standard method. The objective function, Ψ, defined below is minimized,

$$\Psi = \|\mathbf{\Gamma}_{cal} - \mathbf{\Gamma}_{mes}\|^2 + \lambda\|\mathbf{x}\|^2, \tag{49}$$

where $\mathbf{\Gamma}_{cal}$ and $\mathbf{\Gamma}_{mes}$ are the vectors of the measured and calculated (featured) TD data with the components of $\mathbf{\Gamma}_{cal}(\mathbf{r}_{sm}, \mathbf{r}_{dm} : \mu_a(\mathbf{r}_n), D(\mathbf{r}_n))$, and $\mathbf{\Gamma}_{mes}(\mathbf{r}_{sm}, \mathbf{r}_{dm})$, respectively, the index m (=1, ..., M) specifies the TD data number with M being the number of the TD data, the index n (=1, ..., N) specifies the voxel number with N being the number of voxels in the medium, respectively, $\mathbf{x} = [\mu_a(\mathbf{r}_n), D(\mathbf{r}_n)]^T$ is a vector consisting of the optical properties at the n-th voxel, and λ is the regularization parameter. Here, the measurement errors are neglected for simplicity. The Levenberg–Marquardt algorithm, which is a standard method in the Newton-type minimization process, leads to the following inversion matrix equation for the updates at the k-th iteration, $\delta\mathbf{x}^{(k)}$ [69,166],

$$\mathbf{W}^{(k)}\delta\mathbf{x}^{(k)} = \Delta\mathbf{\Gamma}^{(k)} = \mathbf{\Gamma}_{cal}^{(k)} - \mathbf{\Gamma}_{mes}, \tag{50}$$

where $\mathbf{W}$ is the sensitivity (weight or Jacobian) matrix of the featured TD data to the changes in the optical properties, $\delta\mathbf{x}$. If the time-gated intensity is chosen for the featured TD data, $\mathbf{W}$ will be given by equations similar to Equations (17) and (18). Then the optical properties are updated by use of the Tikhonv regularization as follows,

$$\delta\mathbf{x}^{(k)} = (\mathbf{W}^{(k)T}\mathbf{W}^{(k)} + \lambda\mathbf{I})^{-1}\mathbf{W}^{(k)T}\Delta\mathbf{\Gamma}^{(k)}, \tag{51}$$

$$\mathbf{x}^{(k+1)} = \mathbf{x}^{(k)} + \delta\mathbf{x}^{(k)}. \tag{52}$$

Some other various schemes have been employed for image reconstruction, and please refer to References [4,167,168] for examples.

Schweiger and Arridge evaluated qualities of the reconstructed images with varying featured TD data such as the CW intensity $M_{DC(0,\infty)} = E$, first temporal moment $M_1 = <t>$, 3rd central moment c_3, normalized Laplace transform $M_{ML}(0,0.001)/E$, normalized Mellin–Laplace transform set $M_{ML}(1,0.01)/E + M_{ML}(3,0.001)/E$, and combination of $c_3 + M_{ML}(0,0.001)/E$ [169]. Here, $c_n =$

$E^{-1} \int_0^\infty (t- <t>)^n \Gamma(t)dt$. Among these six featured TD data, the combination of $c_3 + M_{ML}(0,0.001)/E$ provided the best reconstruction results.

Gao et al. proved that using full TOF distributions provides better images over the featured TD data, although the computation cost was as high as two orders of magnitude of that using featured TD data such as E (CW), $<t>$, $<t> + c_2$, and $<t> + c_2 + c_3$ even for 2D reconstruction [170]. Figure 16 illustrates that the μ_a image reconstructed using the full TOF distributions shows qualities better than those using the featured TD data of E and $<t>$.

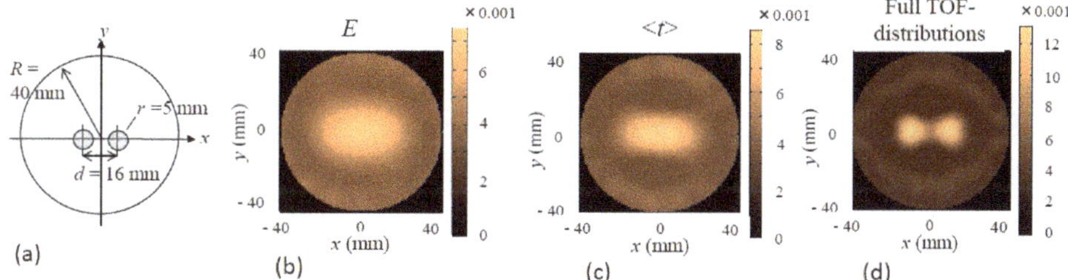

Figure 16. Effects of the type of featured data on the quality of the reconstructed DOT images of μ_a. (**a**) Geometry of a 2D object; reconstructed DOT images using (**b**) E (CW), (**c**) $<t>$, and (**d**) full TOF-distributions normalized by E. The 2D object has a diameter of 80 mm and contains two small absorbers with a diameter of 10 mm and a separation of 16 mm. Modified from Reference [170].

5.3.2. Modified Generalized Pulse Spectrum Technique for TD-DOT

Gao et al. [171] proposed a modified generalized pulse spectrum technique (GPST) for image reconstruction of TD-DOT. The modified GPST is based on the Laplace transformed RTE and employs a ratio of the Laplace transformed signals at two Laplace parameters (denoted p_1 and p_2), $R = \Gamma(p_2)/\Gamma(p_1)$, as the featured TD data. Then, with some assumptions, the sensitivities of R to the changes both in μ_a and in μ_s' are expressed using the product of the Green's functions of the Laplace transformed fluence rate and flux as follows,

$$W_a^{(R)}(\mathbf{r}_d, \mathbf{r}_s, t) = \int_{V_p} G^{(R)}(r_d, r\prime, p)G(r\prime, r_s, p)dr\prime, \tag{53}$$

$$W_s^{(R)}(\mathbf{r}_d, \mathbf{r}_s, t) = \frac{\mu_a + p}{\mu_s\prime} W_a^{(R)}(\mathbf{r}_d, \mathbf{r}_s, t). \tag{54}$$

Note that the sensitivity to change in μ_s', $W_s^{(R)}(\mathbf{r}_d, \mathbf{r}_s, t)$, is easily calculated while the sensitivities of other featured TD data to the change in μ_s' are usually expressed by the spatial gradients of both the Green's functions of the fluence rate and flux as Equation (18), which makes computation ineffective and leads to crosstalk between μ_a and μ_s' images. This difference in calculating the sensitivity to the change in μ_s' is the biggest advantage of the modified GPST over the other featured TD data approaches. In addition, taking positive and negative Laplace parameters for p_1 and p_2 is interpreted as weighting the early and late times in the TOF distributions, thus covering the key features of the TOF distributions. With these advantages, the modified GPST makes the simultaneous reconstruction of μ_a and μ_s' images possible with less crosstalk between them. μ_s' images provide anatomical information while μ_a images provide physiological information. The modified GPST was successfully extended from a 2D case to a semi-3D case with numerical simulations and phantom experiments [171–173].

A numerical simulation and phantom experimental studies using the modified GPST were performed for imaging brain activation from TR reflectance measurements in a two-layered model [174]. Comparisons of the reconstructed images among various featured data types such as E, $<t>$, $<t> + c2$, $E + <t>$, modified GPST and full TOF distributions were made, and it was found that the full TOF distributions provided the best images at the cost of long computation times [175].

5.3.3. Other Techniques for TD-DOT

Lyubimov et al. [176] applied the concept of the photon average trajectory to image reconstruction algorithm of TD-DOT to speed up computation.

Hervé et al., Puszka et al., and Puszka et al. studied the performances of TD-DOT using the Mellin–Laplace transform with parameters of n and p in Equation (47) by numerical simulations and phantom experiments [177–179]. TD-DOT images reconstructed from the featured TD data using the Mellin–Laplace transform were found to be more robust to measurement noises than those using <t>, and the larger the value of n, the deeper the sensitivities extended. Also, the feasibility of short SD distances of 10 mm and 15 mm was studied for application of the null SD distance technique using an experimental setup employing a fast-gated SPAD [133,179].

For saving computation time and memory, Naser and Deen [180] developed a recursive equation to calculate the sensitivity (Jacobian) at a specific time step from the calculated fluence rates at all the previous time steps.

5.3.4. Brain Imaging

A TD-DOT system developed by Benaron's group was applied to neonatal heads to reconstruct 2D-DOT images of hemoglobin saturation to reveal the existence of hemorrhages in neonatal brains [181]. This system was also applied to adult heads to reveal the focal areas of neuronal activity in the brains [182]. The DOT images were reconstructed using a unique and heuristic curvilinear back-projection algorithm which was very fast in computation at the cost of a low accuracy. Their TD-DOT system, built in the early 1990s, required about 6 h to acquire one dataset for one image.

Hebden et al. reported 3D-DOT images of hemodynamics in premature infant brains with hemorrhage using MONSTIR and custom-made helmets with optodes [49,183]. They employed a non-linear inversion algorithm for image reconstruction (TOAST) [164]. The reconstructed images of $C_{total-HB}$ and S_{O2} indicated the existence of hemorrhage with physiologically reasonable values. Hebden et al. also obtained DOT images of ventilated infant brains to show the hemodynamic responses to changes in the oxygen and carbon dioxide partial pressures in the ventilating air [184]. The changes in $C_{total-HB}$ and S_{O2} revealed by the DOT images were in qualitative agreement with physiologically expected changes.

Ueda et al. used a 16-channel multi-wavelength reflectance TD-DOT system to obtain DOT images of human brain activity using a reconstruction algorithm based on the microscopic Beer–Lambert law [9,50]. The algorithm requires calculation of the photon path distribution only for a non-absorbing medium, and the changes in μ_a are easily reconstructed. Two sets of optodes were attached onto the right and left of the forehead of a subject performing a video game task. The DOT images of ΔC_{oxy-HB}, $\Delta C_{deoxy-HB}$, and $\Delta C_{total-HB}$ overlaid on the MR images of his brain clearly showed the localized activity of the prefrontal cortex as shown in Figure 17, although the hemodynamic changes appeared in the dura mater due to localization errors.

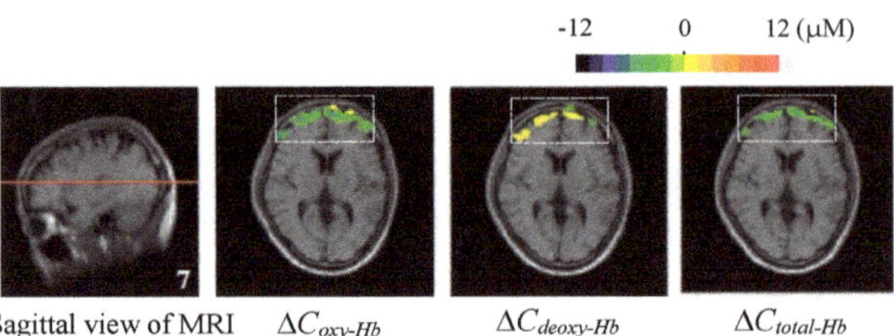

Figure 17. Hemodynamic changes in the prefrontal cortex observed by TD-DOT images during a video game task. Modified from Reference [50]. Copyright (c) 2005 The Japan Society of Applied Physics.

Using MONSTIR [49], Gibson et al. obtained 3D-DOT images showing the response of a neonate motor cortex to the task of arm movement using a linear algorithm for image reconstruction [185]. It took about 2 h to acquire one dataset for one image, which is too long to obtain images of fast neuronal responses.

Ueno et al. obtained DOT images of ventilated premature infant brains with the help of reducing the artifacts induced by the relative movements of the optodes against the scalp surface [46,48]. Fukuzawa et al. improved the image quality by introducing the coupling coefficients of the optodes as the unknowns in the reconstruction algorithm [46,47,186]. Although the quality of these DOT images for infant brains still needs to be improved for clinical use, these results demonstrate the feasibility of the whole-head imaging of neonates by taking advantage of their small head size and skull thickness.

Imaging of neuronal activities by NIRS technology have recently been greatly advanced using the high-density reflectance CW-domain DOT having a data acquisition speed of 12 frames/s which is fast enough to visualize the hemodynamic response of neuronal activities [187]. Current TD-DOT systems are too slow for that purpose, and advancements in hardware for TD-DOT are highly desired for further applications in neuronal imaging.

5.3.5. Breast Imaging

Breasts are recognized to have smaller μ_a and μ_s' than other tissues so that transmittance measurements are available. Ntziachristos et al. [188] developed a transmittance TD-DOT system to image the optical properties of breast phantoms and reported the potentiality of TD-DOT for breast tumor diagnosis. The same group used the TD system with an exogenous agent, indocyanine green (ICG), for image enhancement in in vivo measurements on patients [189]. They obtained better localization of target tumors, observed the differences in the ICG distributions among malignant tumors, benign tumors, and healthy tissues, and compared the DOT images with the concurrently acquired gadolinium-enhanced MRI images. However, the use of exogenous agents for frequent measurements may cause complications for patients and may thus be avoided.

Intes et al. [190] reported a four-wavelength (760, 780, 830, and 850 nm) TD-DOT system for breast tumor measurement using a commercially available instrument. Tomographic images reconstructed by a non-linear inversion algorithm showed the distributions of C_{oxy-HB}, $C_{deoxy-HB}$, water content, and lipid content in compressed breasts from 49 patients, and differences in the blood content and water fraction between malignant and benign tumors were found.

For the purposes of screening of breast cancers and of evaluating the effectiveness of chemotherapy for breast cancers, Ueda et al. [52] used a 48-channel TD-DOT system. In the DOT images of the breasts, cancer tissues were imaged with higher absorption than healthy surrounding tissues. In the case of breast cancers which had positively responded to chemotherapy, μ_a of the tumor after chemotherapy drastically decreased by 34% from 0.0076 mm^{-1} before chemotherapy, indicating the effectiveness of quantitative evaluation of chemotherapy. Yoshimoto et al. [53] used a 12-channel TD-NIRS system with a hand-held probe for use at bed side and indicated the usefulness of the TD-DOT system for evaluation of the effectiveness of chemotherapy to breast cancers.

Enfield et al. [191] investigated the response to hormone therapy of breast cancers using a 3D TD-DOT system. They found that the images of $\Delta C_{total-HB}$ strongly correlated with the clinical assessment of response to hormone treatment of breast cancers.

Zhang et al. [192] demonstrated the feasibility of improving the breast tumor diagnosis using combined TD-DOT and TD-FT. They used two wavelengths of 780 nm and 830 nm for the source and measured the TOF-distributions at 32 sites around cylindrical phantoms. The modified GPST was employed to reconstruct the images of (μ_a, μ_s') and (y_f, τ_f) where y_f and τ_f are the fluorescence yield and lifetime, respectively. The images of (μ_a, μ_s') and (y_f, τ_f) were reconstructed independently with a reasonable quantitativeness, but when the images of (y_f, τ_f) were used to provide the locations of targets as prior information for reconstructing (μ_a, μ_s') images (fluorescence guided DOT), the quantitativeness of the (μ_a, μ_s') images was highly improved; for example, the ratios of the reconstructed to correct values of (μ_a, μ_s') were (23.5%, 31.0%) for separate reconstruction and (76.6%, 86.2%) for fluorescence guided reconstruction.

5.3.6. Muscle Imaging

For muscle imaging, Hillman et al. reported DOT images of μ_a and μ_s' inside a human adult forearm during hand grip exercises with transmittance measurements using MONSTIR [49,193]. The DOT images at wavelengths of 790 and 820 nm showed the responses of μ_a and μ_s' to the exercises.

Using another multi-channel TD-DOT system [45], Zhao et al. reconstructed DOT images of μ_a and μ_s' in human lower legs and forearms [47,194]. For the case of the forearms, DOT images of μ_a, μ_s', $\Delta C_{oxy\text{-}HB}$, $\Delta C_{deoxy\text{-}HB}$, and $\Delta C_{total\text{-}HB}$ in human forearms during excises were obtained as shown in Figure 18.

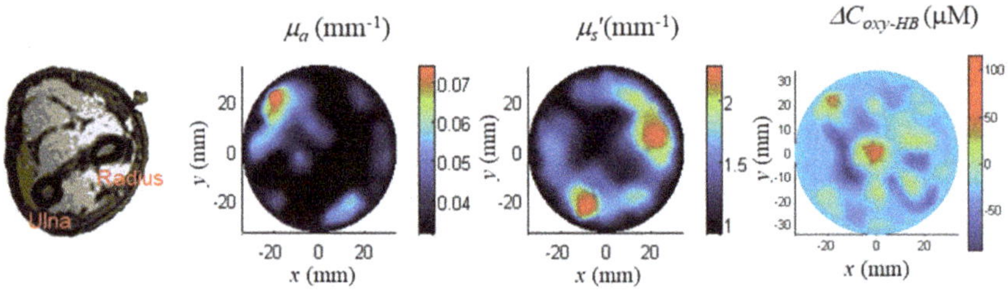

Figure 18. TD-DOT images of the human forearm. (**Left**) MRI of the forearm showing the ulna and radius. (**Center Two**) Reconstructed μ_a and μ_s' images for 759 nm at rest state. μ_s' image reflects the positions of the two bones. (**Right**) Change in $C_{oxy\text{-}HB}$ from rest to task states. Red areas indicate the positions of thick blood vessels while deep-blue areas indicate muscles. Modified from Reference [47].

5.4. Time-Domain Fluorescence Diffuse Optical Spectroscopy (TD-FS) and Tomography (TD-FT)

5.4.1. Fundamental Equations for TD-FS and TD-FT

Fluorescence diffuse optical spectroscopy (FS) and tomography (FT) have been investigated as one of the modalities for molecular imaging [195,196], and TD-FT is an extension of TD-DOT to the coupled fluorescence excitation and emission phenomena of light propagation. Fundamental equations of light propagation for TD-FS and TD-FT are usually described by the coupled TD-DEs for excitation and emission light,

$$\left[\frac{1}{c}\frac{\partial}{\partial t} - \nabla \cdot D_x(\mathbf{r})\nabla + \mu_{ax}(\mathbf{r}) + \varepsilon N(\mathbf{r})\right]\phi_x(\mathbf{r},t) = q_x(\mathbf{r},t), \tag{55}$$

$$\left[\frac{1}{c}\frac{\partial}{\partial t} - \nabla \cdot D_m(\mathbf{r})\nabla + \mu_{am}(\mathbf{r})\right]\phi_m(\mathbf{r},t) = \frac{\gamma(\mathbf{r})\varepsilon N(\mathbf{r})}{\tau(r)}\int_0^t \phi_x(r,t\prime)\exp(\frac{t-t\prime}{\tau(r)})dt\prime, \tag{56}$$

where subscripts x and m refer to excitation and emission light, respectively, $N(\mathbf{r})$, $\gamma(\mathbf{r})$, and $\tau(\mathbf{r})$ are the spatial distributions of the fluorescence properties, i.e., concentration, quantum efficiency, and lifetime of the fluorophore, respectively, and ε is the extinction coefficient of the fluorophore at the excitation wavelength. The goal of FT is to reconstruct the distributions of fluorescence properties, and $N(\mathbf{r})$ and $\tau(\mathbf{r})$ are the main targets for molecular imaging. In general, the image reconstruction algorithm for TD-DOT is extended for TD-FT, which uses the emitted fluorescence light intensities measured at the object surface as the input data in addition to the measured reemitted excitation light intensities. The optical and fluorescence properties are assumed first, and Equations (55) and (56) are solved as the forward problem to obtain the calculated excitation and fluorescence light intensities, which are compared with the measured ones. If they do not agree, the optical and fluorescence properties are upgraded by an optimization procedure introduced in Section 5.3.1, and Equations (55) and (56) are solved again. This process is repeated until convergence is reached. Many studies have been conducted on TD-FS and TD-FT, and some studies are reviewed in the following.

5.4.2. Analytical Solutions of the Equations for TD-FS

Patterson and Pogue derived analytical formulations of fluorescence emission light intensities measured at surfaces of homogeneously fluorescing semi-infinite media for TD-FS and FD-FS [197]. The emission light intensities are analytically given using the product of the fluence rate in the medium and the escape function, which is equivalent to the sensitivity of reflectance to absorption changes expressed by Equation (17). Sadoqi et al. provided analytical solutions of Equations (55) and (56) for cuboidal, spherical, and cylindrical media using eigen functions, and TOF distributions of the fluorescence emission were compared with those obtained by phantom experiments [198].

5.4.3. Clinical Applications of TD-FS

Many studies have been done toward clinical applications of TD-FS, but only a few studies are reviewed in this article. For evaluating excised tumor specimens, Butte et al. induced tissue autofluorescence with a pulsed (1.2 ns) nitrogen laser at 337 nm and measured its decay profiles in a wavelength range from 370 nm to 500 nm using a fast oscilloscope and a photomultiplier [199]. Obtaining data at two times in the decay profiles (times at the intensity of $1/e$ and 10% of the maximum intensity) at six wavelengths, a linear discrimination analysis enabled discrimination of meningioma tissue from normal tissue with a sensitivity larger than 89% and specificity of 100%. The feasibility of using TD-FS for evaluating tumor tissue was discussed.

Milej et al. tried a method to measure the inflow and washout of an optical contrast agent (ICG) for the purpose of providing information on the blood supply to the brain by TR measurements of fluorescence at the head surface [200]. In order to provide information for interpretation of the measured fluorescence signals, they performed MC simulations and in vivo TR measurements to find the effects of the SD distances, positions of the SDs, and doses of injected ICG. Furthermore, Milej et al. made MC simulations of light propagation of the excitation and emission light in a two-layered medium mimicking intra- and extracerebral tissue compartments after injecting ICG [201]. They found that the knowledge of the absorption properties of the medium is essential for interpretation of the TR fluorescence signals measured at the head surface.

5.4.4. TD-FT Using Full TOF-Distributions and Effects of Featured Data Types

For TD-FT, Gao et al. reported FT images of the fluorescence yield ($\gamma \varepsilon N$) and lifetime (τ) using full TOF distributions in numerical simulations [202]. Reconstructed FT images using featured TD data can be evaluated by comparing with those using the full TOF distributions. To improve the quality of the images reconstructed from full TOF distributions, Li et al. proposed an overlap time-gating approach which simultaneously achieves high time-resolution and high signal-to-noise ratio for input data extracted from the measured TOF distributions [203]. Riley et al. discussed the effect of featured data types on reconstructions of the depth and lifetime of the fluorophore, and concluded that local data types such as the peak values, peak times, FWHMs (full widths at half maximum), and slopes of decaying period of measured TOF distributions have advantages over global data types such as the CW intensities, mean TOFs, variances, and skews [204].

5.4.5. TD-FT Using Early Arriving Photons

Many studies of TD-FT have been done using early arriving photons. Some of them to name a few are reviewed below.

Wu et al. used a streak camera for TR measurements of early photons, TOF distributions of which were Laplace transformed with an optimum parameter, and the fluorophore targets were accurately localized under the presence of the noise of background fluorescence [205].

Niedre et al. reconstructed TD-FT images of lung carcinoma in mice in vivo from early photons detected by a high-speed gated ICCD with a gate time of 100 ps to obtain 360° projections [206]. Light propagation was described by a cumulant solution of the TD-RTE to generate three-point Jacobians, and a standard inversion process was employed to reconstruct 3D images of the fluorescence targets.

Patwardhan and Culver developed a TD-DOT system to quantitatively reconstruct μ_a and μ_s' images inside small animals such as mice for the purpose of improvement of TD-FT images [207]. They used high-speed time-gated ICCD with a time gate less than 300 ps to obtain TR transmission data which were converted to FD data for image reconstruction of μ_a and μ_s' using a non-linear iterative inversion algorithm.

Leblond et al. studied early-photon TD-FT to explain improved spatial resolutions of the images reconstructed from time-gated measurements with 200~400 ps time-gates by mathematical formulations and in silico phantoms [208]. They used the TD-DE for the forward problem and singular-value analysis for the inverse problem. Zhu et al. reported that using not only early photons but also late photons improve both the spatial resolution and image contrast in 3D TD-FT on a heterogeneous mouse model [209]. They employed the normalized Born-ratios as the featured data type, derived the Jacobians of the featured data to the changes in the absorption coefficient of fluorophores, and solved the non-linear iterative inversion problem to reconstruct the TD-FT images.

Cheng et al. developed a reconstruction algorithm of TD-FT for small animals using the first order time-derivative (or slope) of the rising portion of the measured TOF distributions in the early time regime [210]. Using the first derivative provided robustness to measurement noises and high-spatial resolutions in the reconstructed TD-FT images. As the forward model, they employed the telegraph equation (TE) of Equation (5) having the term, $\partial^2 \phi(\mathbf{r}, t)/\partial t^2$, because the TD-DE was derived by assuming $\partial^2 \phi(\mathbf{r}, t)/\partial t^2 << \partial \phi(\mathbf{r}, t)/\partial t$ which may breakdown for early photons.

5.4.6. TD-FT Using the GPST Algorithm

Gao et al. reconstructed TD-FT images using the GPST algorithm which was developed in TD-DOT [171,211]. From phantom experiments using a four-channel TCSPC system and GPST algorithm, combined DOT and FT provided images of μ_a, μ_s', N, and τ, and hemoglobin images were obtained from μ_a images at two wavelengths of 780 nm and 830 nm [192].

5.4.7. Total Light Approach in TD-FS and TD-FT

Marjono et al. proposed a concept of the total light to calculate light propagation of fluorescence emission [212]. The total light is defined as $\phi_t = \phi_x + \phi_m^*/\gamma$, where ϕ_m^* is the fluence rate of the emission light when the fluorescence life time is zero, i.e., $\phi_m(\mathbf{r}, t) = \phi_m^*(\mathbf{r}, t) \otimes (1/\tau)\exp(-t/\tau)$ with $\otimes$ denoting the convolution operator. With additional assumption of $\mu_{ax} = \mu_{am} = \mu_a$ and $D_x = D_m = D$, the DE for ϕ_t is given as,

$$\left[\frac{1}{c}\frac{\partial}{\partial t} - \nabla \cdot D(\mathbf{r})\nabla + \mu_a(\mathbf{r}) \right] \phi_t(\mathbf{r}, t) = q_x(\mathbf{r}, t), \tag{57}$$

which is almost the same form as Equation (55) except the absorption by the fluorophore. So, computation is simplified. The total light approach was applied to TD-FT for reconstructing μ_a and $\varepsilon N(\mathbf{r})$ in a 2D circular medium with a single and double fluorophore targets using <t> as a featured TD-data [213]. Nishimura et al. applied the total light approach to estimate the life-time function, $(1/\tau)\exp(-t/\tau)$, in heterogeneous scattering media with a more general scheme using the TD-RTE [214].

5.4.8. Transformation of TD-FT to FD-FT

Transformation of TD-FT to FD-FT have often been employed because integration of the right-hand side of Equation (56) can be avoided in FD [215–217]. For cases of a single fluorophore target, fluorescence signals in FD can be simply written as the product of the fluence rate, the Green's function of the excitation light and life-time function.

5.4.9. Application of MC Method for TD-FT

Monte Carlo simulations were also employed for forward calculations of TD-FT [218,219]. The perturbation MC (pMC) and adjoint MC were found to be appropriate for generating time-gated TR-Jacobians, and the positions and lifetimes of fluorophore targets were reconstructed using data of multiple time gates.

6. Clinical Applications of Commercially Available TD-NIRS Systems by Japanese Researchers

In this section, results of clinical applications of commercially available TD-NIRS systems by some Japanese clinician groups are briefly reviewed. All of them used TRS-10 or TRS-20 provided by Hamamatsu Photonics as described in Section 3.2.2 under the assumption of homogeneous semi-infinite medium. The applications of these instruments to multi-layered tissues may lead to some unreasonable results with overestimation or underestimation of physiological changes but may provide new findings and insights in the fields of hemodynamics, physiology, and neuroscience. These studies are still at the initial stage of clinical applications for TD-NIRS and may lead to a large leap in the near future.

6.1. Group from Kagawa Medical University

After some preliminary studies by Kusaka et al. and Ijichi et al. using hypoxia models of piglets, they measured μ_s', DPF, cerebral blood volume (CBV), etc., in 22 neonate heads, and found a significant correlation between the postconceptional age, μ_s', and CBV [220–222]. Ogawa et al. measured the changes in $C_{oxy\text{-}Hb}$, $C_{deoxy\text{-}Hb}$, $C_{total\text{-}Hb}$, and S_{O2} of breast tissue during breastfeeding and found that all four parameters decreased during breastfeeding [223]. Koyano et al. studied the effect of blood transfusion on cerebral hemodynamics in preterm infants and found that cerebral S_{O2} decreased after transfusion while CBV increased [224]. Nakamura et al. tried to use the TRS-10 to evaluate the seriousness of asphyxiated neonates and found CBV and cerebral S_{O2} were significantly higher for neonates with adverse outcomes of hypothermic therapy [225]. Kusaka et al. reviewed the usefulness of TD-NIRS for cerebral hemodynamic treatments in neonates [226].

6.2. Group from Kagoshima University Hospital

A group from the Kagoshima University Hospital collaborating with Hamamatsu Photonics used TD-NIRS (TRS-10) to monitor cerebral hemodynamics ($C_{oxy\text{-}Hb}$, $C_{deoxy\text{-}Hb}$, $C_{total\text{-}Hb}$ and S_{O2}) of patients during cardiopulmonary bypass surgery [227]. They found that $C_{total\text{-}Hb}$ correlated well with hematocrit measured by a blood gas analyzer. Then, Kakihana et al. evaluated the occurrence of postoperative cognitive dysfunction (POCD) after cardiopulmonary bypass surgery with hypothermic treatment by S_{O2} measured by TRS-10 and internal jugular vein oxygen saturation (S_{jvO2}) measured by a conventional method [228]. In the patients with POCD, S_{jvO2} were found to be significantly larger than S_{O2}. The same group also tried to monitor hepatic oxygenation by TRS-10 and found that abdominal $C_{total\text{-}Hb}$ were significantly higher in the liver area than other area in healthy people and that hepatic oxygenation measured by TRS-10 may work for early detection of intestinal ischemia [229]. Recently, monitoring of post-resuscitation encephalopathy by TRS-10 was experimentally tried for pigs with cardiac arrest induced by electrical stimuli for preliminary study toward clinical applications [230].

6.3. Group from Nihon University School of Medicine

Sakatani et al. collaborated with Hamamatsu Photonics in using the 16-channel TCSPC system (exceptionally in this section) to measure the changes in the mean optical pathlengths and cerebral blood oxygenation at the adult prefrontal cortex during verbal fluency and driving simulation tasks [50,231]. During the verbal fluency task, $C_{oxy\text{-}Hb}$ and $C_{total\text{-}Hb}$ increased while $C_{deoxy\text{-}Hb}$ decreased. On the other hand, $C_{oxy\text{-}Hb}$ and $C_{total\text{-}Hb}$ decreased while $C_{deoxy\text{-}Hb}$ increased during the driving simulation task. The mean optical pathlengths did not change by the tasks. Yokose et al used the TRS-10 to detect vasospasm of the patients with subarachnoid hemorrhage by measuring the cortical oxygen saturation [232]. Tanida et al. used the TRS-10 to conclude that the change in $C_{oxy\text{-}Hb}$ at the lateral prefrontal cortex during a working memory task correlated with working memory performance [233]. Sakatani et al. examined the effect of the extract of *Ginko biloba* leaves on the performance of working memory with TRS-10 probes attached on the prefrontal cortex and found that the right laterality score of $C_{oxy\text{-}Hb}$ increased by administration of *Ginko biloba* leaves [234]. Machida et al. showed that cosmetic therapy to elderly women with mild cognitive impairment was effective to improve the activities of the prefrontal cortex with increased $C_{oxy\text{-}Hb}$ and $C_{total\text{-}Hb}$ measured by the TRS-20 [235]. Tanida et al. also used TRS-20 to evaluate the difference in the women's emotion between pleasure and displeasure induced by the difference in lipsticks and found that the difference in the lipsticks resulted in different activities in the left and right prefrontal cortex [236]. Murayama et al. studied the relationship between cognitive function and cerebral blood oxygenation measured with the TRS-20 in the prefrontal cortexes of 113 elderly people and found strong correlations between working memory function and $C_{oxy\text{-}Hb}$, $C_{total\text{-}Hb}$, and S_{O2} [237].

6.4. Group of Professor Hamaoka (Tokyo Medical University)

In 2000, Hamaoka et al. used TD-NIRS to study the changes in $C_{oxy\text{-}Hb}$, $C_{deoxy\text{-}Hb}$, $C_{total\text{-}Hb}$, and S_{O2} in radial digitorum extensor muscles with arterial occlusion and found reasonable agreements of S_{O2} measured by TD-NIRS and a blood gas analyzer [238]. Later, the group used TD-NIRS to examine the activity of brown adipose tissue (BAT), which can be a counter-measure of obesity and obesity-induced metabolic disorders. Brown adipose tissue has more capillary and mitochondria than other tissues, and the densities of capillaries and mitochondria were found to be strongly correlated with μ_a and μ_s', respectively, which are measurable with TD-NIRS [239]. Using TD-NIRS, Nirengi et al. showed that daily ingestion of capsinoids (thermogenic capsisin analog) for eight weeks increased the BAT density by 46%, resulting in the same results with the conventional method using ^{18}F-fluoro-deoy-glucose positron emission tomography (PET) combined with X-ray CT [240]. Fuse et al. performed a study with 423 Japanese subjects to confirm the relationship between the BAT density and $C_{total\text{-}Hb}$ measured with TD-NIRS [241].

6.5. Other Groups in Japan

In 2006, Ohmae et al. made measurements for six healthy human subjects with pharmacologically perturbed cerebral hemodynamics using the TRS-10 and PET simultaneously, and reported a correlation between the hemodynamic changes obtained by the TRS-10 and the PET-parameter changes in gray matter regions [242]. Sato et al. used the TRS-10 with eight adult male heads during carotid endarterectomy and reported that TD-NIRS measurements were useful for monitoring the cerebral blood circulation and oxygenation during neurosurgical operations [243]. Yamazaki et al. used the TRS-20 on the foreheads of pregnant women during caesarean section [244]. In the case of massive bleeding due to placenta previa, S_{O2} measured by the TRS-20 showed clear decreases from 67% to 54%, while Sp_{O2} measured by pulse oximetry did not change. The TD-NIRS systems were found useful for monitoring bleeding during caesarean section. Ueda et al. evaluated the usefulness of TD-DOS imaging of primary breast cancers using the TRS-20 [245]. Lesions with $C_{total\text{-}Hb}$ 20% higher than normal tissue exhibited more advanced cancer stage, higher mitotic counts, and higher ^{18}Fluoro-deoxy-glucose uptake in PET, and TD-DOS imaging of $C_{total\text{-}Hb}$ was found useful for prediction of patient prognosis and potential response to treatment.

7. Summary

We reviewed the development of TD-NIRS from the views of its theoretical background, instruments, advanced theories and methods, the literature on future clinical applications, and current clinical applications of commercially available TD-NIRS systems. At the beginning of the summary section, we identified the major/key developments in the instruments as well as the theories, methods, and applications of TD-NIRS and imaging from the previous sections, and they are listed in Tables 1 and 2 in a chronological style. In addition, the major developments in Tables 1 and 2 are schematically and chronologically shown in Figures 19 and 20, where the relationship between the major developments are indicated by vertical arrows to help us view the whole picture of TD-NIRS and imaging technology. From these tables and figures, it can be said that the theories, methods, and applications of TD-NIRS developed steadily in the 1990s and 2000s, while the development of TD-NIRS instruments was rather slow during 1990s and 2000s. However, the development of such instruments began to accelerate during the 2010s, probably owing to the advent of the SiPM (MPPC), as seen from Figure 19.

Table 1. Chronology of major events in instruments of TD-NIRS (Abbreviations: "Ch" for channel and "pLD" for pulsed laser diode).

Year	Event	Ref.
1988	TD-NIRS system using a streak camera or TCSPC	[34,36]
1999	2-wavelength multi Ch TD NIRS oximeter using pLDs and multi-anode PMTs	[39]
	Commercial 1-Ch TD-NIRS system using TCSPC for research use: "TRS-10"	[40]
	1-wavelength 1-Ch TD optical mammography using a pLD and PMT-TCSPC	[41]
	64-channel TD-DOT system using pLDs and PMT-TCSPCs	[45]
2000	32-channel TD-DOT system (MONSTIR)	[49]
2003	TD-NIRS system using the spread spectrum technique and pseudo-random bit sequences	[63]
2005	TD-NIRS system using time-gated ICCD	[65]
	16-channel TD-DOT system using pLDs and PMT-TCSPCs	[50]
2009	Commercial 2-Ch TD-NIRS system using TCSPC for research use: "TRS-20"	[43]
2011	48-Ch 3-wavelength TD-NIRS optical mammography	[52]
2013	TD-DOT system incorporating SPADs into TCSPC	[58]
2014	Commercial 2-Ch TD-NIRS system using MPPCs for medical use: "tNIRS-1"	[55]
	MONSTIR II employing an SC laser with an AOTF for 4 wavelengths	[51]
2016	TD-NIRS mammography for imaging the contents of water, lipid, collagen, oxy-Hb and deoxy-Hb using 7 wavelengths	[54]
	Compact 2-wavelength TD-NIRS system and detector probe using SiPM	[56,57]
	TD-NIRS system using an SC laser and SPADs for non-contact measurements	[60]
2017	12-Ch TD-NIRS mammography with a hand-held probe	[53]
2018	TD-DOT system using an SC laser and SPAD camera	[62]
	Compact TD-NIRS system for measuring the contents of water, lipid, oxy-Hb, and deoxy-Hb using 6 wavelengths	[68]
	Compact 1-Ch TD-NIRS system using telecommunication devices	[3]

Table 2. Chronology of major events in theories, methods and applications of TD-NIRS.

Year	Event	Ref.
1983	Monte Carlo method applied to photon migration	[22]
1988	TD measurement of optical pathlength	[34]
	TD-NIRS of hemoglobin and myoglobin in muscle	[36]
1989	Analytical solutions of the TD-DE for semi-infinite and slab media	[18]
1991~1995	TD sensitivity functions	[26–29]
1991, 1993	Method of TD-DOT image reconstruction including forward and inverse models	[69,164]
1992	Analytical solutions of the TD-DE for various simple geometries	[19]
	Monte Carlo code for multi-layered tissue, MCML	[23]
1993, 1995	TOF and absorbance imaging of biological media and neonates	[156,181]
1994	Mathematical model for TD-FT	[197]
1994, 2006	Diffusion coefficient independent of the absorption coefficient	[8,17]
1996	TD imaging based on the perturbation model	[87]
1996, 1998	Perturbation Monte Carlo simulation	[105,107]
1997	Light propagation in a model of the adult head	[126]
	TD-FT using early-arriving photons	[205]
1998	TR reflectance from two-layered media	[98]
	Simultaneous MR and TD-NIRS mammography	[191
2000, 2014	Open source software for TD-DOT: TOAST and TOAST++	[49,165]
2001	TD-DOT of human forearm	[193]
	Photon path distribution based on the microscopic Beer–Lambert law	[9]
2002	GPST and full TR algorithms for TD-DOT	[170,171]
	3D TD-DOT of premature infant brain	[185]
2005	Perturbation model for layered media	[91]
	TR reflectance at null space SD distance	[131]
	Measurements of optical properties in neonates using a commercial TD-NIRS system: TRS-10	[222]
2010	(Monograph) Light propagation through biological tissue	[32]
2014	Hybrid TD-RTE and TD-DE	[119,120]
2014, 2016	Estimate of tissue composition in breasts using TR reflectance at 7 wavelengths	[54,139]

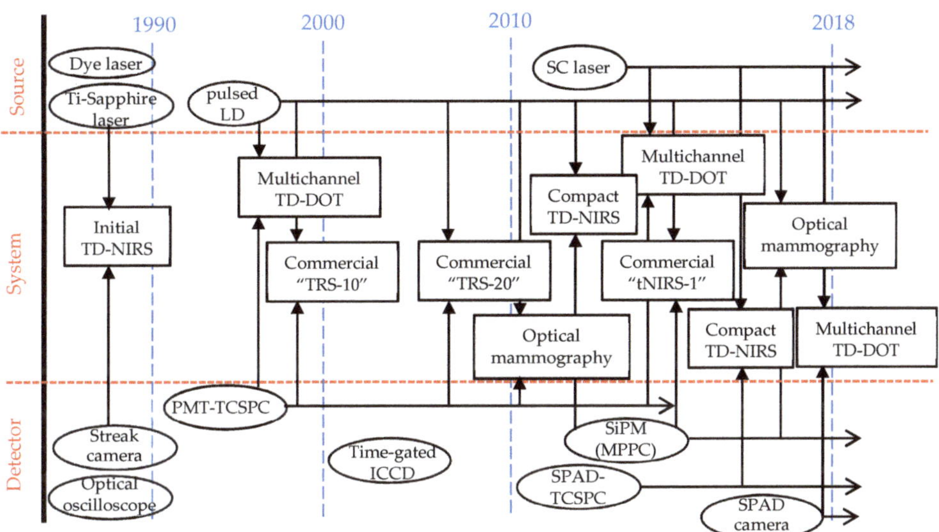

Figure 19. Schematic view of the chronological development of the TD-NIRS instruments classifying the sources, detectors, and systems, indicating incorporation of the sources and detectors into the systems by vertical arrows (abbreviation: pLD—pulsed laser diode).

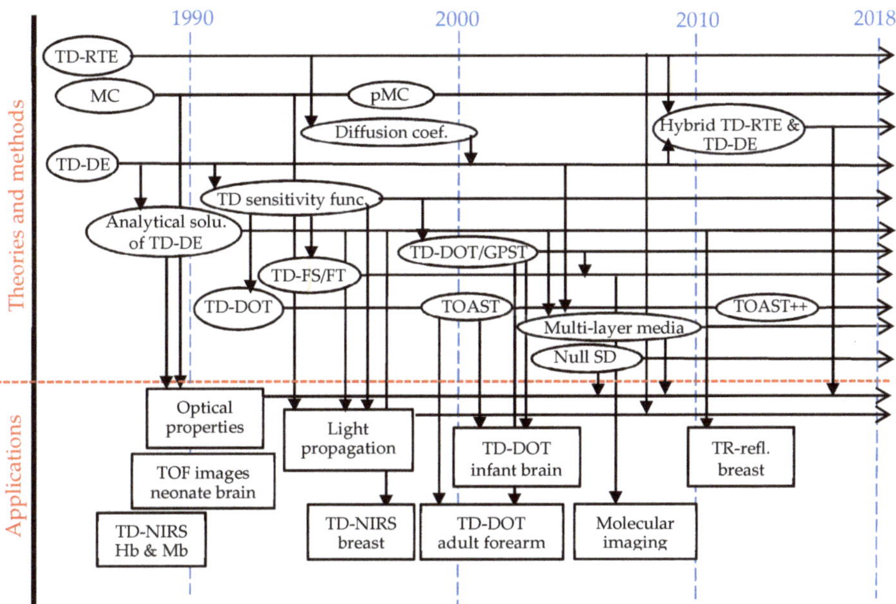

Figure 20. Schematic view of the chronological development of TD-NIRS theories, methods, and applications, indicating their relationship with vertical arrows.

The reputation of being cumbersome, bulky, and very expensive has prevented TD-NIRS systems from being used widely in clinical applications, while the theoretical studies of TD-NIRS have antecedently developed many features toward clinical applications.

In this situation, slow but steady progresses in TD-NIRS systems over the conventional TD-NIRS techniques have recently shown the possibility and feasibility of solving the drawbacks. As described in Section 3.3, many new systems without using the PMTs have been developed such as the compact, low-power consuming and dual-channel or dual-wavelength TD-NIRS systems using the SiPMs (MPPCs) [55,56], the systems incorporating fast-gated SPADs [58,60,61], the systems based on a spread spectrum approach using laser diodes with pseudo-random-bit sequences [63,64], the systems using time-gated ICCD cameras [65,66], the system incorporating a low-cost commercially available optical transceiver module widely used in telecommunications [3], the system using a supercontinuum laser [67], and the system using state-of-the-art SPAD camera chips [62]. In particular, the SiPM (MPPC) that is a new photon counting semiconductor device may replace the conventional PMTs. Resultantly, it is highly expected that TD-NIRS systems may be greatly improved to meet the requirements in clinical applications.

In concert with the recent developments of the instruments, the theories, methods, and applications in TD-NIRS will be further developed in the future. In particular, the calculation method of the TD-RTE and the hybrid TD-RTE and TD-DE techniques will be developed so that the developments in the TD-NIRS instruments may fully exhibit their performances in the picosecond and nanosecond time-resolved measurements.

If the advanced theoretical methods and algorithms are installed even in the existing commercially available TD-NIRS systems, clinicians will be able to obtain more useful information for diagnostics. For example, if the methods to determine the optical properties of two-layered media instead of assuming the homogeneous media are installed into the commercially available TD-NIRS systems (such as TRS-1 and tNIRS-1 sold by Hamamatsu), the provided information will be greatly extended to become more useful for clinicians. When many TD-NIRS or TD-DOT results are accumulated, clinical applications for new and unforeseen diagnostic methods will be discovered.

Combining the improvements of the hardware and software as stated above is also expected to make it possible to transfer the sophisticated technologies of TD-DOT and TD-FT from current bench-top to bed-side instruments toward wide clinical applications in the near future.

Acknowledgments: This article was partly supported by JSPS KAKENHI, Grant Number 18K18448. Authors acknowledge Yukio Ueda at Hamamatsu Photonics for his advice and Takashi Kusaka at Kagawa Medical University, Yasuyuki Kakihana at Kagoshima University Hospital, and Kaoru Sakatani at Nihon University School of Medicine for providing information.

Conflicts of Interest: The authors declare no conflict of interest.

Abbreviations

(In alphabetical order): A/D (analog to digital), AOTF (acoustic-optic tunable filter), BAT (brown adipose tissue), CBV (cerebral blood volume), CFD (constant fraction discriminator), CPU (central processing unit), CSF (cerebro-spinal fluid), CT (computed tomography), CW (continuous wave), DE (diffusion equation), DOM (discrete ordinate method), DOT (diffuse optical tomography), DPF (differential pathlength factor), FD (frequency-domain), FEM (finite element method), FS (fluorescence diffuse optical spectroscopy), FT (fluorescence diffuse optical tomography), GPST (generalized pulse spectrum technique), GPU (graphic processing unit), ICCD (intensified charge coupled device), ICG (indocyanine green), IRF (instrumental response function), MC (Monte Carlo), MOBIIR (model-based iterative image reconstruction), MPPC (multi-pixel photon counter), MRI (magnetic resonance imaging), NIRS (near-infrared spectroscopy), NN (neural network), PET (positron emission tomography), pMC (perturbation Monte Carlo), PMT (photomultiplier tube), POCD (postoperative cognitive dysfunction), RTE (radiative transfer equation), SC (supercontinuum), SD (source-detector), SiPM (silicon photomultiplier), SPAD (single-photon avalanche diode), TAC (time-to-amplitude converter), TCSPC (time-correlated single photon counting), TD (time-domain), TDC (time-to-digital converter), TE (telegraph equation), TOF (time-of-flight), TR (time-resolved), 2D (two-dimensional), 3D (three-dimensional).

References

1. Ferrari, M.; Quaresima, V. Near infrared brain and muscle oximetry: From the discovery to current applications. *J. Near Infrared Spectrosc.* **2012**, *20*, 1–14. [CrossRef]

2.   Torricelli, A.; Contini, D.; Pifferi, A.; Caffini, M.; Re, R.; Zucchelli, L.; Spinelli, L. Time domain functional NIRS imaging for human brain mapping. *NeuroImage* **2014**, *85*, 28–50. [CrossRef]

3.   Papadimitriou, K.I.; Dempsey, L.A.; Hebden, J.C.; Arridge, S.R.; Powell, S. A spread spectrum approach to time-domain near-infrared diffuse optical imaging using inexpensive optical transceiver modules. *Biomed. Opt. Express* **2018**, *9*, 2648–2663. [CrossRef] [PubMed]

4.   Arridge, S.R. Optical tomography in medical imaging. *Inverse Prob.* **1999**, *15*, R41–R93. [CrossRef]

5.   Duderstadt, J.J.; Hamilton, L.J. *Nuclear Reactor Analysis*; Wiley: New York, NY, USA, 1976; ISBN 978-0-471-22363-4.

6.   Kaltenbach, J.M.; Kascheke, M. Frequency- and time-domain modelling of light transport in random media. In *Medical Optical Tomography: Functional Imaging and Monitoring*; Muller, G.J., Ed.; SPIE Optical Engineering Press: Bellingham, WA, USA, 1993; pp. 65–86. ISBN 0819413801, 0819413798.

7.   Bigio, I.J.; Fantini, S. *Quantitative Biomedical Optics; Theory, Methods, and Applications*; Cambridge University Press: Cambridge, UK, 2016; ISBN 978-0-521-87656-8.

8.   Furutsu, K.; Yamada, Y. Diffusion approximation for a dissipative random media and the applications. *Phys. Rev. E* **1994**, *50*, 3634–3640.

9.   Tsuchiya, Y. Photon path distribution and optical responses of turbid media: Theoretical analysis based on the microscopic Beer–Lambert law. *Phys. Med. Biol.* **2001**, *46*, 2067–2084. [CrossRef] [PubMed]

10.   Graaff, R.; Ten Bosch, J.J. Diffusion coefficient in photon diffusion theory. *Opt. Lett.* **2000**, *25*, 43–45. [CrossRef] [PubMed]

11.   Aronson, R.; Corngold, N. Photon diffusion coefficient in an absorbing medium. *J. Opt. Soc. Am. A* **1999**, *16*, 1066–1071. [CrossRef]

12.   Elaloufi, R.; Carminati, R.; Greffet, J.-J. Definition of the diffusion coefficient in scattering and absorbing media. *J. Opt. Soc. Am. A* **2003**, *20*, 678–685. [CrossRef]

13.   Bassani, M.; Martelli, F.; Zaccanti, G.; Contini, D. Independence of the diffusion coefficient from absorption: Experimental and numerical evidence. *Opt. Lett.* **1997**, *22*, 853–855. [CrossRef]

14.   Nakai, T.; Nishimura, G.; Yamamoto, K. Tamura, Expression of the optical diffusion coefficient in high-absorption turbid media. *Phys. Med. Biol.* **1997**, *42*, 2541–2549. [CrossRef]

15.   Durduran, T.; Chance, B.; Yodh, A.G.; Boas, D.A. Does the photon diffusion coefficient depend on absorption? *J. Opt. Soc. Am. A* **1997**, *14*, 3358–3365. [CrossRef]

16.   Furutsu, K. Pulse wave scattering by an absorber and integrated attenuation in the diffusion approximation. *J. Opt. Soc. Am. A* **1997**, *14*, 267–274. [CrossRef]

17.   Pierrat, R.; Greffet, J.-J.; Carminati, R. Photon diffusion coefficient in scattering and absorbing media. *J. Opt. Soc. Am. A* **2006**, *23*, 1106–1110. [CrossRef]

18.   Patterson, M.S.; Chance, B.; Wilson, B.C. Time resolved reflectance and transmittance for the noninvasive measurement of tissue optical properties. *Appl. Opt.* **1989**, *28*, 2331–2336. [CrossRef] [PubMed]

19.   Arridge, S.R.; Cope, M.; Delpy, D.T. The theoretical basis for the determination of optical pathlengths in tissue: Temporal and frequency analysis. *Phys. Med. Biol.* **1992**, *37*, 1531–1560. [CrossRef]

20.   Kienle, A.; Foschum, F.; Hohmann, A. Light propagation in structural anisotropic media in the steady-state and time domains. *Phys. Med. Biol.* **2013**, *58*, 6205–6223. [CrossRef]

21.   Bonner, R.F.; Nossal, R.; Havlin, S.; Weiss, G.H. Model for photon migration in turbid biological media. *J. Opt. Soc. Am. A* **1987**, *4*, 423–432. [CrossRef]

22.   Wilson, B.C.; Adam, G. A Monte Carlo model for the absorption and flux distributions of light in tissue. *Med. Phys.* **1983**, *10*, 824–830. [CrossRef]

23.   Wang, L.; Jacques, S.L.; Zeng, L. MCML—Monte Carlo modeling of light transport in multi-layered tissues. *Comput. Method. Prog. Biomed.* **1992**, *47*, 131–146. [CrossRef]

24.   Monte Carlo Light Scattering Programs. Available online: https://omlc.org/software/mc/ (accessed on 26 August 2018).

25.   Boas, D.A.; Culver, J.P.; Stott, J.J.; Dunn, A.K. Three dimensional Monte Carlo code for photon migration through complex heterogeneous media including the adult human head. *Opt. Express* **2002**, *10*, 159–170. [CrossRef] [PubMed]

26.   Cui, W.; Wang, N.; Chance, B. Study of photon migration depths with time-resolved spectroscopy. *Opt. Lett.* **1991**, *16*, 1632–1634. [CrossRef] [PubMed]

27. Schotland, J.C.; Haselgrove, J.C.; Leigh, J.S. Photon hitting density. *Appl. Opt.* **1993**, *32*, 448–453. [CrossRef] [PubMed]

28. Arridge, S.R. Photon-measurement density functions. Part I: Analytical forms. *Appl. Opt.* **1995**, *34*, 7395–7409. [CrossRef] [PubMed]

29. Patterson, M.S.; Andersson-Engels, S.; Wilson, B.C.; Osei, E.K. Absorption spectroscopy in tissue-simulating materials: A theoretical and experimental study of photon paths. *Appl. Opt.* **1995**, *34*, 22–30. [CrossRef]

30. Sawosz, P.; Kacprzak, M.; Weigl, W.; Borowska-Solonynko, A.; Krajewski, P.; Zolek, N.; Ciszek, B.; Maniewski, R.; Liebert, A. Experimental estimation of the photons visiting probability profiles in time-resolved diffuse reflectance measurement. *Phys. Med. Biol.* **2012**, *57*, 7973–7981. [CrossRef]

31. Choi, J.H.; Wolf, M.; Toronov, V.; Wolf, U.; Polzonetti, C.; Hueber, D.; Safonova, L.P.; Gupta, R.; Michalos, A.; Mantulin, W.; et al. Noninvasive determination of the optical properties of adult brain: Near-infrared spectroscopy approach. *J. Biomed. Opt.* **2004**, *9*, 221–229. [CrossRef]

32. Martelli, F.; Del Bianco, S.; Ismaelli, A.; Zaccanti, G. *Light Propagation through BIOLOGICAL Tissue and Other Diffusive Media*; SPIE Press: Bellingham, WA, USA, 2010; ISBN 978-0-8194-7658-6.

33. Brazy, J.E.; Lewis, D.V.; Mitnick, M.H.; Jöbsis vander Vliet, F.F. Noninvasive Monitoring of Cerebral Oxygenation in Preterm Infants: Preliminary Observations. *Pediatrics* **1985**, *75*, 217–225.

34. Delpy, D.T.; Cope, M.; van der Zee, P.; Arridge, S.R.; Wray, S.; Wyatt, J. Estimation of optical pathlength through tissue from direct time of flight measurement. *Phys. Med. Biol.* **1988**, *33*, 1433–1442. [CrossRef]

35. van der Zee, P.; Cope, M.; Arridge, S.R.; Essenpreis, M.; Potter, L.A.; Edwards, A.D.; Wyatt, J.S.; McCormick, D.C.; Roth, S.C.; Reynolds, E.O.R.; et al. Experimentally measured optical pathlengths for the adult head, calf and forearm and the head of the newborn infant as a function of inter optode spacing. *Adv. Exp. Med. Biol.* **1992**, *316*, 143–153.

36. Chance, B.; Nioka, S.; Kent, J.; McCully, K.; Fountain, M.; Greenfeld, R.; Holtom, G. Time-resolved spectroscopy of hemoglobin and myoglobin in resting and ischemic muscle. *Anal. Biochem.* **1988**, *174*, 698–707. [CrossRef]

37. O'connor, D.V.; Phillips, D. *Time Correlated Single Photon Counting*; Academic Press: London, UK, 1984; ISBN 9780323141444.

38. Nomura, Y.; Hazeki, O.; Tamura, M. Exponential attenuation of light along nonlinear path through the biological model. *Adv. Exp. Med. Biol.* **1989**, *248*, 77–80. [PubMed]

39. Cubeddu, R.; Pifferi, A.; Taroni, P.; Torricelli, A.; Valentini, G. Compact tissue oximeter based on dual-wavelength multichannel time-resolved reflectance. *Appl. Opt.* **1999**, *38*, 3670–3680. [CrossRef] [PubMed]

40. Oda, M.; Yamashita, Y.; Nakano, T.; Suzuki, A.; Shimizu, K.; Hirano, I.; Shimomura, F.; Ohmae, E.; Suzuki, T.; Tsuchiya, Y. Near infrared time-resolved spectroscopy system for tissue oxygenation monitor. *Proc. SPIE* **1999**, *3597*, 611–617.

41. Grosenick, D.; Wabnitz, H.; Rinneberg, H.H.; Moesta, K.T.; Schlag, P.M. Development of a time-domain optical mammograph and first in vivo applications. *Appl. Opt.* **1999**, *38*, 2927–2943. [CrossRef]

42. Ida, T.; Iwata, Y. Correction for counting losses in X-ray diffractometry. *J. Appl. Cryst.* **2005**, *38*, 426–432. [CrossRef]

43. Oda, M.; Ohmae-Yamaki, E.; Suzuki, H.; Suzuki, T.; Yamashita, Y. Tissue oxygenation measurements using near-infrared time-resolved spectroscopy. *J. Jpn. Coll. Angiol.* **2009**, *49*, 131–137. (In Japanese)

44. Oda, M.; Yamashita, Y.; Nakano, T.; Suzuki, A.; Shimizu, K.; Hirano, I.; Shimomura, F.; Ohmae, E.; Suzuki, T.; Tsuchiya, Y. Near-infrared time-resolved spectroscopy system for tissue oxygenation monitor. *Proc. SPIE* **2000**, *4160*, 204–210.

45. Eda, H.; Oda, I.; Ito, Y.; Wada, Y.; Oikawa, Y.; Tsunazawa, Y.; Takada, M.; Tsuchiya, Y.; Yamashita, Y.; Oda, M.; et al. Multichannel time-resolved optical tomographic imaging system. *Rev. Sci. Instrum.* **1999**, *70*, 3595–3602. [CrossRef]

46. Fukuzawa, R.; Okawa, S.; Matsuhashi, S.; Kusaka, T.; Tanikawa, Y.; Hoshi, Y.; Gao, F.; Yamada, Y. Reduction of image artifacts induced by change in the optode coupling in time-resolved diffuse optical tomography. *J. Biomed. Opt.* **2011**, *16*, 116022. [CrossRef]

47. Zhao, H.; Gao, F.; Tanikawa, Y.; Yamada, Y. Time-resolved diffuse optical tomography and its application to in vitro and in vivo imaging. *J. Biomed. Opt.* **2007**, *12*, 062107. [CrossRef] [PubMed]

48. Ueno, M.; Fukuzawa, R.; Okawa, S.; Yamada, Y.; Kusaka, T.; Nishida, T.; Isobe, K.; Tanikawa, Y.; Gao, F.; Sato, C.; et al. In vivo measurement of premature neonate head using diffuse optical tomography. Book of Abstracts, 3rd Asian and Pacific Rim Symp. In Proceedings of the Biophotonics (APBP2007), Cairns, Australia, 9–11 July 2007; pp. 43–44.

49. Schmidt, F.E.W.; Fry, M.E.; Hillman, E.M.C.; Hebden, J.C.; Delpy, D.T. A 32-channel time-resolved instrument for medical optical tomography. *Rev. Sci. Instrum.* **2000**, *71*, 256–265. [CrossRef]

50. Ueda, Y.; Yamanaka, T.; Yamashita, D.; Suzuki, T.; Ohmae, E.; Oda, M.; Yamashita, Y. Reflectance Diffuse Optical Tomography: Its Application to Human Brain Mapping. *Jpn. J. Appl. Phys.* **2005**, *44*, L1203–L1206. [CrossRef]

51. Cooper, R.J.; Magee, E.; Everdell, N.; Magazov, S.; Varela, M.; Airantzis, D.; Gibson, A.P.; Hebden, J.C. MONSTIR II: A 32-channel, multispectral, time-resolved optical tomography system for neonatal brain imaging. *Rev. Sci. Instrum.* **2014**, *85*, 053105. [CrossRef] [PubMed]

52. Ueda, Y.; Yoshimoto, K.; Ohmae, E.; Suzuki, T.; Yamanaka, T.; Yamashita, D.; Ogura, H.; Teruya, C.; Nasu, H.; Imi, E.; et al. Time-Resolved Optical Mammography and Its Preliminary Clinical Results. *Tech. Cancer Res. Treat.* **2011**, *10*, 393–401. [CrossRef]

53. Yoshimoto, K.; Ohmae, E.; Yamashita, D.; Suzuki, H.; Homma, S.; Mimura, T.; Wada, H.; Suzuki, T.; Yoshizawa, N.; Nasu, H.; et al. Development of time-resolved reflectance diffuse optical tomography for breast cancer monitoring. *Proc. Spie* **2017**, *10059*, 100590M.

54. Taroni, P.; Paganoni, A.M.; Ieva, F.; Pifferi, A.; Quarto, G.; Abbate, F.; Cassano, E.; Cubeddu, R. Non-invasive optical estimate of tissue composition to differentiate malignant from benign breast lesions: A pilot study. *Sci. Rep.* **2016**, *7*, 40683. [CrossRef]

55. Fujisaka, S.; Ozaki, T.; Suzuki, T.; Kamada, T.; Kitazawa, K.; Nishizawa, M.; Takahashi, A.; Suzuki, S. A clinical tissue oximeter using NIR time-resolved spectroscopy. *Adv. Exp. Med. Biol.* **2016**, *876*, 427–433.

56. Buttafava, M.; Martinenghi, E.; Tamborini, D.; Contini, D.; Dalla Mora, A.; Renna, M.; Torricelli, A.; Pifferi, A.; Zappa, F.; Tosi, A. A Compact Two-Wavelength Time-Domain NIRS System Based on SiPM and Pulsed Diode Lasers. *IEEE Photonics J.* **2016**, *9*, 7800114. [CrossRef]

57. Re, R.; Martinenghi, E.; Dalla Mora, A.; Contini, D.; Pifferi, A.; Torricellia, A. Probe-hosted silicon photomultipliers for time-domain functional near-infrared spectroscopy: Phantom and in vivo tests. *Neurophotonics* **2016**, *3*, 045004. [CrossRef]

58. Puszka, A.; Di Sieno, L.; Dalla Mora, A.; Pifferi, A.; Contini, D.; Boso, G.; Tosi, A.; Hervé, L.; Planat-Chrétien, A.; Koenig, A.; et al. Time-resolved diffuse optical tomography using fast-gated single-photon avalanche diodes. *Biomed. Opt. Express* **2013**, *4*, 1351–1365. [CrossRef]

59. Dalla Mora, A.; Contini, D.; Arridge, S.; Martelli, F.; Tosi, A.; Boso, G.; Farina, A.; Durduran, T.; Martinenghi, E.; Torricelli, A.; et al. Towards next-generation time-domain diffuse optics for extreme depth penetration and sensitivity. *Biomed. Opt. Express* **2015**, *6*, 1749–1760. [CrossRef] [PubMed]

60. Di Sieno, L.; Wabnitz, H.; Pifferi, A.; Mazurenka, M.; Hoshi, Y.; Dalla Mora, A.; Contini, D.; Boso, G.; Becker, W.; Martelli, F.; et al. Characterization of a time-resolved non-contact scanning diffuse optical imaging system exploiting fast-gated single-photon avalanche diode detection. *Rev. Sci. Instrum.* **2016**, *87*, 035118. [CrossRef]

61. Sinha, L.; Brankov, J.G.; Tichauer, K.M. Enhanced detection of early photons in time-domain optical imaging by running in the "dead-time" regime. *Opt. Lett.* **2016**, *41*, 3225–3228. [CrossRef] [PubMed]

62. Kalyanov, A.; Jiang, J.-J.; Lindner, S.; Ahnen, L.; di Costanzo, A.; Mata Pavia, J.; Sanchez Majos, S.; Wolf, M. Time Domain Near-Infrared Optical Tomography with Time-of-Flight SPAD Camera: The New Generation. In Proceedings of the Optical Tomography and Spectroscopy, Biophotonics Congress: Biomedical Optics Congress 2018, OSA, FL, USA, 3–6 April 2018.

63. Chen, N.G.; Zhu, Q. Time-resolved diffusive optical imaging using pseudo-random bit sequences. *Opt. Express* **2003**, *11*, 3445–3454. [CrossRef] [PubMed]

64. Mo, W.R.; Chen, N.G. Fast time-domain diffuse optical tomography using pseudorandom bit sequences. *Opt. Express* **2008**, *16*, 13643–13650. [CrossRef]

65. Selb, J.; Stott, J.J.; Franceschini, M.A.; Sorensen, A.G.; Boas, D.A. Improved sensitivity to cerebral hemodynamics during brain activation with a time-gated optical system: Analytical model and experimental validation. *J. Biomed. Opt.* **2005**, *10*, 011013. [CrossRef]

66. Zhao, Q.; Spinelli, L.; Bassi, A.; Valentini, G.; Contini, D.; Torricelli, A.; Cubeddu, R.; Zaccanti, G.; Martelli, F.; Pifferi, A. Functional tomography using a time-gated ICCD camera. *Biomed. Opt. Express* **2011**, *2*, 705–716. [CrossRef] [PubMed]

67. Lange, F.; Peyrin, F.; Montcel, B. Broadband time-resolved multi-channel functional near-infrared spectroscopy system to monitor in vivo physiological changes of human brain activity. *Appl. Opt.* **2018**, *57*, 6417–6429. [CrossRef] [PubMed]

68. Ohmae, E.; Yoshizawa, N.; Yoshimoto, K.; Hayashi, M.; Wada, H.; Mimura, T.; Suzuki, H.; Homma, S.; Suzuki, N.; Ogura, H.; et al. Stable tissue-simulating phantoms with various water and lipid contents for diffuse optical spectroscopy. *Biomed. Opt. Exp.* **2018**, *9*, 5792–5808. [CrossRef]

69. Schweiger, M.; Arridge, S.R.; Delpy, D.T. Application of the Finite-Element Method for the Forward and Inverse Models in Optical Tomography. *J. Math. Imag. Vis.* **1993**, *3*, 263–283. [CrossRef]

70. Lewis, E.E.; Miller, W.F., Jr. *Computational Methods of Neutron Transport*; American Nuclear Society, Inc.: La Grande Park, IL, USA, 1993; ISBN 0-89448-452-4.

71. Das, C.; Trivedi, A.; Mitra, K.; Vo-Dinh, T. Experimental and numerical analysis of short-pulse laser interaction with tissue phantoms containing inhomogeneities. *Appl. Opt.* **2003**, *42*, 5173–5180. [CrossRef] [PubMed]

72. Pal, G.; Basu, S.; Mitra, K.; Vo-Dinh, T. Time-resolved optical tomography using short-pulse laser for tumor detection. *Appl. Opt.* **2006**, *45*, 6270–6282. [CrossRef] [PubMed]

73. Fujii, H.; Yamada, Y.; Kobayashi, K.; Watanabe, M.; Hoshi, Y. Modeling of light propagation in the human neck for diagnoses of thyroid cancers by diffuse optical tomography. *Int. J. Numer. Meth. Biomed. Eng.* **2017**, *33*, e2826. [CrossRef] [PubMed]

74. Asllanaj, F.; Fumeron, S. Applying a new computational method for biological tissue optics based on the time-dependent two-dimensional radiative transfer equation. *J. Biomed. Opt.* **2012**, *17*, 075007.

75. Fujii, H.; Yamada, Y.; Chiba, G.; Hoshi, Y.; Kobayashi, K.; Watanabe, M. Accurate and efficient computation of the 3D radiative transfer equation in highly forward-peaked scattering media using a renormalization approach. *J. Comput. Phys.* **2018**, *374*, 591–604. [CrossRef]

76. Paasschens, J.C.J. Solution of the time-dependent Boltzmann equation. *Phys. Rev. E* **1997**, *56*, 1135–1141. [CrossRef]

77. Martelli, F.; Sassaroli, A.; Pifferi, A.; Torricelli, A.; Spinelli, L.; Zaccanti, G. Heuristic Green's function of the time dependent radiative transfer equation for a semi-infinite medium. *Opt Express* **2007**, *26*, 18168–18175. [CrossRef]

78. Liemert, A.; Kienle, A. Infinite space Green's function of the time-dependent radiative transfer equation. *Biomed. Opt. Express* **2012**, *3*, 543–551. [CrossRef]

79. Liemert, A.; Kienle, A. Green's function of the time-dependent radiative transport equation in terms of rotated spherical harmonics. *Phys. Rev. E* **2012**, *86*, 036603. [CrossRef]

80. Simon, E.; Foschum, F.; Kienle, A. Hybrid Green's function of the time-dependent radiative transfer equation for anisotropically scattering semi-infinite media. *J. Biomed. Opt.* **2013**, *18*, 015001. [CrossRef] [PubMed]

81. Ferwerda, H.A. The radiative transfer equation for scattering media with a spatially varying refractive index. *J. Opt. A Pure Appl. Opt.* **1999**, *1*, L1–L2. [CrossRef]

82. Khan, T.; Jiang, H. A new diffusion approximation to the radiative transfer equation for scattering media with spatially varying refractive indices. *J. Opt. A Pure Appl. Opt.* **2003**, *5*, 137–141. [CrossRef]

83. Tualle, J.-M.; Tinet, E. Derivation of the radiative transfer equation for scattering media with a spatially varying refractive index. *Opt. Commun.* **2003**, *228*, 33–38. [CrossRef]

84. Kumar, S.; Mitra, K.; Yamada, Y. Hyperbolic damped-wave models for transient light-pulse propagation in scattering media. *Appl. Opt.* **1996**, *36*, 3372–3378. [CrossRef] [PubMed]

85. Durian, D.J.; Rudnick, J. Photon migration at short times and distances and in cases of strong absorption. *J. Opt. Soc. Am. A* **1997**, *14*, 235–245. [CrossRef]

86. Polonsky, I.N.; Box, M.A. General perturbation technique for the calculation of radiative effects in scattering and absorbing media. *J. Opt. Soc. Am. A* **2002**, *19*, 2281–2292. [CrossRef]

87. Hebden, J.C.; Arridge, S.R. Imaging through scattering media by the use of an analytical model of perturbation amplitudes in the time domain. *Appl. Opt.* **1996**, *35*, 6788–6796. [CrossRef] [PubMed]

88. Morin, M.; Verreault, S.; Mailloux, A.; Fre'chette, J.; Chatigny, S.; Painchaud, Y.; Beaudry, P. Inclusion characterization in a scattering slab with time-resolved transmittance measurements: Perturbation analysis. *Appl. Opt.* **2000**, *39*, 2840–2852. [CrossRef]

89. Carraresi, S.; Mohamed Shatir, T.S.; Martelli, F.; Zaccanti, G. Accuracy of a perturbation model to predict the effect of scattering and absorbing inhomogeneities on photon migration. *Appl. Opt.* **2001**, *40*, 4622–4632. [CrossRef]

90. Spinelli, L.; Torricelli, A.; Pifferi, A.; Taroni, P.; Cubeddu, R. Experimental test of a perturbation model for time-resolved imaging in diffusive media. *Appl. Opt.* **2003**, *42*, 3145–3153. [CrossRef] [PubMed]

91. Martelli, F.; Del Bianco, S.; Zaccanti, G. Perturbation model for light propagation through diffusive layered media. *Phys. Med. Biol.* **2005**, *50*, 2159–2166. [CrossRef] [PubMed]

92. Wassermann, B. Limits of high-order perturbation theory in time-domain optical mammography. *Phys. Rev. E* **2006**, *74*, 031908. [CrossRef]

93. Grosenick, D.; Kummrow, A.; Macdonald, R.; Schlag, P.M.; Rinneberg, H. Evaluation of higher-order time-domain perturbation theory of photon diffusion on breast-equivalent phantoms and optical mammograms. *Phys. Rev. E* **2007**, *76*, 061908. [CrossRef] [PubMed]

94. Sassaroli, A.; Martelli, F.; Fantini, S. Perturbation theory for the diffusion equation by use of the moments of the generalized temporal point-spread function. I. Theory. *J. Opt. Soc. Am. A* **2006**, *23*, 2105–2118. [CrossRef]

95. Sassaroli, A.; Martelli, F.; Fantini, S. Perturbation theory for the diffusion equation by use of the moments of the generalized temporal point-spread function. III. Frequency-domain and time-domain results. *J. Opt. Soc. Am. A* **2010**, *27*, 1723–1742. [CrossRef]

96. Martelli, F.; Sassaroli, A.; Del Bianco, S.; Yamada, Y.; Zaccanti, G. Solution of the time-dependent diffusion equation for layered diffusive media by the eigenfunction method. *Phys. Rev. E* **2003**, *67*, 056623. [CrossRef]

97. Martelli, F.; Sassaroli, A.; Del Bianco, S.; Zaccanti, G. Effect of the refractive index mismatch on light propagation through diffusive layered media. *Phys. Rev. E* **2004**, *70*, 011907. [CrossRef]

98. Kienle, A.; Glanzmann, T.; Wagnières, G.; van den Bergh, H. Investigation of two-layered turbid media with time-resolved reflectance. *Appl. Opt.* **1998**, *37*, 6852–6862. [CrossRef]

99. Tualle, J.-M.; Prat, J.; Tinet, E.; Avrillier, S. Real-space Green's function calculation for the solution of the diffusion equation in stratified turbid media. *J. Opt. Soc. Am. A* **2000**, *17*, 2046–2055. [CrossRef]

100. Tualle, J.-M.; Nghiem, H.L.; Ettori, D.; Sablong, R.; Tinet, E.; Avrillier, S. Asymptotic behavior and inverse problem in layered scattering media. *J. Opt. Soc. Am. A* **2004**, *21*, 24–34. [CrossRef]

101. Liemert, A.; Kienle, A. Light diffusion in *N*-layered turbid media: Frequency and time domains. *J. Biomed. Opt.* **2010**, *15*, 025002. [CrossRef] [PubMed]

102. Martelli, F.; Sassaroli, A.; Yamada, Y.; Zaccanti, G. Analytical approximate solutions of the time-domain diffusion equation in layered slabs. *J. Opt. Soc. Am. A* **2002**, *19*, 71–80. [CrossRef]

103. Martelli, F.; Sassaroli, A.; Del Bianco, S.; Zaccanti, G. Solution of the time-dependent diffusion equation for a three-layer medium: Application to study photon migration through a simplified adult head model. *Phys. Med. Biol.* **2007**, *52*, 2827–2843. [CrossRef]

104. Barnett, A.H. A fast numerical method for time-resolved photon diffusion in general stratified turbid media. *J. Comput. Phys.* **2004**, *201*, 771–797. [CrossRef]

105. Kienle, A.; Patterson, M.S. Determination of the optical properties of turbid media from a single Monte Carlo simulation. *Phys. Med. Biol.* **1996**, *41*, 2221–2227. [CrossRef] [PubMed]

106. Alerstam, E.; Andersson-Engels, S.; Svensson, T. White Monte Carlo for time-resolved photon migration. *J. Biomed. Opt.* **2008**, *13*, 041304. [CrossRef]

107. Sassaroli, A.; Blumetti, C.; Martelli, F.; Alianelli, L.; Contini, D.; Ismaelli, A.; Zaccanti, G. Monte Carlo procedure for investigating light propagation and imaging of highly scattering media. *Appl. Opt.* **1998**, *37*, 7392–7400. [CrossRef]

108. Hayakawa, C.K.; Spanier, J.; Bevilacqua, F.; Dunn, A.K.; You, J.S.; Tromberg, B.J.; Venugopalan, V. Perturbation Monte Carlo methods to solve inverse photon migration problems in heterogeneous tissues. *Opt. Lett.* **2001**, *26*, 1335–1337. [CrossRef]

109. Sassaroli, A. Fast perturbation Monte Carlo method for photon migration in heterogeneous turbid media. *Opt. Lett.* **2011**, *36*, 2095–2097. [CrossRef]

110. Kumar, Y.P.; Vasu, R.M. Reconstruction of optical properties of low-scattering tissue using derivative estimated through perturbation Monte-Carlo method. *J. Biomed. Opt.* **2004**, *9*, 1002–1012. [CrossRef]

111. Chen, J.; Intes, X. Time-gated perturbation Monte Carlo for whole body functional imaging in small animals. *Opt. Express* **2009**, *17*, 19566–19570. [CrossRef]

112. Yao, R.; Intes, X.; Fang, Q.Q. Direct approach to compute Jacobians for diffuse optical tomography using perturbation Monte Carlo-based photon "replay". *Biomed. Opt. Express* **2018**, *9*, 4588–4602. [CrossRef]

113. Fang, Q.Q.; Boas, D.A. Monte Carlo simulation of photon migration in 3D turbid media accelerated by graphics processing units. *Opt. Express* **2009**, *17*, 20178–20190. [CrossRef]

114. Zoller, C.J.; Hohmann, A.; Foschum, F.; Geiger, S.; Geiger, M.; Ertl, T.P.; Kienle, A. Parallelized Monte Carlo software to efficiently simulate the light propagation in arbitrarily shaped objects and aligned scattering media. *J. Biomed. Opt.* **2018**, *23*, 065004.

115. Voit, F.; Hohmann, A.; Schäfer, J.; Kienle, A. Multiple scattering of polarized light: Comparison of Maxwell theory and radiative transfer theory. *J. Biomed. Opt.* **2012**, *17*, 045003. [CrossRef]

116. Jia, M.; Zhao, H.; Li, J.; Liu, L.; Zhang, L.; Jiang, J.; Gao, F. Coupling between radiative transport and diffusion approximation for enhanced near-field photon-migration modeling based on transient photon kinetics. *J. Biomed. Opt.* **2016**, *21*, 050501. [CrossRef]

117. Lehtikangas, O.; Tarvainen, T. Hybrid forward-peaked-scattering diffusion approximations for light propagation in turbid media with low-scattering regions. *J. Quant. Spectrosc. Radiat. Transf.* **2013**, *116*, 132–144. [CrossRef]

118. Gorpas, D.; Andersson-Engels, S. Evaluation of a radiative transfer equation and diffusion approximation hybrid forward solver for fluorescence molecular imaging. *J. Biomed. Opt.* **2012**, *17*, 126010. [CrossRef]

119. Fujii, H.; Okawa, S.; Yamada, Y.; Hoshi, Y. Hybrid model of light propagation in random media based on the time-dependent radiative transfer and diffusion equations. *J. Quant. Spectrosc. Radiat. Transf.* **2014**, *147*, 145–154. [CrossRef]

120. Roger, M.; Caliot, C.; Crouseilles, N.; Coelho, P.J. A hybrid transport-diffusion model for radiative transfer in absorbing and scattering media. *J. Comput. Phys.* **2014**, *275*, 346–362. [CrossRef]

121. Coelho, P.J.; Crouseilles, N.; Pereira, P.; Roger, M. Multi-scale methods for the solution of the radiative transfer equation. *J. Quant. Spectrosc. Radiat. Transf.* **2015**, *172*, 36–49. [CrossRef]

122. Nickell, S.; Hermann, M.; Essenpreis, M.; Farrell, T.J.; Krämer, U.; Patterson, M.S. Anisotropy of light propagation in human skin. *Phys. Med. Biol.* **2000**, *45*, 2873–2886. [CrossRef] [PubMed]

123. Heino, J.; Arridge, S.; Sikora, J.; Somersalo, E. Anisotropic effects in highly scattering media. *Phys. Rev. E* **2003**, *68*, 031908. [CrossRef] [PubMed]

124. Hasegawa, Y.; Yamada, Y.; Tamura, M.; Nomura, Y. Monte Carlo simulation of light transmission through living tissues. *Appl. Opt.* **1991**, *30*, 4515–4520. [CrossRef] [PubMed]

125. Nomura, Y.; Hazeki, O.; Tamura, M. Relationship between time-resolved and non-time-resolved Beer–Lambert law in turbid media. *Phys. Med. Biol.* **1997**, *42*, 1009–1022. [CrossRef]

126. Okada, E.; Firbank, M.; Schweiger, M.; Arridge, S.R.; Cope, M.; Delpy, D.T. Theoretical and experimental investigation of near-infrared light propagation in a model of the adult head. *Appl. Opt.* **1997**, *36*, 21–31. [CrossRef]

127. Fukui, Y.; Ajichi, Y.; Okada, E. Monte Carlo prediction of near-infrared light propagation in realistic adult and neonatal head models. *Appl. Opt.* **2003**, *42*, 2881–2887. [CrossRef]

128. Firbank, M.; Okada, E.; Delpy, D.T. A Theoretical Study of the Signal Contribution of Regions of the Adult Head to Near-Infrared Spectroscopy Studies of Visual Evoked Responses. *NeuroImage* **1998**, *8*, 69–78. [CrossRef]

129. Steinbrink, J.; Wabnitz, H.; Obrig, H.; Villringer, A.; Rinneberg, H. Determining changes in NIR absorption using a layered model of the human head. *Phys. Med. Biol.* **2001**, *46*, 879–896. [CrossRef]

130. Del Bianco, S.; Martelli, F.; Zaccanti, G. Penetration depth of light re-emitted by a diffusive medium: Theoretical and experimental investigation. *Phys. Med. Biol.* **2002**, *47*, 4131–4144. [CrossRef] [PubMed]

131. Torricelli, A.; Pifferi, A.; Spinelli, L.; Cubeddu, R. Time-Resolved Reflectance at Null Source-Detector Separation: Improving Contrast and Resolution in Diffuse Optical Imaging. *Phys. Rev. Lett.* **2005**, *95*, 078101. [CrossRef]

132. Spinelli, L.; Martelli, F.; Del Bianco, S.; Pifferi, A.; Torricelli, A.; Cubeddu, R.; Zaccanti, G. Absorption and scattering perturbations in homogeneous and layered diffusive media probed by time-resolved reflectance at null source-detector separation. *Phys. Rev. E* **2006**, *74*, 021919. [CrossRef] [PubMed]

133. Mazurenka, M.; Jelzow, A.; Wabnitz, H.; Contini, D.; Spinelli, L.; Pifferi, A.; Cubeddu, R.; Dalla Mora, A.; Tosi, A.; Zappa, F.; et al. Non-contact time-resolved diffuse reflectance imaging at null source-detector separation. *Opt. Express* **2012**, *20*, 283–290. [CrossRef] [PubMed]

134. Zhao, H.; Tanikawa, Y.; Gao, F.; Onodera, Y.; Sassaroli, A.; Tanaka, K.; Yamada, Y. Maps of optical differential pathlength factor of human adult forehead, somatosensory motor, occipital regions at multi-wavelengths in NIR. *Phys. Med. Biol.* **2002**, *47*, 2075–2093. [CrossRef]

135. Bonnéry, C.; Leclerc, P.-O.; Desjardins, M.; Hoge, R.; Bherer, L.; Pouliot, P.; Lesage, F. Changes in diffusion path length with old age in diffuse optical tomography. *J. Biomed. Opt.* **2012**, *17*, 056002. [CrossRef]

136. Jacques, S.L. Time-Resolved Reflectance Spectroscopy in Turbid Tissues. *IEEE Trans. Biomed. Eng.* **1989**, *36*, 1155–1161. [CrossRef]

137. Sassarol1, A.; Martelli, F.; Tanikawa, Y.; Tanaka, K.; Araki, R.; Onodera, Y.; Yamada, Y. Time-Resolved Measurements of in vivo Optical Properties of Piglet Brain. *Opt. Rev.* **2000**, *7*, 420–425. [CrossRef]

138. Comelli, D.; Bassi, A.; Pifferi, A.; Taroni, P.; Torricelli, A.; Cubeddu, R.; Martelli, F.; Zaccanti, G. In vivo time-resolved reflectance spectroscopy of the human forehead. *Appl. Opt.* **2007**, *46*, 1717–1725. [CrossRef]

139. Quarto, G.; Spinelli, L.; Pifferi, A.; Torricelli, A.; Cubeddu, R.; Abbate, F.; Balestreri, N.; Menna, S.; Cassano, E.; Taroni, P. Estimate of tissue composition in malignant and benign breast lesions by time-domain optical mammography. *Biomed. Opt. Express* **2014**, *5*, 3684–3698. [CrossRef]

140. Guggenheim, J.A.; Bargigia, I.; Farina, A.; Pifferi, A.; Dehghani, H. Time resolved diffuse optical spectroscopy with geometrically accurate models for bulk parameter recovery. *Biomed. Opt. Express* **2016**, *7*, 3784–3794. [CrossRef] [PubMed]

141. Kienle, A.; Glanzmann, T. In vivo determination of the optical properties of muscle with time-resolved reflectance using a layered model. *Phys. Med. Biol.* **1999**, *44*, 2689–2702. [CrossRef] [PubMed]

142. Gagnon, L.; Gauthier, C.; Hoge, R.D.; Lesage, F.; Selb, J.; Boas, D.A. Double-layer estimation of intra- and extracerebral hemoglobin concentration with a time-resolved system. *J. Biomed. Opt.* **2008**, *13*, 054019. [CrossRef] [PubMed]

143. Martelli, F.; Del Bianco, S.; Zaccanti, G. Procedure for retrieving the optical properties of a two-layered medium from time-resolved reflectance measurements. *Opt. Lett.* **2003**, *28*, 1236–1238. [CrossRef] [PubMed]

144. Martelli, F.; Del Bianco, S.; Zaccanti, G.; Pifferi, A.; Torricelli, A.; Bassi, A.; Taroni, P.; Cubeddu, R. Phantom validation and in vivo application of an inversion procedure for retrieving the optical properties of diffusive layered media from time-resolved reflectance measurements. *Opt. Lett.* **2004**, *29*, 2037–2039. [CrossRef]

145. Sato, C.; Shimada, M.; Yamada, Y.; Hoshi, Y. Extraction of depth-dependent signals from time-resolved reflectance in layered turbid media. *J. Biomed. Opt.* **2005**, *10*, 064008. [CrossRef]

146. Shimada, M.; Hoshi, Y.; Yamada, Y. Simple algorithm for the measurement of absorption coefficients of a two-layered medium by spatially resolved and time-resolved reflectance. *Appl. Opt.* **2005**, *44*, 7554–7563. [CrossRef]

147. Shimada, M.; Sato, C.; Hoshi, Y.; Yamada, Y. Estimation of the absorption coefficients of two-layered media by a simple method using spatially and time-resolved reflectances. *Phys. Med. Biol.* **2009**, *54*, 5057–5071. [CrossRef]

148. Hoshi, Y.; Shimada, M.; Sato, C.; Iguchi, Y. Reevaluation of near-infrared light propagation in the adult human head: Implications for functional near-infrared spectroscopy. *J. Biomed. Opt.* **2005**, *10*, 064032. [CrossRef]

149. Jäger, M.; Kienle, A. Non-invasive determination of the absorption coefficient of the brain from time-resolved reflectance using a neural network. *Phys. Med. Biol.* **2011**, *56*, N139–N144. [CrossRef]

150. Martelli, F.; Del Bianco, S.; Zaccanti, G. Retrieval procedure for time-resolved near-infrared tissue spectroscopy based on the optimal estimation method. *Phys. Med. Biol.* **2012**, *57*, 2915–2929. [CrossRef]

151. Martelli, F.; Del Bianco, S.; Spinelli, L.; Cavalieri, S.; Di Ninni, P.; Binzoni, T.; Jelzow, A.; Macdonald, R.; Wabnitz, H. Optimal estimation reconstruction of the optical properties of a two-layered tissue phantom from time-resolved single-distance measurements. *J. Biomed. Opt.* **2015**, *20*, 115001. [CrossRef] [PubMed]

152. Zucchelli, L.; Contini, D.; Re, R.; Torricelli, A.; Spinelli, L. Method for the discrimination of superficial and deep absorption variations by time domain fNIRS. *Biomed. Opt. Express* **2013**, *4*, 2893–2910. [CrossRef]

153. Re, R.; Contini, D.; Zucchelli, L.; Torricelli, A.; Spinelli, L. Effect of a thin superficial layer on the estimate of hemodynamic changes in a two-layer medium by time domain NIRS. *Biomed. Opt. Express* **2016**, *7*, 264–277. [CrossRef] [PubMed]

154. Farina, A.; Torricelli, A.; Bargigia, I.; Spinelli, L.; Cubeddu, R.; Foschum, F.; Jäger, M.; Simon, E.; Fugger, O.; Kienle, A.; et al. In-vivo multilaboratory investigation of the optical properties of the human head. *Biomed. Opt. Express* **2015**, *6*, 2609–2623. [CrossRef] [PubMed]

155. Liebert, A.; Wabnitz, H.; Steinbrink, J.; Obrig, H.; Möller, M.; Macdonald, R.; Villringer, A.; Rinneberg, H. Time-resolved multidistance near-infrared spectroscopy of the adult head: Intracerebral and extracerebral absorption changes from moments of distribution of times of flight of photons. *Appl. Opt.* **2004**, *43*, 3037–3047. [CrossRef] [PubMed]

156. Benaron, D.A.; Stevenson, D.K. Optical Time-of-Flight and Absorbance Imaging of Biologic Media. *Science* **1993**, *259*, 1463–1466. [CrossRef] [PubMed]

157. Ntziachristos, V.; Ma, X.H.; Yodh, A.G.; Chance, B. Multichannel photon counting instrument for spatially resolved near infrared spectroscopy. *Rev. Sci. Instrum.* **1999**, *70*, 193–201. [CrossRef]

158. Yamada, Y.; Hasegawa, Y.; Yamashita, Y. Simulation of fan-beam-type optical computed-tomography imaging of strongly scattering and weakly absorbing media. *Appl. Opt.* **1993**, *32*, 4808–4814. [CrossRef] [PubMed]

159. Oda, I.; Eda, H.; Tsunazawa, Y.; Takada, M.; Yamada, Y.; Nishimura, G.; Tamura, M. Optical tomography by the temporally extrapolated absorbance method. *Appl. Opt.* **1996**, *35*, 169–175. [CrossRef]

160. Hebden, J.C.; Wong, K.S. Time-resolved optical tomography. *Appl. Opt.* **1993**, *32*, 372–380. [CrossRef]

161. Zevallos, M.E.; Gayen, S.K.; Das, B.B.; Alrubaiee, M.; Alfano, R.R. Picosecond Electronic Time-Gated Imaging of Bones in Tissues. *IEEE Select. Top. Quant. Electron.* **1999**, *5*, 916–922. [CrossRef]

162. Chen, K.; Perelman, L.T.; Zhang, Q.G.; Dasari, R.R.; Feld, M.S. Optical computed tomography in a turbid medium using early arriving photons. *J. Biomed. Opt.* **2000**, *5*, 144–154. [CrossRef] [PubMed]

163. Turner, G.M.; Zacharakis, G.; Soubret, A.; Ripoll, J.; Ntziachristos, V. Complete-angle projection diffuse optical tomography by use of early photons. *Opt. Lett.* **2005**, *30*, 409–411. [CrossRef]

164. Arridge, S.R.; van der Zee, P.; Cope, M.; Delpy, D.T. Reconstruction methods for infrared absorption imaging. *Proc. Spie* **1991**, *1431*, 204–215.

165. Schweiger, M.; Arridge, S. The Toast++ software suite for forward and inverse modeling in optical tomography. *J. Biomed. Opt.* **2014**, *19*, 040801. [CrossRef] [PubMed]

166. Yamada, Y.; Okawa, S. Diffuse Optical Tomography: Present Status and Its Future. *Opt. Rev.* **2014**, *21*, 185–205. [CrossRef]

167. Arridge, S.R.; Schweiger, M. A gradient-based optimisation scheme for optical tomography. *Opt. Express* **1998**, *2*, 213–226. [CrossRef]

168. Intes, X.; Ntziachristos, V.; Culver, J.P.; Yodh, A.; Chance, B. Projection access order in algebraic reconstruction technique for diffuse optical tomography. *Phys. Med. Biol.* **2002**, *47*, N1–N10. [CrossRef]

169. Schweiger, M.; Arridge, S.R. Application of temporal filters to time resolved data in optical tomography. *Phys. Med. Biol.* **1999**, *44*, 1699–1717. [CrossRef]

170. Gao, F.; Zhao, H.; Yamada, Y. Improvement of image quality in diffuse optical tomography by use of full time-resolved data. *Appl. Opt.* **2002**, *41*, 778–791. [CrossRef]

171. Gao, F.; Zhao, H.; Tanikawa, Y.; Yamada, Y. Time-Resolved Diffuse Optical Tomography Using a Modified Pulse Spectrum Technique. *IEICE Trans. Inf. Syst.* **2002**, *E85-D*, 133–142.

172. Gao, F.; Tanikawa, Y.; Zhao, H.; Yamada, Y. Semi-three-dimensional algorithm for time-resolved diffuse optical tomography by use of the generalized pulse spectrum technique. *Appl. Opt.* **2002**, *41*, 7346–7358. [CrossRef]

173. Zhao, H.; Gao, F.; Tanikawa, Y.; Onodera, Y.; Ohmi, M.; Haruna, M.; Yamada, Y. Imaging of in vitro chicken leg using time-resolved near-infrared optical tomography. *Phys. Med. Biol.* **2002**, *47*, 1979–1993. [CrossRef]

174. Gao, F.; Zhao, H.; Tanikawa, Y.; Yamada, Y. Optical tomographic mapping of cerebral haemodynamics by means of time-domain detection: Methodology and phantom validation. *Phys. Med. Biol.* **2004**, *49*, 1055–1078. [CrossRef]

175. Gao, F.; Zhao, H.; Tanikawa, Y.; Homma, K.; Yamada, Y. Influences of target size and contrast on near infrared diffuse optical tomography—a comparison between featured-data and full time-resolved schemes. *Opt. Quant. Electron.* **2005**, *37*, 1287–1304. [CrossRef]

176. Lyubimov, V.V.; Kalintsev, A.G.; Konovalov, A.B.; Lyamtsev, O.V.; Kravtsenyuk, O.V.; Murzin, A.G.; Golubkina, O.V.; Mordvinov, G.B.; Soms, L.N.; Yavorskaya, L.M. Application of the photon average trajectories method to real-time reconstruction of tissue inhomogeneities in diffuse optical tomography of strongly scattering media. *Phys. Med. Biol.* **2002**, *47*, 2109–2128. [CrossRef]

177. Hervé, L.; Puszka, A.; Planat-Chrétien, A.; Dinten, J.-M. Time-domain diffuse optical tomography processing by using the Mellin–Laplace transform. *Appl. Opt.* **2012**, *51*, 5978–5988. [CrossRef]

178. Puszka, A.; Hervé, L.; Planat-Chrétien, A.; Koenig, A.; Derouard, J.; Dinten, J.-M. Time-domain reflectance diffuse optical tomography with Mellin-Laplace transform for experimental detection and depth localization of a single absorbing inclusion. *Biomed. Opt. Express* **2013**, *4*, 569–583. [CrossRef] [PubMed]

179. Puszka, A.; Di Sieno, L.; Mora, A.D.; Pifferi, A.; Contini, D.; Planat-Chrétien, A.; Koenig, A.; Boso, G.; Tosi, A.; Hervé, L.; et al. Spatial resolution in depth for time-resolved diffuse optical tomography using short source detector separations. *Biomed. Opt. Express* **2015**, *6*, 1–10. [CrossRef]

180. Naser, M.A.; Deen, M.J. Time-domain diffuse optical tomography using recursive direct method of calculating Jacobian at selected temporal points. *Biomed. Phys. Eng. Express* **2015**, *1*, 045207. [CrossRef]

181. Benaron, D.A.; Van Houten, J.P.; Cheong, W.-F.; Kermit, E.L.; King, R.A. Early clinical results of time-of-flight optical tomography in a neonatal intensive care unit. *Proc. Spie* **1995**, *2389*, 582–596.

182. Benaron, D.A.; Hintz, S.R.; Villringer, A.; Boas, D.; Kleinschmidt, A.; Frahm, J.; Hirth, C.; Obrig, H.; van Houten, J.C.; Kermit, E.L.; et al. Noninvasive Functional Imaging of Human Brain Using Light. *J. Cereb. Blood Flow Metab.* **2000**, *20*, 469–477. [CrossRef] [PubMed]

183. Hebden, J.C.; Gibson, A.; Yusof, R.M.; Everdell, N.; Hillman, E.M.C.; Delpy, D.T.; Arridge, S.R.; Austin, T.; Meek, J.H.; Wyatt, J.S. Three-dimensional optical tomography of the premature infant brain. *Phys. Med. Biol.* **2002**, *47*, 4155–4166. [CrossRef] [PubMed]

184. Hebden, J.C.; Gibson, A.; Austin, T.; Yusof, R.M.; Everdell, N.; Delpy, D.T.; Arridge, S.R.; Meek, J.H.; Wyatt, J.S. Imaging changes in blood volume and oxygenation in the newborn infant brain using three-dimensional optical tomography. *Phys. Med. Biol.* **2004**, *49*, 1117–1130. [CrossRef] [PubMed]

185. Gibson, A.P.; Austin, T.; Everdell, N.L.; Schweiger, M.; Arridge, S.R.; Meek, J.H.; Wyatt, J.S.; Delpy, D.T.; Hebden, J.C. Three-dimensional whole-head optical tomography of passive motor evoked responses in the neonate. *NeuroImage* **2006**, *30*, 521–528. [CrossRef] [PubMed]

186. Boas, D.A.; Gaudette, T.; Arridge, S.R. Simultaneous imaging and optode calibration with diffuse optical tomography. *Opt. Express* **2001**, *8*, 263–270. [CrossRef]

187. Zeff, B.W.; White, B.R.; Dehghani, H.; Schlaggar, B.L.; Culver, J.P. Retinotopic mapping of adult human visual cortex with high-density diffuse optical tomography. *PNAS* **2007**, *104*, 12169–12174. [CrossRef]

188. Ntziachristos, V.; Ma, X.H.; Chance, B. Time-correlated single photon counting imager for simultaneous magnetic resonance and near-infrared mammography. *Rev. Sci. Instrum.* **1998**, *69*, 4221–4233. [CrossRef]

189. Ntziachristos, V.; Yodh, A.G.; Schnall, M.; Chance, B. Concurrent MRI and diffuse optical tomography of breast after indocyanine green enhancement. *Proc. Natl. Acad. Sci. USA* **2000**, *97*, 2767–2772. [CrossRef]

190. Intes, X.; Djeziri, S.; Ichalalene, Z.; Mincu, N.; Wang, Y.; St-Jean, P.; Lesage, F.; Hall, D.; Boas, D.A.; Polyzos, M. Time-Domain Optical Mammography Softscan®: Initial Results on Detection and Characterization of Breast Tumors. *Proc. Spie* **2004**, *5578*, 188–197.

191. Enfield, L.; Cantanhede, G.; Douek, M.; Ramalingam, V.; Purushotham, A.; Hebden, J.; Gibson, A. Monitoring the response to neoadjuvant hormone therapy for locally advanced breast cancer using three-dimensional time-resolved optical mammography. *J. Biomed. Opt.* **2013**, *18*, 056012. [CrossRef]

192. Zhang, W.; Wu, L.-H.; Li, J.; Yi, X.; Wang, X.; Lu, Y.-M.; Chen, W.-T.; Zhou, Z.-X.; Zhang, L.-M.; Zhao, H.-J.; et al. Combined hemoglobin and fluorescence diffuse optical tomography for breast tumor diagnosis: A pilot study on time-domain methodology. *Biomed. Opt. Express* **2013**, *4*, 331–348. [CrossRef]

193. Hillman, E.M.C.; Hebden, J.C.; Schweiger, M.; Dehghani, H.; Schmidt, F.E.W.; Delpy, D.T.; Arridge, S.R. Time resolved optical tomography of the human forearm. *Phys. Med. Biol.* **2001**, *46*, 1117–1130. [CrossRef]

194. Zhao, H.; Gao, F.; Tanikawa, Y.; Homma, K.; Yamada, Y. Time-resolved diffuse optical tomographic imaging for the provision of both anatomical and functional information about biological tissue. *Appl. Opt.* **2005**, *44*, 1905–1916. [CrossRef]

195. Massoud, T.F.; Gambhir, S.S. Molecular imaging in living subjects: Seeing fundamental biological processes in a new light. *Genes Dev.* **2003**, *17*, 545–580. [CrossRef]

196. Cherry, S.R. In vivo molecular and genomic imaging: New challenges for imaging physics. *Phys. Med. Biol.* **2004**, *49*, R13–R48. [CrossRef]

197. Patterson, M.S.; Pogue, B.W. Mathematical model for time-resolved and frequency-domain fluorescence spectroscopy in biological tissues. *Appl. Opt.* **1994**, *33*, 1963–1974. [CrossRef]

198. Sadoqi, M.; Riseborough, P.; Kumar, S. Analytical models for time resolved fluorescence spectroscopy in tissues. *Phys. Med. Biol.* **2001**, *46*, 2725–2743. [CrossRef]

199. Butte, P.V.; Pikul, B.K.; Hever, A.; Yong, W.H.; Black, K.L.; Marcu, L. Diagnosis of meningioma by time-resolved fluorescence spectroscopy. *J. Biomed. Opt.* **2005**, *10*, 064026. [CrossRef]

200. Milej, D.; Gerega, A.; Żołek, N.; Weig, W.; Kacprzak, M.; Sawosz, P.; Mączewska, J.; Fronczewska, K.; Mayzner-Zawadzka, E.; Królicki, L.; et al. Time-resolved detection of fluorescent light during inflow of ICG to the brain—a methodological study. *Phys. Med. Biol.* **2012**, *57*, 6725–6742. [CrossRef] [PubMed]

201. Milej, D.; Gerega, A.; Wabnitz, H.; Liebert, A. A Monte Carlo study of fluorescence generation probability in a two-layered tissue model. *Phys. Med. Biol.* **2014**, *59*, 1407–1424. [CrossRef] [PubMed]

202. Gao, F.; Zhao, H.; Zhang, L.; Tanikawa, Y.; Marjono, A.; Yamada, Y. A self-normalized, full time-resolved method for fluorescence diffuse optical tomography. *Opt. Express* **2008**, *16*, 13104–13121. [CrossRef] [PubMed]

203. Li, J.; Yi, X.; Wang, X.; Lu, Y.M.; Zhang, L.M.; Zhao, H.; Gao, F. Overlap time-gating approach for improving time-domain diffuse fluorescence tomography based on the IRF-calibrated Born normalization. *Opt. Lett.* **2013**, *38*, 1841–1843. [CrossRef]

204. Riley, J.; Hassan, M.; Chernomordik, V.; Gandjbakhche, A. Choice of data types in time resolved fluorescence enhanced diffuse optical tomography. *Med. Phys.* **2007**, *34*, 4890–4900. [CrossRef]

205. Wu, J.; Perelman, L.; Dasari, R.R.; Feld, M.S. Fluorescence tomographic imaging in turbid media using early-arriving photons and Laplace transforms. *Proc. Natl. Acad. Sci. USA* **1997**, *94*, 8783–8788. [CrossRef]

206. Niedre, M.J.; de Kleine, R.H.; Aikawa, E.; Kirsch, D.G.; Weissleder, R.; Ntziachristos, V. Early photon tomography allows fluorescence detection of lung carcinomas and disease progression in mice in vivo. *Proc. Natl. Acad. Sci. USA* **2008**, *105*, 19126–19131. [CrossRef] [PubMed]

207. Patwardhan, S.V.; Culver, J.P. Quantitative diffuse optical tomography for small animals using an ultrafast gated image intensifier. *J. Biomed. Opt.* **2008**, *13*, 011009. [CrossRef] [PubMed]

208. Leblond, F.; Dehghani, H.; Kepshire, D.; Pogue, B.W. Early-photon fluorescence tomography: Spatial resolution improvements and noise stability considerations. *J. Opt. Soc. Am. A* **2009**, *26*, 1444–1457. [CrossRef]

209. Zhu, Q.; Dehghani, H.; Tichauer, K.M.; Holt, R.W.; Vishwanath, K.; Leblond, F.; Pogue, B.W. A three-dimensional finite element model and image reconstruction algorithm for time-domain fluorescence imaging in highly scattering media. *Phys. Med. Biol.* **2011**, *56*, 7419–7434. [CrossRef]

210. Cheng, J.; Cai, C.; Luo, J. Reconstruction of high-resolution early-photon tomography based on the first derivative of temporal point spread function. *J. Biomed. Opt.* **2018**, *23*, 060503. [CrossRef] [PubMed]

211. Gao, F.; Zhao, H.; Tanikawa, Y.; Yamada, Y. A linear, featured-data scheme for image reconstruction in time-domain fluorescence molecular tomography. *Opt. Express* **2006**, *14*, 7109–7124. [CrossRef] [PubMed]

212. Marjono, A.; Okawa, S.; Gao, F.; Yamada, Y. Light Propagation for Time-Domain Fluorescence Diffuse Optical Tomography by Convolution Using Lifetime Function. *Opt. Rev.* **2007**, *14*, 131–138.

213. Marjono, A.; Yano, A.; Okawa, S.; Gao, F.; Yamada, Y. Total light approach of time-domain fluorescence diffuse optical tomography. *Opt. Express* **2008**, *16*, 15268–15285. [CrossRef] [PubMed]

214. Nishimura, G.; Awasthi, K.; Furukawa, D. Fluorescence lifetime measurements in heterogeneous scattering medium. *J. Biomed. Opt.* **2016**, *21*, 075013. [CrossRef] [PubMed]

215. Kumar, A.T.N.; Skoch, J.; Hammond, F.L., III; Dunn, A.K.; Boas, D.A.; Bacskai, B.J. Time resolved fluorescence imaging in diffuse media. *Spie Proc.* **2005**, *6009*, 60090Y.

216. Soloviev, V.Y.; D'Andrea, C.; Valentini, G.; Cubeddu, R.; Arridge, S.R. Combined reconstruction of fluorescent and optical parameters using time-resolved data. *Appl. Opt.* **2009**, *48*, 28–36. [CrossRef]

217. Yi, X.; Wang, B.-Y.; Wan, W.-B.; Wang, Y.-H.; Zhang, Y.-Q.; Zhao, H.; Gao, F. Full time-resolved diffuse fluorescence tomography accelerated with parallelized Fourier-series truncated diffusion approximation. *J. Biomed. Opt.* **2015**, *20*, 056003. [CrossRef]

218. Chen, J.; Intes, X. Comparison of Monte Carlo methods for fluorescence molecular tomography—computational efficiency. *Med. Phys.* **2011**, *38*, 5788–5798. [CrossRef]

219. Chen, J.; Venugopal, V.; Intes, X. Monte Carlo based method for fluorescence tomographic imaging with lifetime multiplexing using time gates. *Biomed. Opt. Express* **2011**, *2*, 871–886. [CrossRef]

220. Kusaka, T.; Hisamatsu, Y.; Kawada, K.; Okubo, K.; Okada, H.; Namba, M.; Imai, T.; Isobe, K.; Itoh, S. Measurement of Cerebral Optical Pathlength as a Function of Oxygenation Using Near-Infrared Time-resolved Spectroscopy in a Piglet Model of Hypoxia. *Opt. Rev.* **2003**, *10*, 466–469. [CrossRef]

221. Ijichi, S.; Kusaka, T.; Isobe, K.; Islam, F.; Okubo, K.; Okada, H.; Namba, M.; Kawada, K.; Imai, T.; Itoh, S. Quantification of cerebral hemoglobin as a function of oxygenation using near-infrared time-resolved spectroscopy in a piglet model of hypoxia. *J. Biomed. Opt.* **2005**, *10*, 024026. [CrossRef] [PubMed]

222. Ijichi, S.; Kusaka, T.; Isobe, K.; Okubo, K.; Kawada, K.; Namba, M.; Okada, H.; Nishida, T.; Imai, T.; Itoh, S. Developmental Changes of Optical Properties in Neonates Determined by Near-Infrared Time-Resolved Spectroscopy. *Pediatric Res.* **2005**, *58*, 568–573. [CrossRef] [PubMed]

223. Ogawa, K.; Kusaka, T.; Tanimoto, K.; Nishida, T.; Isobe, K.; Itoh, S. Changes in Breast Hemodynamics in Breastfeeding Mothers. *J. Hum. Lat.* **2008**, *24*, 415–421. [CrossRef]

224. Koyano, K.; Kusaka, T.; Nakamura, S.; Nakamura, M.; Konishi, Y.; Miki, T.; Ueno, M.; Yasuda, S.; Okada, H.; Nishida, T.; et al. The effect of blood transfusion on cerebral hemodynamics in preterm infants. *Transfusion* **2013**, *53*, 1459–1467. [CrossRef] [PubMed]

225. Nakamura, S.; Koyano, K.; Jinnai, W.; Hamano, S.; Yasuda, S.; Konishi, Y.; Kuboi, T.; Kanenishi, K.; Nishida, T.; Kusaka, T. Simultaneous measurement of cerebral hemoglobin oxygen saturation and blood volume in asphyxiated neonates by near-infrared time-resolved spectroscopy. *Brain Dev.* **2015**, *37*, 925–932. [CrossRef]

226. Kusaka, T.; Isobe, K.; Yasuda, S.; Koyano, K.; Nakamura, S.; Nakamura, M.; Ueno, M.; Miki, T.; Itoh, S. Evaluation of cerebral circulation and oxygen metabolism in infants using near-infrared light. *Brain Dev.* **2014**, *36*, 277–283. [CrossRef] [PubMed]

227. Ohmae, E.; Oda, M.; Suzuki, T.; Yamashita, Y.; Kakihana, Y.; Matsunaga, A.; Kanmura, Y.; Tamura, M. Clinical evaluation of time-resolved spectroscopy by measuring cerebral hemodynamics during cardiopulmonary bypass surgery. *J. Biomed. Opt.* **2007**, *12*, 062112. [CrossRef]

228. Kakihana, Y.; Okayama, N.; Matsunaga, A.; Yasuda, T.; Imabayashi, M.; Nakahara, M.; Kiyonaga, N.; Ikoma, K.; Kikuchi, T.; Kanmura, Y.; et al. Cerebral Monitoring Using Near-Infrared Time-Resolved Spectroscopy and Postoperative Cognitive Dysfunction. *Oxyg. Transp. Tissue Xxxiiiadv. Exp. Med. Biol.* **2012**, *737*, 19–24.

229. Yasuda, T.; Yamaguchi, K.; Futatsuki, T.; Furubeppu, H.; Nakahara, M.; Eguchi, T.; Miyamoto, S.; Madokoro, Y.; Terada, S.; Nakamura, K.; et al. Non-invasive Monitoring of Hepatic Oxygenation Using Time-Resolved Spectroscopy. *Oxyg. Transp. Tissue Xxxviiadv. Exp. Med. Biol.* **2016**, *876*, 407–412.

230. Kakihana, Y.; Kamikokuryo, C.; Furubeppu, H.; Madokoro, Y.; Futatsuki, T.; Miyamoto, S.; Haraura, H.; Hatanaka, K.; Eguchi, T.; Saitoh, Y.; et al. Monitoring of Brain Oxygenation During and After Cardiopulmonary Resuscitation: A Prospective Porcine Study. *Oxyg. Transp. Tissue Xladv. Exp. Med. Biol.* **2018**, *1072*, 83–87.

231. Sakatani, K.; Yamashita, D.; Yamanaka, T.; Oda, M.; Yamashita, Y.; Hoshino, T.; Fujiwara, N.; Murata, Y.; Katayama, Y. Changes of cerebral blood oxygenation and optical pathlength during activation and deactivation in the prefrontal cortex measured by time-resolved near infrared spectroscopy. *Life Sci.* **2006**, *78*, 2734–2741. [CrossRef] [PubMed]

232. Yokose, N.; Sakatani, K.; Murata, Y.; Awano, T.; Igarashi, T.; Nakamura, S.; Hoshino, T.; Katayama, Y. Bedside Monitoring of Cerebral Blood Oxygenation and Hemodynamics after Aneurysmal Subarachnoid Hemorrhage by Quantitative Time-Resolved Near-Infrared Spectroscopy. *World Neurosurg.* **2010**, *73*, 508–513. [CrossRef]

233. Tanida, M.; Sakatani, K.; Tsujii, T.; Katayama, Y. Relation between working memory performance and evoked cerebral blood oxygenation changes in the prefrontal cortex evaluated by quantitative time-resolved near-infrared spectroscopy. *Neurolog. Res.* **2012**, *34*, 114–119. [CrossRef] [PubMed]

234. Sakatani, K.; Tanida, M.; Hirao, N.; Takemura, N. *Ginkobiloba* Extract Improves Working Memory Performance in Middle-Aged Women: Role of Asymmetry of Prefrontal Cortex Activity during a Working Memory Task. *Oxyg. Transp. Tissue Xxxviadv. Exp. Med. Biol.* **2014**, *812*, 295–301.

235. Machida, A.; Shirato, M.; Tanida, M.; Kanemaru, C.; Nagai, S.; Sakatani, K. Effects of Cosmetic Therapy on Cognitive Function in Elderly Women Evaluated by Time-Resolved Spectroscopy Study. *Oxyg. Transp. Tissue Xxxviiadv. Exp. Med. Biol.* **2016**, *876*, 289–295.

236. Tanida, M.; Okabe, M.; Tagai, K.; Sakatani, K. Evaluation of Pleasure-Displeasure Induced by Use of Lipsticks with Near-Infrared Spectroscopy (NIRS): Usefulness of 2-Channel NIRS in Neuromarketing. *Oxyg. Transp. Tissue Xxxixadv. Exp. Med. Biol.* **2017**, *977*, 215–220.

237. Murayama, Y.; Sato, Y.; Hu, L.; Brugnera, A.; Compare, A.; Sakatani, K. Relation Between Cognitive Function and Baseline Concentrations of Hemoglobin in Prefrontal Cortex of Elderly People Measured by Time-Resolved Near-Infrared Spectroscopy. *Oxyg. Transp. Tissue Xxxixadv. Exp. Med. Biol.* **2017**, *977*, 269–276.

238. Hamaoka, T.; Katsumura, T.; Murase, N.; Nishio, S.; Osada, T.; Sako, T.; Higuchi, H.; Kurosawa, Y.; Shimomitsu, T.; Miwa, M.; et al. Quantification of ischemic muscle deoxygenation by near infrared time-resolved spectroscopy. *J. Biomed. Opt.* **2000**, *5*, 102–105. [CrossRef] [PubMed]

239. Nirengi, S.; Yoneshiro, T.; Sugie, H.; Saito, M.; Hamaoka, T. Human Brown Adipose Tissue Assessed by Simple, Noninvasive Near-Infrared Time-Resolved Spectroscopy. *Obesity* **2015**, *23*, 973–980. [CrossRef]

240. Nirengi, S.; Homma, T.; Inoue, N.; Sato, H.; Yoneshiro, T.; Matsushita, M.; Kameya, T.; Sugie, H.; Tsuzaki, K.; Saito, M.; et al. Assessment of human brown adipose tissue density during daily ingestion of thermogenic capsinoids using near-infrared time-resolved spectroscopy. *J. Biomed. Opt.* **2016**, *21*, 091305. [CrossRef] [PubMed]

241. Fuse, S.; Nirengi, S.; Amagasa, S.; Homma, T.; Kime, R.; Endo, T.; Sakane, N.; Matsushita, M.; Saito, M.; Yoneshiro, T.; et al. Brown adipose tissue density measured by near-infrared time-resolved spectroscopy in Japanese, across a wide age range. *J. Biomed. Opt.* **2018**, *23*, 065002. [CrossRef] [PubMed]

242. Ohmae, E.; Ouchi, Y.; Oda, M.; Suzuki, T.; Nobesawa, S.; Kanno, T.; Yoshikawa, E.; Futatsubashi, M.; Ueda, Y.; Okada, H.; et al. Cerebral hemodynamics evaluation by near-infrared time-resolved spectroscopy: Correlation with simultaneous positron emission tomography measurements. *Neuroimage* **2006**, *29*, 697–705. [CrossRef]

243. Sato, C.; Yamaguchi, T.; Seida, M.; Ota, Y.; Yu, I.; Iguchi, Y.; Nemoto, M.; Hoshi, Y. Intraoperative monitoring of depth-dependent hemoglobin concentration changes during carotid endarterectomy by time-resolved spectroscopy. *Appl. Opt.* **2007**, *46*, 2785–2792. [CrossRef] [PubMed]

244. Yamazaki, K.; Suzuki, K.; Itoh, H.; Muramatsu, K.; Nagahashi, K.; Tamura, N.; Uchida, T.; Sugihara, K.; Maeda, H.; Kanayama, N. Cerebral oxygen saturation evaluated by near-infrared time-resolved spectroscopy (TRS) in pregnant women during caesarean section—a promising new method of maternal monitoring. *Clin. Physiol. Funct. Imaging* **2013**, *33*, 109–116. [CrossRef] [PubMed]

245. Ueda, S.; Nakamiya, N.; Matsuura, K.; Shigekawa, T.; Sano, H.; Hirokawa, E.; Shimada, H.; Suzuki, H.; Oda, M.; Yamashita, Y.; et al. Optical imaging of tumor vascularity associated with proliferation and glucose metabolism in early breast cancer: Clinical application of total hemoglobin measurements in the breast. *BMC Cancer* **2013**, *13*, 514. [CrossRef] [PubMed]

 *applied sciences*

Review

Clinical Brain Monitoring with Time Domain NIRS: A Review and Future Perspectives

Frédéric Lange * and Ilias Tachtsidis

Biomedical Optics Research Laboratory, Department of Medical Physics and Biomedical Engineering, University College London, London WC1E 6BT, UK; i.tachtsidis@ucl.ac.uk
* Correspondence: f.lange@ucl.ac.uk

Received: 28 February 2019; Accepted: 1 April 2019; Published: 18 April 2019

Abstract: Near-infrared spectroscopy (NIRS) is an optical technique that can measure brain tissue oxygenation and haemodynamics in real-time and at the patient bedside allowing medical doctors to access important physiological information. However, despite this, the use of NIRS in a clinical environment is hindered due to limitations, such as poor reproducibility, lack of depth sensitivity and poor brain-specificity. Time domain NIRS (or TD-NIRS) can resolve these issues and offer detailed information of the optical properties of the tissue, allowing better physiological information to be retrieved. This is achieved at the cost of increased instrument complexity, operation complexity and price. In this review, we focus on brain monitoring clinical applications of TD-NIRS. A total of 52 publications were identified, spanning the fields of neonatal imaging, stroke assessment, traumatic brain injury (TBI) assessment, brain death assessment, psychiatry, peroperative care, neuronal disorders assessment and communication with patient with locked-in syndrome. In all the publications, the advantages of the TD-NIRS measurement to (1) extract absolute values of haemoglobin concentration and tissue oxygen saturation, (2) assess the reduced scattering coefficient, and (3) separate between extra-cerebral and cerebral tissues, are highlighted; and emphasize the utility of TD-NIRS in a clinical context. In the last sections of this review, we explore the recent developments of TD-NIRS, in terms of instrumentation and methodologies that might impact and broaden its use in the hospital.

Keywords: NIRS; diffuse optics; time-domain; time-resolved; brain oxygenation; tissue saturation; scattering; absorption

---

1. Introduction

Over the last two decades, the near-infrared spectroscopy (NIRS) field has gained lots of attention as more and more instruments are used in patients within the hospitals [1]. NIRS relies on the fact that light, in the range of 600 to 1000 nm, can penetrate deep into the biological tissue, because in that wavelength range absorption is low, and scattering is the dominant interaction process. Moreover, each absorber, called chromophore, has a specific absorption or extinction coefficient [2]. It is thus possible to quantify the contribution of each chromophore and to resolve its concentration in the tissue. One of the main chromophores in tissue that is oxygen dependent is haemoglobin; NIRS can quantify the in vivo concentrations of oxygenated (HbO_2) and deoxygenated (HHb) haemoglobin and hence monitor tissue oxygenation non-invasively, and in real-time. Since its first introduction by Jobsis in 1977 [3], this technique has been used to probe various types of tissue, such as brain and muscle [4] and breast [5].

From a methodological point of view, a NIRS measurement can be achieved in three different detection modes: continuous wave (CW), frequency domain (FD), and time domain (TD). These modes differ by their level of complexity and the amount of information that they can retrieve. We discuss the details of these techniques in Section 2, and Table 1 lists their main characteristics. Of those three techniques, the CW-NIRS mode is the most widespread and has already been used extensively in the hospitals. Indeed, several reviews can be found in the literature that discuss the use of NIRS in the context of clinical brain monitoring. In particular, a recent review by Obrig discusses the use of NIRS in neuroscience, and the clinical areas of cerebrovascular disease, epileptic disorders, and in the evaluation of the functional activation of the diseased brain [6]. The use of NIRS for monitoring, traumatic brain injury (TBI) [7] and clinical interventions, including the use of the technique in anaesthesiology [8,9], neonatology [10] and psychiatry [11–14], have also been reviewed.

While these reviews demonstrate the potential and popularity of the NIRS technique; there is still debate on whether NIRS can be used reliably in a clinical context. Most NIRS instruments used in hospitals are CW devices due to their commercial availability and simplicity of operation; however, these instruments have several limitations, mainly their lack of brain-specificity and the large variability in absolute quantification of oxygenation [15]. TD-NIRS can resolve these issues and enable an in-depth measurement of absolutes values (see next section). TD-NIRS technology is often viewed as more complicated when compared to CW-NIRS and is not widely available. Indeed, the society for functional Near Infrared Spectroscopy website lists companies providing NIRS instruments [16] and out of the 14 companies listed, only Hamamatsu Photonics (Hamamatsu Photonics K.K., Hamamatsu, Japan) provides commercial TD-NIRS instruments

The goal of the present work is to review the current use of TD-NIRS for clinical applications, with a particular focus on brain monitoring. We will begin with a brief description of the NIRS basics in terms of methodology and technology. Then, we will review the reported work on clinical brain monitoring with TD-NIRS. Finally, we will highlight some recent TD-NIRS developments that we believe will have a direct impact on the clinical popularity of the technique.

Table 1. Characteristic of the three NIRS methodologies.

Main Characteristics	Methodology		
	CW	FD	TD
Practicalities			
Instrument cost	Depends on number of channels, from low to high	High	Very high
Instrument size	Depends on number of channels, wearable systems available	Bulky but transportable	Bulky but transportable
Availability of commercial systems	Yes, multiple companies	Yes, but only one company	Yes, but only one company
Technical considerations			
Number of channels	Can be very high (more than 1000) with the latest HD-NIRS systems. Commercial systems with up to 100 channels	Can be high. Commercial system with up to 500 channels	Lowest. Usually between 1 and 30 channels. Commercial system with up to 2 channels
Sampling rate	Highest: 100 Hz	Middle: 50 Hz	Lowest: 1–10 Hz (can be higher, depends on number of channels, wavelengths)
Non-contact scanning possibilities	Yes	No	Yes
Depth sensitivity	Low, depends on source detector distance	Deep, depends on source detector distance	Deep, do not depends on source detector distance
Measurement parameters			
[HbO$_2$], [HHb], [HbT]	Yes, changes	Yes, absolute but requires multi-distance	Yes, absolute, even with a single channel
Absolute values of absorption and scattering coefficients, and path length	No	Yes, but requires multi-distance	Yes
StO$_2$	Yes, but requires multi-distance	Yes, but requires multi-distance	Yes
DOT possible	Yes	Yes	Yes

Abbreviations: CW: continuous-wave; FD: frequency-domain; TD: time domain; StO$_2$: tissue saturation in oxygen; DOT: diffuse optical tomography; HD-NIRS: high-density NIRS.

2. Basics of NIRS

In the wavelength range used in NIRS (i.e., 600–1000 nm), two main physical phenomena affect the photon trajectory in tissues: scattering and absorption [17]. The scattering is the dominant effect in biological tissue and is caused by microscopic refractive index changes inside the tissue. The resulting effect is that the initial trajectory of the photon is lost, and the photon is deviated to another direction. In NIRS, this quantity is represented by the reduced scattering coefficient μ'_s, which represents the scattering probability per unit length together with the average scattering direction. This parameter can give an insight to the tissue structure [18]. On the other hand, the absorption is the consequence of the loss of a photon which is due to the presence of particular chromophores inside the tissue that convert light intensity into other kinds of energy, either radiative (i.e., fluorescence) or nonradiative (i.e., vibrational). In NIRS, this quantity is represented by the absorption coefficient μ_a, which represents the absorption probability per unit length. As each chromophore has a specific spectral shape, they will each contribute differently to the overall absorption. Then, using a multiwavelength light source with well-chosen wavelengths s us to separate the contribution of each chromophore in order to quantify its concentration. A more detailed explanation of those processes can be found in [17,19].

The effects of absorption and scattering contribute non-linearly to the global attenuation of the light by the tissue. Hence, it is not straightforward to disentangle them. Indeed, an inverse problem must be solved in order to have access to the absorption and scattering independently [20]. Basically, light transport in tissue can be modelled accurately by the radiative transport equation (RTE) or by the simpler diffusion equation (DE), which is derived from the RTE given certain assumptions [19]. Usually, the RTE is solved numerically using Monte Carlo (MC) approaches and gives the most accurate description of light propagation. This accuracy comes at the cost of long calculation time [21]. The DE can also be solved numerically, by using the finite element method (FEM). Software platforms are available to facilitate this process include TOAST++ and NIRFAST [22,23]. However, the real strength of the DE is that it can be solved analytically in various simple geometries [17]. Thus, the calculation time is fast, and real-time methods can be implemented.

As mentioned in the introduction, data acquisition can be performed in three modes: CW, FD, and TD. The principle of those acquisition modes is summarized in Figure 1. Below we describe the basics of each technique, so the reader can appreciate their advantages and disadvantages. For further, more detailed information we refer the reader to [24]. In all the three modes, the basic idea is to shine light onto the tissue, and to collect the transmitted/reflected light few centimetres away from the source point. The single point CW technique is the simplest one; a continuous light is emitted onto the tissue, and the transmitted attenuated light is collected few centimetres away. CW-NIRS only measures the changes in light attenuation, which is defined as the variation and reduction of the transmitted/reflected light intensity from the emitted light. Indeed, using the change in light attenuation at several wavelengths and assuming that light scattering has not changed between measurement time points, CW-NIRS can calculate, using the modified Beer–Lambert law (MBLL) [25], the changes in [HbO$_2$] and [HHb] concentrations, which are assumed to be the sole chromophores that contributes to the contrast. The MBLL considers the extra path length between the source and the detector due to the scattering with a parameter named the differential path length factor (DPF). As this parameter cannot be measured by CW systems, since they cannot distinguish absorption and scattering, this DPF is generally extracted from tabulated values [26] or using a modelling approach [27]. This limits the quantitative accuracy of the measurement since the overall scattering coefficients of the probed tissues is subject-dependent. This is also the reason why single point CW-NIRS only reports changes in haemoglobin concentration. However, some extensions of the CW technique enable us to access some absolute parameters. Indeed, using the spatially resolved spectroscopy (SRS) technique, based on the measurement of the light attenuation at several source/detector separations, the absolute μ_a of the tissue can be estimated by fitting the distribution of the spatially resolved light attenuation measured, to the solution of the DE in the CW regime [28]. Then, measuring the spatial distribution of the light attenuation at several wavelengths, and assuming (1) that [HbO$_2$] and [HHb] are the sole chromophores contributing to the absorption and (2) an a-priori distribution of μ'_s as a function of wavelength, then one can extract information about the absolute tissue oxygen saturation (referred as tissue oxygenation index (TOI) or tissue saturation (StO$_2$) in the literature) [29]. The tissue saturation reflects the ratio between the concentration of [HbO$_2$] and the concentration of total haemoglobin ([HbT] = [HbO$_2$] + [HHb]). This has been widely exploited by commercial brain oximeters [30]. Moreover, some recent developments in CW system methodology employing a broadband spectrum can be used in order to extract absolute optical properties of the tissue [31]. The use of a broadband spectra can also unlock information on a third chromophore, cytochrome-c-oxidase (CCO), which is a marker of the metabolism. Therefore, the detection of this third chromophore can refine the picture of the tissue status obtained by NIRS, through integration of the metabolic and hemodynamic information [32].

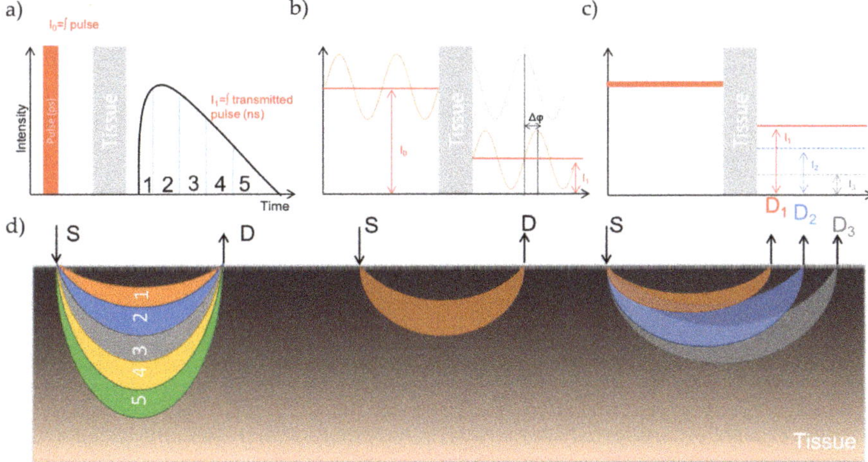

Figure 1. Schematic of the three NIRS detection modes. (**a**) TD-NIRS mode, an ultra-short light impulse is shined onto the tissue and the broadened and attenuated re-emitted pulse is measured after passing through the tissue. (**b**) FD-NIRS, a modulated (MHz) continuous light source is shined onto the tissues and the attenuated, phase shifted reemitted light is measured. (**c**) CW-NIRS, a continuous light is shined onto the tissues and the attenuated re-emitted light is measured. (**d**) Representation of the photon path in tissues for each technique. S: source, D: detector.

One of the drawbacks of NIRS is its poor spatial resolution due to the inherent properties of light transport in tissue. This drawback can be reduced when using not only one, but several detection channels. Indeed, more complex arrangements have been developed to give access to a topographic (2D) or tomographic (3D) maps of brain haemoglobin concentration (referred as DOT: diffuse optical tomo/topography). This approach has largely been exploited in functional NIRS (fNIRS), where brain activity is being monitored [33]. Recently, the development of high-density DOT proved to have a resolution comparable to MRI [34], marking a new cornerstone in the development of the NIRS technology.

The second NIRS technique is frequency domain NIRS (FD-NIRS). In FD-NIRS the light shined onto the tissue is modulated at frequencies in the MHz range. The tissue will attenuate and shift the phase of the reflected light. Analysis of the phase-shift provides information on the total distance that the light has travelled through tissue. To solve the DE and obtain measurements of absorption and scattering, the intensity and phase at two or more, source-detector distances is required [35]. FD-NIRS enables the estimation of absolute brain tissue oxygen saturation and enhances depth resolution comparable to TD-NIRS [36]. However, Davies and colleagues [37] have recently compared CW-NIRS with FD-NIRS in monitoring clinical hypoxia, reporting that there were no advantageous difference observed in parameters recovered from the FD-NIRS compared with those detected by CW-NIRS.

2.1. TD-NIRS Methodology

In TD-NIRS an ultra-short light impulse (typically in the picosecond range) illuminates the tissue. Tissue has the effect of attenuating and broadening the re-emitted pulse (the resulting pulse-width is typically few nanoseconds wide). The measured signal is typically called a temporal point spread function (TPSF) or distribution of time of flights (DTOFs) of photons. Indeed, the measurand here is the histogram of the arrival time of photons. Thus, the amount of information acquired by time resolved systems outstrips the two previous techniques since they only acquire intensity (CW-NIRS) or intensity and phase (FD-NIRS) dimensions.

The DTOF acquired by the TD-NIRS systems contains information on the optical properties of the tissue and enables various data processing workflows to be used. Each method has its own strengths and drawbacks, in terms of computational cost accuracy, or need for a priori, as depicted in Figure 2. There are three main approaches for the data processing of TD-NIRS: the fitting approach, the moment calculation approach, and the gating approach (also called time gating or time windowing).

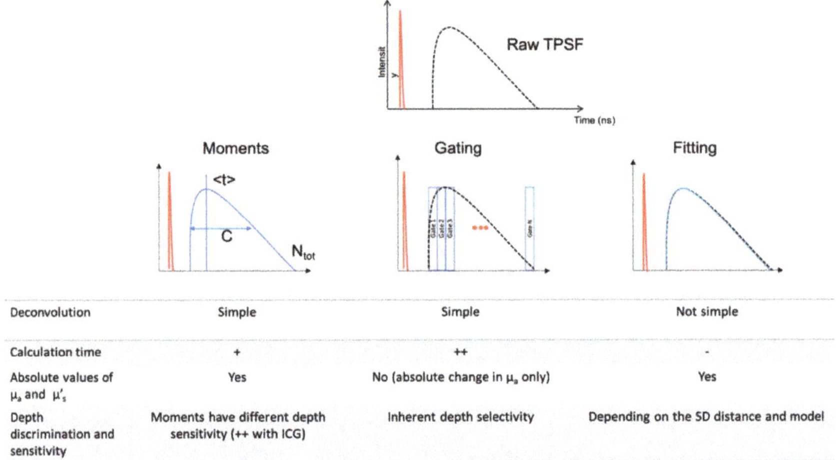

	Moments	Gating	Fitting
Deconvolution	Simple	Simple	Not simple
Calculation time	+	++	-
Absolute values of μ_a and μ'_s	Yes	No (absolute change in μ_a only)	Yes
Depth discrimination and sensitivity	Moments have different depth sensitivity (++ with ICG)	Inherent depth selectivity	Depending on the SD distance and model

Figure 2. Schematic of the three-data processing approach in TD-NIRS.

The fitting approach. This approach was the first to be used to process TPSF data [38]. It consists of minimising the different between an analytical solution of the DE for a particular geometry, and the measured TPSF. Typically, a non-linear fitting procedure based on the classical Levenberg–Marquardt approach is used in order to determine tissue μ_a and μ'_s [39]. Different geometry can be assumed, the most commonly used being the homogeneous semi-infinite one, notably for its simplicity [38]. However, assuming that the head is completely homogeneous might appear an over simplification. More accurate tissue geometries can be considered, such as the multi-layered geometry, that represents more accurately the brain geometry [40]. For example, one could considerer an extra-cerebral and an intra-cerebral layer. As mentioned earlier, FEM and MC methods can also be used to model-specific anatomies which provides the best accuracy. Once the absorption coefficient is known at several wavelengths, the absolute concentration of each chromophore considered can be calculated using the Beer–Lambert Law. Typically, the considered chromophores are HbO_2, HHb and water. The water content is usually set to 80% and the free fitting parameters are [HbO_2] and [HHb] [41]. The tissue saturation can then be extracted from those concentrations. It worth mentioning that no assumption of μ'_s is needed here as it is calculated.

It worth mentioning that in all TD-NIRS measurements, the contribution of the instrument to the DTOF has to be taken into account by measuring what is called the instrument response function (IRF). The ability to deconvolute the IRF from the tissue response function is crucial in order to retrieve accurate data [42]. This deconvolution step is non-trivial and the standard approach when fitting the data is to minimise the error between the TPSF and the model convoluted by the IRF [43].

The moments calculation. This approach reduces the amount of information contained in the entire DTOF by calculating 3 metrics, the 0th order moment corresponding to the total number of photons (equivalent to CW), the 1st order moment corresponding to the mean arrival time of photon (i.e., used to calculate the DPF (see refs. [26,44])), and the second-order moment corresponding to the variance of the TPSF [45]. Thus, it reduces the computational cost of the calculation. Moreover, the deconvolution of the TPSF is made easier since the deconvolution process only requires subtraction of the moments of the IRF from the moments of the TPSF.

Analytical expressions based on these moments are available in order to determine the absorption and scattering coefficients [45]. Once the optical properties are known, the same process as described for the fitting approach can be applied to the haemoglobin concentrations and tissue saturation. Finally, it has been shown that this approach can enhance the depth sensitivity of the measurements, since these three metrics have different depth-sensitivity [46].

The gating (or time gated approach). This approach is based on the separation of the TPSF in different time gates, representing different arrival time of photons. Usually, one can separate an early and a late gate, the early gate being more sensitive to the superficial layers, and the late gate to the deep tissues [47]. This effect is illustrated in supplementary material 1, which shows a video of the change of absorption sensitivity (x, y) as function as the position of the gate in the TPSF. It shows clearly that the late gates are more sensitive to deep tissues [48]. This technique has been used with systems based on ICCD cameras [49,50] or, more recently, SPADs [51].

2.2. Measure of Brain Perfusion

Finally, it worth mentioning that all the approaches described above rely on the fact that the changes in absorption are due to endogenous oxygen-dependent contrast (i.e., haemoglobin). This principle is at the heart of the NIRS technique as it enables us to follow the blood volume (i.e., via HbT), and StO_2, which reflects the balance between the oxygen delivery and the oxygen consumption. This implies that NIRS cannot directly measure the cerebral perfusion. However, this parameter can be assessed by using alternative methodologies and instruments.

Indeed, the use of indocyanine green (ICG), a contrast enhancing agent, makes the assessment of cerebral perfusion with NIRS possible. ICG is a non-toxic light-absorbing dye that has a high absorption and fluorescence emission in the NIR [52]. By following the transit time of an ICG bolus injected intravenously, one can estimate the cerebral blood flow (CBF) [53]. Although, this transit time can be monitored in depth with DOT-CW-NIRS systems [52], it has been showed that when using only one channel, the use of TD-NIRS enhances the depth sensitivity of the ICG bolus tracking and enables us to distinguish between intra- and extra-cerebral layers. Moreover, the use of TD-NIRS gives also access to the measurement of the fluorescence signal of the ICG, which also increase the depth sensitivity [54]. The measured signal of fluorescence is often referred to as the distributions of times of arrival of fluorescence (DTAs). The main advantage of this technique is that it can provide a CBF value in standard physiological units (i.e., mL of blood/100 g/min) [55]. The main drawback of this technique is that it cannot continuously monitor the CBF as the injection of ICG is limited by a maximum recommended daily dose. We will then refer to this technique as DCE-TD-NIRS, where DCE stands for dynamic contrast enhanced.

On the other hand, diffuse correlation spectroscopy (DCS) is a technique that enables continuous monitoring of a CBF index called blood flow index (BFI) without the injection of a contrast agent. Briefly, in DCS, the tissue of interest is illuminated by coherent near-infrared light, which causes a speckle interference pattern to form after the light scatters multiple times through the tissue. Thus, the changes in light scattering due to the moving red blood cells causes the speckle pattern to fluctuate rapidly. Those speckle patterns are then analysed to extract blood flow information. In order to quantify the fluctuation of the speckle pattern, the optical properties of the tissue need to be assumed or measured. Therefore, the use of TD-NIRS is well-suited to provide such information, especially since the two techniques are compatible and can be combined. A more detailed explanation of DCS principles can be found in [56].

3. Clinical Brain Monitoring with TD-NIRS

The focus if the present review was to evaluate the use of TR-NIRS in a clinical context regarding brain monitoring. Therefore, papers were identified using PubMed and Scopus, searching for a combination of keywords including (near-infrared spectroscopy | near infrared | optical | tomography) and (time domain | time resolved) and (brain | cerebral). Moreover, a manual search from articles' references was performed and authors related to the subject were contacted. Papers were rejected if only CW-NIRS or FD-NIRS were used, if the application was not focused on the brain, and if no clinical application could be identified (i.e., applications on healthy volunteers or animals only). A total of 52 publications covering a variety of applications were identified. We have grouped the applications into seven subgroups that we will discuss below: neonatal applications (19 studies), stroke assessment (8 studies), traumatic brain injury (TBI) assessment (four studies), psychiatry (nine studies), brain death assessment (three studies), perioperative care assessment (seven studies), and other applications (two studies).

Table 2 provides a summary of the main characteristics of the instrument used in the studies reported in this review, and Figure 3 shows pictures of the instrumentation typically used. A more detailed list of TD-NIRS instruments is available in the review by Torricelli and colleagues [57]. Moreover, a summary of the studies included in this review is presented in Table 3 and includes the system used, the number of patients scanned, the methodology used to process the data, and finally states the measurand reported in these studies.

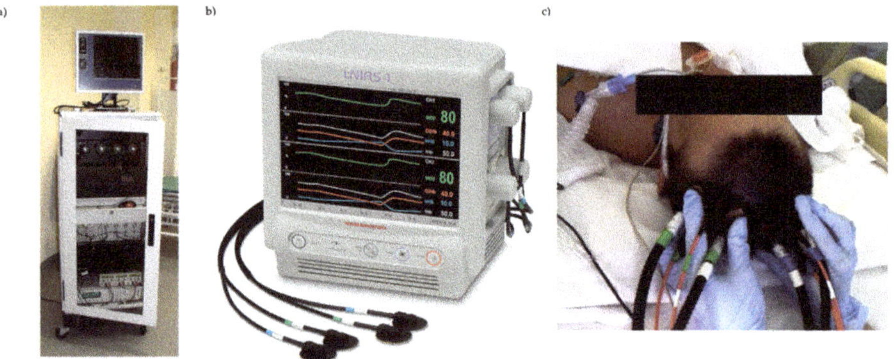

Figure 3. Example of homemade and commercial TD-NIRS instrumentation. (**a**) Homemade TD-NIRS system used in [58], used with ICG. (**b**) The tNIRS-1 commercial system, from Hamamatsu (Japan), courtesy of Hamamatsu. (**c**) Example of probe positioning over both cerebral hemispheres of a patient (the system is the one illustrated in (**a**)). Extracted from [58].

Finally, Figure 4 summarizes the main characteristic of the TD-NIRS systems used in the reviewed publications and shows the number of publications over the last 20 years that report the use of TD-NIRS in a clinical context.

Table 2. List of the system used in the publication reviewed, with their principle characteristics.

SYSTEM ID	GROUP	SYSTEM NAME	NB OF WAVELENGTH USED	WAVELENGTH USED	REPETITION RATE (MHZ)	SOURCE TYPE	DETECTOR TYPE	NB OF CHANNELS	ACQ. FREQ. (HZ) **
1	Hamamatsu Photonics K.K.	TRS-10	3	761, 795, 835 nm	5	Laser diode	PMT	1	0.2
2	Hamamatsu Photonics K.K.	TRS-20	3	759,797,833	5	Laser diode	PMT	2	0.2
3	Hamamatsu Photonics K.K.	TRS-21	6	762, 802, 838, 908, 936, 976 nm	5	Laser diode	PMT	2	0.2
4	Hamamatsu Photonics K.K.	tNIRS-1	3	755, 826, 850 nm	9	Laser diode	MPPC	2	0.2
5	Hamamatsu Photonics K.K.	NIRO-TRS 1	3	755, 826, 850 nm	9	Laser diode	MPPC	2	0.2
6	Standford University, Standford, USA	NA	2	785, 850 nm	NA	Laser diode	PMT	34	>0.05
7	UCL, London, UK	MONSTIR	2	780, 815 nm	40	Ti:sapphire laser	MCP-PMT	32	< 0.05
8	UCL, London, UK	MONSTIR II	4	690, 750, 800, 850 nm	40	Supercontinuum laser	PMT	32	≈ 0.01
9	Politecnico di Milano, Milan, Italy	NA	2	690, 820 nm	80	Laser diode	PMT	24	1
10 *	BabyLux project	BabyLux	3	685, 760, 822 nm	20	Laser diode	hybrid PMT	1	1
11	Physikalisch-Technische Bundesanstalt, Berlin, Germany	NA	2	803, 807nm	60	Laser diode	PMT	4	20
12	Physikalisch-Technische Bundesanstalt, Berlin, Germany	NA	1	785 nm	50	Laser diode	PMT	8	20
13	Physikalisch-Technische Bundesanstalt, Berlin, Germany	NA	3	687, 803, 826 nm	45	Laser diode	PMT	4	0.4
14	Politecnico di Milano, Milan, Italy	NA	3	690, 785, 830 nm	80	Laser diode	hybrid PMT	1	0.03
15	Institute of Biocybernetics and Biomedical Engineering, Warsaw, Poland	NA	2	687, 832 nm	80	Laser diode	PMT	32	10
16*	University of Pennsylvania, Philadelphia, Pennsylvania, USA	NA	7	730, 750, 786, 810, 830, 850, 808 nm	78	Supercontinuum laser	hybrid PMT	1	0.9
17	Institute of Biocybernetics and Biomedical Engineering, Warsaw, Poland	NA	1	760 nm	80	Laser diode	PMT	8	10
18	Western University, London, Ontario, Canada	NA	2	760, 830 nm	80	Laser diode	hybrid PMT	4	3.33

* Systems 10 and 16 also have DCS instrumentation. Here only the TD-NIRS part is reported. Refer to [59,60] respectively for complete instrumental details. ** Typical values than can vary depending on the instrument settings. See individual references for more details. Abbreviations: PMT: photomultiplier tube; MCP-PMT: microchannel plate PMT; MPPC: multi-pixel photon counters.

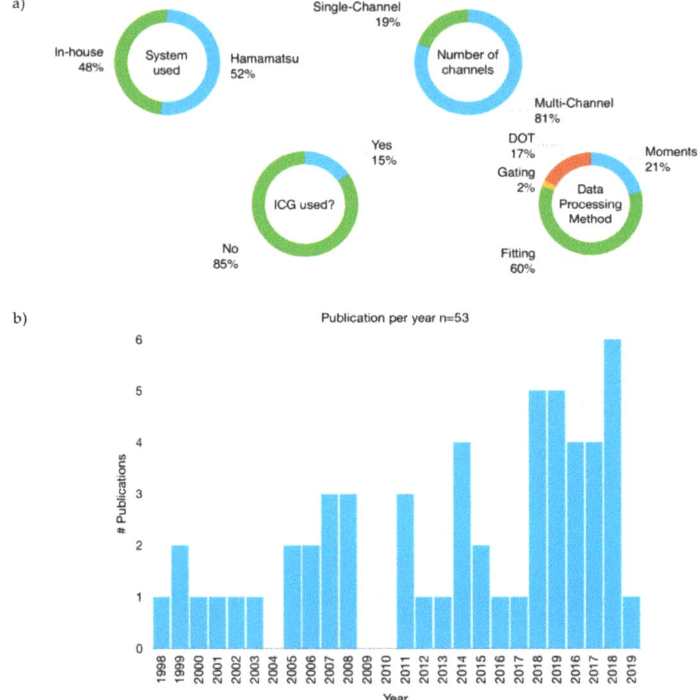

Figure 4. Summary of the literature review results. (**a**) Summary of the principle TD-NIRS component used in the 52 reviewed publications. (**b**) Graph showing the number of publications using TD-NIRS in brain clinical application since the first study in 1998.

3.1. Neonatology

The use of TD-NIRS in neonatology is the area that has been one of the most explored, starting from the very early days of TD-NIRS. This particular population is well-suited to NIRS, allowing DOT to be performed [61]. Indeed, the first papers reporting the use of an in-house developed TD-NIRS instrument (ID: 6 in Table 2) on neonates dates from the late 1990s, by the group of Dr. David Benaron at Stanford University, USA. In a series of papers [62–65], they reported the first tomographic images of neonates using TD-NIRS. This team then successfully tomographically-identified intracranial haemorrhage [62,63] and focal regions of low oxygenation after acute stroke [65]. Hintz and colleagues [66] also retrieved a topographic image of the activation of the motor cortex of one infant measured at six days postnatal age.

At University College London (UCL), UK, an in-house 32 channel TD-NIRS system was developed. This system was called MONSTIR (for Multi-channel Opto-electronic Near-infrared System for Time-resolved Image Reconstruction) [67,68]. With this system, tomographic images of healthy infants [69] and infants with intraventricular haemorrhage (IVH) [69,70] were reconstructed. An example of their reconstructed images is provided in Figure 5a. The image reconstruction required a comparison of the in-vivo data with the data acquired on a phantom with well-known optical properties, producing an image of the differences. The natural extension of this work was to be able to reconstruct images without the necessity to image a reference phantom. Hebden and colleagues [71] first demonstrated this possibility on a severely brain-injured 38-week-old newborn. The changes in absorption and scattering coefficient, and in [HbO$_2$] and [HHb], resulting from small alterations in ventilator settings could be detected during an imaging session of three hours at two wavelengths. Following that, Gibson and colleagues [72], reported the first 3D optical images of entire pre-term babies' heads during motor-evoked response. More recently, the UCL team reported the development of the second generation of the TD-DOT-NIRS MONSTIR, called MONSTIR II [68]. Dempsey and colleagues [73] reported to have successfully scanned a healthy infant with MONSTIR II. This work was largely exploratory to demonstrate the quality of the recording of the data acquired with MONSTIR II in a clinical environment. Moreover, in her PhD thesis [74], Dempsey reported the use of moments to reconstruct images of babies presenting with perinatal arterial ischemic stroke (PAIS). Here, she explored the possibility of using the calculated moments (intensity, time of flight, variance) as a metric to locate the stroke area.

In the literature there is a significant amount of publications using commercial TD-NIRS systems developed by Hamamatsu in neonates. Koyano and colleagues [75] investigated the effect of blood transfusion on cerebral haemodynamic in preterm infants using a TRS-10 system, suggesting that TD-NIRS can be used for determining the need for transfusion in very-low-birth-weight infants. Ishii and colleagues [76] used a TRS-20 system to compare changes in cerebral and systemic perfusion between appropriate- and small-for-gestational-age (AGA and SGA, respectively) infants during the first three days after birth. Also using a TRS-20, Fujioka and colleagues [77] reported the difference in brain oxygenation between term and preterm infants during the first three days of life.

Nakamura and colleagues [78] used a TRS-10 system to measure StO_2 and [HbT] in babies with hypoxic-ischemic encephalopathy (HIE), reporting that neonates with adverse outcomes had higher values within 24 h after birth compared with neonates with a favourable outcome. Additionally, using a TRS-10 system, Hoshino and colleagues [79] reported the changes in oxygenation concentration and tissue saturation before and after an endovascular treatment in a 10 months old patient.

Lastly, some work has been done in order to evaluate normal values of optical properties and physiological parameters in neonates. Ijichi and colleagues [80] used a TRS-10 to investigate both the optical properties and relevant physiological parameters of a population of 22 neonates. They reported the absorption and reduced scattering coefficient, DPF, total haemoglobin concentration and tissue saturation, together with the blood volume. They also showed the relationship between those parameters and the gestational age. The use of the TD-NIRS is highlighted here, with the potential to give detailed information about μ_a and μ'_s, and physiological parameters. Spinelli and colleagues [81] used an in-house developed instrument (ID: 9 in Table 2) to evaluate the optical properties and haemoglobin parameters of 33 neonates. They found consistencies in the measures of blood oxygenation but variations in the μ_a, similar to adult population [82]. They concluded that more work was needed to properly define standard values. Finally, preliminary results from a combined TD-NIRS/DCS system reported the evaluation of μ_a, μ'_s and StO_2, together with the BFI of 21 babies, using the BabyLux system [59].

3.2. Stroke Assessment

The utility of using NIRS technology in the context of stroke has been reviewed by Obrig and Steinbrink in 2011 [83]. In that review, they concluded that using TD-NIRS is mandatory for the assessment of pathophysiological state-dependent variables such as perfusion or absolute values of tissue oxygenation. The first works reporting the use of TD-NIRS on stroke patients was by Liebert and colleagues [84,85]. In these studies, the cerebral perfusion was measured by using the method of DCE-TD-NIRS. The bolus transit time through intra- and extra-cerebral tissue was measured in both hemispheres of two stroke patients, using an in-house developed multichannel TD-NIRS system (ID: 11 in Table 2). Steinkellner and colleagues [86] also reported the use of DCE-TD-NIRS to measure the cerebral perfusion in 10 acute stroke patients using an in-house developed multichannel TD-NIRS system (ID: 12 in Table 2). The authors reported to have used the kinetics of the moments to follow the kinetics of the ICG bolus. The kinetic difference between the affect and unaffected hemisphere was compared. The analysis of the variance enabled them to substantially reduce the movement artefacts and to follow the intracerebral compartment, compared to the traditional signals acquired with CW systems, which have been simulated here by time integrating the TPSF (illustrated in Figure 5b). Using the variance, the authors reported a clear difference in perfusion between the affected and non-affected hemisphere.

Other groups reported the use of an in-house developed system to follow the brain oxygenation of stroke patients. Leistner and colleagues [87] compared the neurovascular coupling, induced by a simple motor test, between six patients with subacute ischemic stroke and four healthy volunteers. In this study, the neurovascular coupling was measured using a DC-MEG and a TD-NIRS system (ID: 13 in Table 2) simultaneously. Another in-house developed TD-NIRS system (ID: 14 in Table 2) was used by Giacalone and colleagues [88] to measure haemoglobin concentrations and StO_2 of the anterior circulation in stroke patients. In this study, 47 stroke patients were monitored in the stroke unit at the bedside. Multiple areas of the brain were probed (i.e., frontal, parietal, occipital) and CT-scans images were used to categorize brain regions below each TD-NIRS probe. Moreover, the results were compared with a group of 35 controls selected from a previous study reported in [82]. Significant differences of haemoglobin concentrations and StO_2 were found between patients and controls, and among patients according to recanalization status.

Finally, commercial Hamamatsu TD-NIRS systems were also used on stroke patients. Yokose and colleges [89,90] reported the detection of vasospasm in patients with aneurismal subarachnoid haemorrhage (SAH). Another study by Sato and colleagues [91] used a TRS-20 system to measure haemoglobin concentrations, StO_2, and optical path length of five patients suffering from chronic stroke. They compared the results between the unaffected and affected hemisphere. Significant differences between the unaffected and affected hemisphere were found for StO_2 (higher in the affected side), [HHb] (higher in the affected side) and optical path length (longer in the affected side).

3.3. Traumatic Brain Injury (TBI) Assessment

The use of NIRS technology in traumatic brain injury (TBI) has been reviewed by Davies and colleagues in 2015 [92]. Here we explicitly review and discuss the use of TD-NIRS systems in the same patient population. We found three recent studies investigating the brain status of TBI patients. Weigl and colleagues [58], used the DCE-TD-NIRS method to assess cerebral perfusion in 26 TBI patients and nine controls, using an in-house developed multichannel TD-NIRS system (ID: 15 in Table 2). In this study, the moments of the DTOFs and DTAs were calculated, and their time courses were compared, to track the ICG bolus. Statistically significant differences between the time courses of the ICG bolus calculated using the different moment parameters obtained in healthy subjects and patients with brain haematoma and brain oedema were observed. Moreover, it was noted that the best optical parameter to differentiate between patients and controls was the variance of DTOFs and DTAs.

Highton and colleagues [93] focused on the changes in light scattering related to TBI. In this study, a total of 21 TBI patients were monitored with a TRS-20 system. In 10 patients suffering from unilateral infarction, even though no significant difference in cerebral saturation or haemoglobin concentration were found, a significant reduction of μ'_s was related to the injury severity.

Finally, He and colleagues [60] used an in-house developed TD-NIRS and DCS system (ID: 16 in Table 2) to continuously monitor the absolute blood flow (i.e., using physiological units, see Section 2) of seven patients with acute brain injury. This absolute assessment was possible by coupling the DCE-TD-NIRS, to extract absolute values, and DCS, to continuously assess the blood flow. Here patients were monitored 8 h a day, four days. This illustrates the use of state-of-the-art TD-NIRS technology and instrumentation to collect information with comprehensive units for a physician.

3.4. Brain Assessment in Death

TD-NIRS could also be used to assess the adequacy of resuscitation. In a pilot study, Lanks and colleagues [94] used a NIRO-TRS1 Hamamatsu system to monitor the cerebral saturation and haemoglobin concentration in 11 patients with septic shock, and compared those measurements with traditional markers of perfusion adequacy. In another study using the same system, Lanks and colleagues [95] reported the case of a young man who suffered a pulseless electrical activity (PEA) arrest (i.e., cardiac arrest) while cortical oxygenation was monitored. The onset of cortical deoxygenation preceded the loss of palpable pulses for 15 min, despite otherwise stable measures of perfusion, reflecting falling cortical microvascular haemoglobin concentration and oxygenation minutes before PEA arrest. This result is presented on Figure 5b. Even though it is a single subject study, this suggest that TD-NIRS might provide the means of detecting PEA arrest, by monitoring cortical oxygenation.

Finally, Liebert and colleagues [96] and Weigl and colleagues [97] investigated the possibility to use DCE-TD-NIRS to confirm brain death. Using an in-house developed multichannel TD-NIRS systems (ID: 17 in Table 2), different stages of cerebral perfusion disturbances were compared in three groups that included controls, patients with post-traumatic cerebral oedema, and patients with brain death. The authors reported that DCE-TD-NIRS was feasible in the demanding ICU environment and had promising initial results. However, at its current stage of development, the accuracy of the method did not reach the high standards required for brain death confirmation.

3.5. Psychiatry and Cognitive Impairement

Another field of clinical application of TD-NIRS is psychiatry, in which NIRS is used to evaluate functional brain activation in different neuropsychiatric disorders, notably in schizophrenic disorders, and normal and pathological aging [12].

In the context of schizophrenia, we found two publications that used Hamamatsu TD-NIRS systems to provide baseline measurement of haemoglobin concentration, tissue saturation and optical path length in this group of patients. In these studies, fNIRS was also performed but using a CW-NIRS system. We will therefore only report the TD-NIRS results. Firstly, Shinba and colleagues [98] used TD-NIRS to investigate the frontal lobe dysfunction in schizophrenia. The authors reported no significant difference in path length between the control group and schizophrenic group. However, in some subjects, the path length was different from the mean value of the group by more than 20%, highlighting the importance of a subject specific path length in order to retrieve accurate quantitative values of the change of the haemoglobin concentrations. Hoshi and colleagues [99] investigated the resting hypofrontality in patients with schizophrenia. They reported that the [HbT] decrease was related to the duration of illness, and it was not observed in patients whose duration of illness was less than 10 years. The authors concluded that despite the fact that resting hypofrontality is a chronically developed feature of schizophrenia, this does not necessarily represent frontal dysfunction, but may reflect anatomical and/or functional changes in frontal microcirculation.

On the other hand, several studies addressed the issue of cognitive impairments and aging using Hamamatsu systems. Harada and colleagues [100] compared the haemoglobin concentration measured by a TRS-10 system, and the cerebral activity, measured by a CW-NIRS system, of the prefrontal cortex between 14 young adults and 14 elderly adults while driving. The main finding regarding the TD-NIRS results for this study was that the concentrations of all forms of haemoglobin in elderly adults were lower compared to young adults.

Two papers [101,102] reported the use of a TRS-20 system to evaluate the effect of cosmetic therapy on prefrontal cortex (PFC) activity in 61 elderly females. The [HbT] and StO_2 were evaluated before and after a three-month program of cosmetic therapy. Murayama and colleagues [103] used a TRS-21 system to investigate the relationship between cognitive function and haemoglobin concentration and tissue saturation of the prefrontal cortex (PFC) during rest in 113 elderly adults. The authors quantified cognitive functions using the Mini-Mental State Examination (MMSE) score and the Touch M score, which evaluates working memory function semi-automatically on a touchscreen. A significant positive correlation between the MMSE scores and StO_2 was found together with a significant positive correlation between Touch M scores and baseline concentrations of [HbO_2], [HbT] and StO_2. It is worth noting that the authors reported that the main advantage of TD-NIRS in that study was to measure absolute information at rest; allowing them, therefore, to assess cognitive function without any task (i.e., fNIRS using CW-NIRS). Oyama and colleagues [104] investigated the possibility of using the haemoglobin concentration and StO_2 information acquired with a TRS-20 system to predict the MMSE score of a group of 202 elderly people. The MMSE was conducted first and then TD-NIRS measurement was performed. The authors found using a deep neural network (DNN) that the StO_2 and the optical path lengths contributed to the prediction of the MMSE score.

Finally, two publications by Sakatani and colleagues [105] and Nakano and colleagues [106] investigated the effects of two drugs on the improvement of cognitive function using a TRS-20 system. In these two studies, the ability of TD-NIRS to measure absolute haemoglobin concentration and StO_2 were crucial, since the measurements were taken several weeks apart.

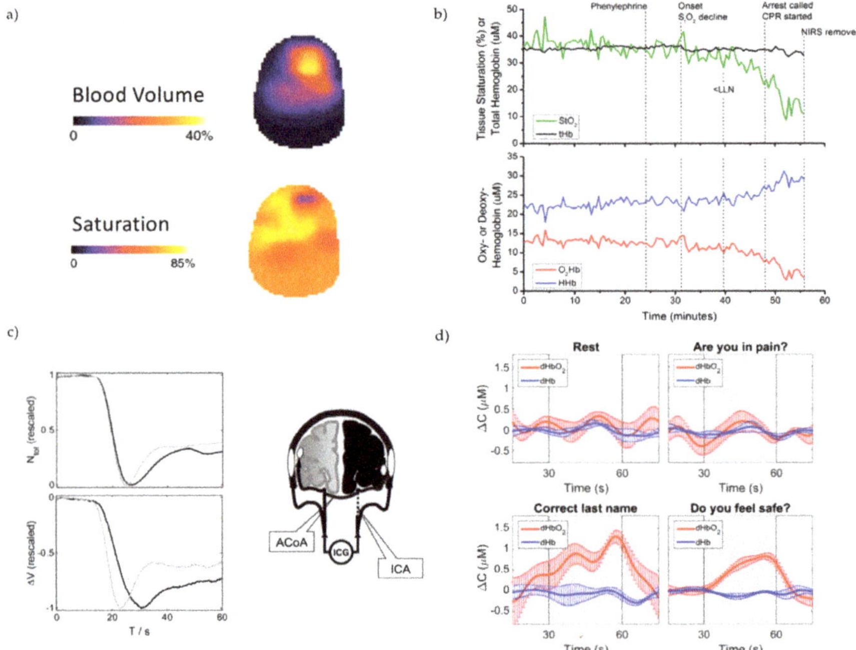

Figure 5. Example of the data produced by TD-NIRS systems. (**a**) Coronal slices of tomographic images of the blood volume and the StO_2 of a baby who had a left ventricular haemorrhage. The haemorrhage can be seen as an increase in blood volume and a decrease in oxygen saturation. Extracted from [69]. (**b**) StO_2 and haemoglobin concentration ([tHb]) in a patient with septic shock during the hour prior to a pulseless electrical activity arrest and cardiopulmonary resuscitation (CPR). The decrease in StO_2 can be seen 15 min prior to the arrest. Extracted from [95]. (**c**) Illustration of a measurement from [86], on a patient with an internal carotid artery (ICA) occlusion on the affected side and a crossflow over the anterior communicating artery (ACoA). The anatomical situation and the influence on the bolus passage are shown schematically on the right diagram, with the affected hemisphere shown in black colour and the unaffected one shown in light grey. The left side of the figure compares the conventional CW-NIRS based approach, which is represented by the time-integrated signal N_{tot} to the moment TD-NIRS approach based on the analysis of variance. The colours of the bolus curves correspond the diagram (black: affected hemisphere, light grey: unaffected hemisphere). In the CW-NIRS approach, the bolus curves differed only slightly. In contrast, in the TD-NIRS approach, the variance signals showed a significant difference between the hemispheres. (**d**) Example of brain activity of a patient with locked-in syndrome during mental imagery (MI), from [107]. The "rest" period corresponds to data acquired without MI activation and is presented as a reference for the contrast observed during the question periods. If the answer was "Yes" the patient was asked to perform a MI task (i.e., play tennis). If the answer was "No", the patient was asked to rest. The grey boxes indicate the response period. One can see that the patient responded "yes" to two questions. At the time of this study, the patient had regained limited eye movement, so those responses could be confirmed.

3.6. Peroperative Care

It has been demonstrated that the use of NIRS to monitor the level of anaesthesia could lead to better postoperative cognitive performance of the patient [108].

In a peroperative context, Kacprzak and colleagues [109] evaluated the application of TD-NIRS to monitor cerebral oxygenation during carotid surgery. The authors used an in-house developed multichannel TD-NIRS system (ID: 15 in Table 2) on a group of 16 patients with atherosclerotic disease and monitored them during routine carotid endarterectomy surgery. The authors reported that the changes in the haemoglobin signals, as estimated from intracerebral tissue, are very sensitive to clamping of the internal carotid artery, whereas its sensitivity to clamping of the external carotid artery is limited. This is a good indicator that the TD-NIRS method used in this paper can achieve in-depth selectivity.

Furthermore, seven studies using Hamamatsu systems were reported in a peroperative context. Ohmae and colleagues [110] reported the monitoring of cerebral circulation during cardiopulmonary bypass surgery using a TRS-10 system. In this study, 23 patients who underwent coronary-artery bypass surgery were evaluated reporting that TD-NIRS can be used to follow brain oxygenation during cardiac bypass surgery. Moreover, the authors reported a large fluctuation of the mean optical path length among patients. The authors also reported that the inter-patient variability of saturation was low compared to that of the haemoglobin concentrations. This finding was also recently reported in healthy volunteers [82]. In another study, Kakihana and colleagues [111] also monitored the cerebral tissue saturation of patients undergoing a cardiac surgery requiring a cardiopulmonary bypass using a TRS-10 system. The jugular venous oxygen saturation ($SjvO_2$) was also monitored. Moreover, postoperative cognitive dysfunction (POCD) was evaluated using the MMSE before and seven days after the operation. It was found that there was no significant difference between the $SjvO_2$ and StO_2 in patients without POCD, but that those two values were significantly different in patients with POCD. Therefore, the combination of $SjvO_2$ and cerebral StO_2 might help predict the occurrence of POCD. Lastly, Satao and colleagues [112] also used a TRS-10 system to investigate the depth-dependence of haemoglobin concentration changes during carotid endarterectomy (CEA) in eight male patients with symptomatic carotid stenosis during CEA surgery reporting brain specificity of the TD-NIRS.

Additionally, two studies using TRS-20 systems investigated the cerebral oxygenation of pregnant women during child delivery. Yamazaki and colleagues [113] reported cerebral oxygenation, on the PFC, of eighteen pregnant women during caesarean section, reporting that in a case of placenta previa, massive bleeding immediately decreased cerebral StO_2. Suzuki and colleagues [114] reported the cerebral [HbT] in pregnant women in delivery. They showed difference between normotensive and hypertensive women with or without epidural anaesthesia. Moreover, a decrease in cerebral [HbT] in the hemi-prefrontal lobe in a woman 2 h after the onset of haemorrhagic stroke in labour could also be detected. This reinforces the possibilities of using TD-NIRS in the context of child delivery.

Finally, Fujisaka and colleagues [115] using the latest TD-NIRS device developed by Hamamatsu (a two-channel tNIRS-1) reported in one patient a decrease in brain [HbT] and StO_2 during circulatory arrest in cardiac surgery.

Table 3. Table of the literature review of the use of TD-NIRS for clinical brain monitoring.

Publication	Year	Patients N = P (C) P: Patients, C: Controls	Apparatus (ID)	Method	Perfusion DCS	ICG	[HbO$_2$]	[HHb]	[HbT]	StO$_2$	μ_a	μ'_s	Path Length	Other
Nenonates (number of studies = 19)														
Hintz et al. [62]	1998	N = 1 (0)	In-house (6)	DOT							x			
Hintz et al. [63]	1999	N = 6 (0)	In-house (6)	DOT							x			
Hintz et al. [64]	1999	N = 6 (0)	In-house (6)	DOT							x			
Benaron et al. [65]	2000	N = 2 (0)	In-house (6)	DOT						x	x			
Hintz et al. [64]	2001	N = 2 (0)	In-house (6)	DOT			x	x			x			
Hebden et al. [70]	2002	N = 3 (0)	In-house (7)	DOT					x	x	x	x		
Hebden et al. [71]	2004	N = 1 (0)	In-house (7)	DOT			x	x			x	x		
Ijichi et al. [80]	2005	N = 22 (0)	TRS-10 (1)	Fitting			x	x	x	x	x	x	x	
Austin et al. [72]	2006	N = 24 (0)	In-house (7)	DOT					x	x	x			
Gibson et al. [69]	2006	N = 6 (0)	In-house (7)	DOT			x	x	x		x	x		
Hoshino et al. [79]	2010	N = 1 (0)	TRS-10 (1)	Fitting			x	x	x	x				
Koyano et al. [75]	2013	N = 19 (0)	TRS-10 (1)	Fitting			x	x	x	x	x	x	x	
Fujioka et al. [77]	2014	N = 72 (0)	TRS-20 (1)	Fitting					x	x				
Ishii et al. [76]	2014	N = 87 (0)	TRS-20 (1)	Fitting					x	x				
Nakamura et al. [78]	2015	N = 11 (0)	TRS-10 (1)	Fitting					x	x				
Dempsey et al. [73]	2015	N = 1 (0)	In-house (8)	Moments										x
Spinelli et al. [81]	2017	N = 33 (0)	In-house (9)	Fitting			x	x	x	x	x	x		
Giovannella et al. [59]	2018	N = 21 (0)	In-house (10)	Fitting	x					x	x	x		
Dempsey (PhD thesis) [74]	2018	N = 5 (0)	In-house (8)	Moments										x
Stoke assessment (number of studies = 8)														
Liebert et al. [84]	2005	N = 2(2)	In-house (11)	Moments		x								
Yokose et al. (a) [90]	2010	N= 14 (11)	TRS-10 (1)	Fitting			x	x	x	x				
Yokose et al. (b) [89]	2010	N= 10 (0)	TRS-10 (1)	Fitting			x	x	x	x				
Leistner et al. [87]	2011	N =11 (0)	In-house (13)	Moments		x								
Steinkellner et al. [86]	2012	N =10 (0)	In-house (12)	Moments		x								
Liebert et al. [85]	2012	N = 2 (2)	In-house (11)	Moments		x								
Sato et al. [91]	2018	N = 5 (0)	TRS-20 (2)	Fitting			x	x	x	x			x	
Giacalone et al. [88]	2019	N = 47 (35)	In-house (14)	Fitting			x	x	x	x				
Traumatic brain injury (TBI) assessment (number of studies = 3)														
Weigl et al. [58]	2014	N = 26 (9)	In-house (15)	Moments		x								
Highton et al. [93]	2016	N = 21 (0)	TRS-20 (2)	Fitting					x	x		x		
He et al. [60]	2018	N = 7 (0)	In-house (16)	Fitting	x	x								
Brain assessment in death (number of studies = 4)														
Liebert et al. [96]	2014	N = 35(NA)	In-house (17)	Moments		x								
Lanks et al. [94]	2017	N = 11 (0)	NIRO-TRS1 (5)	Fitting					x	x				
Lanks et al. [95]	2018	N = 1 (0)	NIRO-TRS1 (5)	Fitting			x	x	x	x				
Weigl et al. [97]	2018	N = 28 (9)	In-house (17)	Moments		x								

Table 3. *Cont.*

Publication	Year	Patients N = P (C) P: Patients, C: Controls	Apparatus (ID)	Method	Perfusion DCS	Perfusion ICG	[HbO$_2$]	[HHb]	[HbT]	StO$_2$	μ_a	μ'_s	Path Length	Other
Psychiatry (number of studies = 9)														
Shinba et al. [98]	2004	N = 13 (10)	TRS-10 (1)	Fitting					x				x	
Hoshi et al. [99]	2006	N = 14 (16)	TRS-10 (1)	Fitting					x					
Harada et al. [100]	2007	N = 14 (14)	TRS-10 (1)	Fitting			x	x	x					
Sakatani et al. [105]	2014	N = 34 (0)	TRS-20 (2)	Fitting			x		x					
Sakatani et al. [101]	2015	N = 61 (0)	TRS-20 (2)	Fitting			x	x	x					
Machida et al. [102]	2016	N = 61 (0)	TRS-20 (2)	Fitting			x	x	x					
Nakano et al. [106]	2016	N = 20 (0)	TRS-20 (2)	Fitting			x	x	x	x				
Murayama et al. [103]	2017	N = 113 (0)	TRS-21 (3)	Fitting			x	x	x					
Oyama et al. [104]	2018	N = 202 (0)	TRS-21 (3)	Fitting			x	x	x	x			x	
Peroperative care (number of studies = 7)														
Ohmae et al. [110]	2007	N = 33 (0)	TRS-10 (1)	Fitting			x	x	x	x	x	x	x	
Sato et al. [112]	2007	N = 7 (0)	TRS-10 (1)	Fitting			x	x	x		x			
Kakihana et al. [111]	2010	N = 10 (0)	TRS-10 (1)	Fitting						x				
Kacprzak et al. [109]	2012	N = 16 (0)	in-house (15)	Moments			x	x						
Yamazaki et al. [113]	2013	N = 18 (0)	TRS-20 (2)	Fitting						x				
Suzuki et al. [114]	2015	N = 80 (0)	TRS-20 (2)	Fitting			x	x	x	x				
Fujisaka et al. [115]	2016	N = 1 (0)	tNIRS-1 (4)	Fitting					x	x				
Others (number of studies = 2)														
Visani et al. [116]	2015	N = 12 (10)	In-house (9)	Gating			x	x						
Abdalmalak et al. [107]	2017	N = 1 (0)	In-house (18)	Moments			x	x						

3.7. Others

Visani and colleagues [116] reported the haemodynamic and neuronal time course to unilateral hand movement in 10 patients with cortical myoclonus, a pathology that triggers sudden, brief, involuntary muscle jerks, compared with 12 healthy volunteers. This particular study was highly multimodal with simultaneous EEG-fMRI and EEG-TD-NIRS measurements. An in-house developed multimodal TD-NIRS system (ID: 9 in Table 2) was used together with the time-gated approach in order to enhance that brain sensitivity. The authors reported a good agreement between the TD-NIRS and fMRI findings, both for controls and patients, and showed reduced hemodynamic changes in patients compared to controls.

Finally, it is known that NIRS can be used within a brain computer interface framework to allow the use of brain activity to control computers or other external devices [117]. In a clinical context, this can be used to communicate with patients with locked-in syndrome. Abdalmalak and colleagues [107] reported the ability to successfully communicate with a functionally locked-in patient suffering from Guillain–Barré syndrome. At the time of the study, the patient had regained limited eye movement, which could be used to confirm the TD-NIRS findings. The hemodynamic response to a motor imagery (MI) activity was measured using an in-house developed multichannel TD-NIRS systems (ID: 18 in Table 2) and using the methods of moments to quantify the changes in μ_a to extract the changes in [HHb] and [HbO$_2$]. The subject was asked to answer three questions: (1) confirming his last name, (2) whether he was in pain, and (3) if he felt safe. He was instructed to stay relaxed if he wanted to answer "no" or to perform a tennis imaginary task if the answer was "yes". The brain activation pattern successfully predicted the correct answers, evaluated by eye movement, to the three questions (see Figure 5d). Moreover, the metrics presented in this paper were in accordance with a previous study, using the same system, on healthy volunteers [118].

4. Discussion

We have summarized the clinical application of TD-NIRS and reported its use (1) in the neonatology unit to follow neonates with various conditions, (2) in the ICU to monitor stroke or TBI patients, or to communicate with patients with locked-in syndrome, (3) in the operating room to follow brain oxygenation during surgery, and (4) in more "standard" neuroscience environments for studies focusing on patients with psychiatric disorders.

Historically, the first clinical application of TD-NIRS was in imaging the brain of premature infants; in particular obtaining 3D tomographical images of their brains in the neonatal intensive care unit. The 17% of TD-NIRS publications exploring TD-DOT are only focused on neonates. However, optical brain tomographs require resolving significant hardware and software challenges, and after a prolific number of publications in early 2000, this field has been less active on the clinical side even though methodological developments are still being made. The reduction in activity might also be explained by the significant improvements made by CW-NIRS-DOT [61].

Most of the TD-NIRS instruments used in clinical applications are multi-channel (81% of the TD-NIRS). Multichannel systems can probe both hemispheres at the same time, which is very important when a comparison is needed between hemispheres. On the other hand, some specific clinical applications, such as in peroperative care where the brain responses are global, do not require monitoring of several brain locations.

The other interesting metric of this review is that the distribution between commercial and in-house developed system was fairly balanced, with 52% of papers reporting commercial system use. However, it is worth mentioning that only Hamamatsu can provide commercial TD-NIRS systems, and that most of the work reported with these systems was originating from Japan. The use of three different versions of their research systems has been reported here (TRS-10/20/21). The use of a NIRO-TRS1 have also been reported, although it is worth noting that it is a research device which is not available for sale. Lastly, we can note that the tNIRS-1 has been approved for use as a medical device in Japan but has not obtained approval outside Japan as of March 2019 (Hamamatsu personal communication).

Nevertheless, this approval in Japan is a great step towards the translation of the technique in the clinic. Indeed, the limited availability of these clinically-approved commercial systems is undoubtfully one of the main pitfalls of TD-NIRS, hindering its use in the hospital.

Regarding the data produced by TD-NIRS systems, it has been seen that a variety of information can be retrieved, as illustrated in Figure 5. Information could be extracted about (1) the absolute optical properties of tissue and path length; (2) the absolute value of haemoglobin and tissue saturation, either as a snapshot tomographic image, or as a time-course at specific locations to follow spontaneous changes, or changes due to a brain activation; (3) the in-depth kinetics of an ICG bolus time to assess perfusion; and (4) the perfusion measured by a DCS system and calibrated with a TD-NIRS system. In order to produce this information, different processing methods of the TPSF were used. It seems that the fitting approach is the most widespread in order to extract the optical properties of tissues. This is partly because this method is the one implemented in the Hamamatsu systems. When using in-house instruments, the moment method tends to be used, especially when ICG is used. Indeed, in the case of bolus tracking, the moments can be used directly to follow the kinetics of the contrast agent. Moreover, the different depth-sensitivity of the moments enables the bolus to be tracked in-depth. To date, there is no consensus on what the best method to extract the TPSF information content is, and it might also depend on the application. This open issue still needs to be evaluated but we can report that new methodological developments might help in this (see next section).

All of these different clinical applications require us to take advantages of the possibility of TD-NIRS to measure absolute values of brain tissue saturation and haemoglobin concentration, to compare either patients with controls, injured and less injured hemispheres, or the same patient over time. In all situations, TD-NIRS showed itself relatively easy to use and functionally robust. Even though TD-NIRS has the reputation of requiring bulky instruments, most of the instruments are transportable and useable at the patient bedside, without disrupting patient care.

Indeed, even though a large number of these clinical applications can be done using CW-NIRS systems, the inability of such instruments to interrogate the absorption and scattering properties of the brain tissue will affect their accuracy. This has been for example demonstrated in [93], where a modification of the scattering properties of TBI patients were reported, or in [91] where modification of the optical path length between the affected and unaffected hemispheres of stroke patients were noted. The use of TD-NIRS to acquire insight in anatomical differences of patients is, thus, an interesting path to explore and could justify their use even further. Moreover, the optical properties information extracted from TD-NIRS system could also be used to strengthen the application of the CW-NIRS, by providing more accurate information of path length, as has been seen in the applications in psychiatry. Lastly, it is worth mentioning that even though TD-NIRS can provide absolute information about HHb and HbO_2 concentrations, the extraction of the absolute properties and then of the concentration of the chromophore are also subjected to assumptions that also limit the accuracy of TD-NIRS. Indeed, the fitting approach is usually based on the solution of the DE in simple homogeneous geometries which do not reflect the complexity of the head architecture [119]. Moreover, one major assumption that needs to be made to calculate the chromophore concentration from the absorption coefficient, is the number of chromophores considered. Indeed, it is usually assumed than HHb, HbO_2 and water are the only significant absorbers, with the water content most often fixed to a certain value (i.e., usually 80% for adults) [41]. However, it is likely that this water content is highly subject-dependent, notably due to the variability of the thickness of skin, skull and CSF and that other chromophores might need to be considered [120]. New methodological and hardware developments are addressing these issues, and these will be detailed in the next section.

We could also see that by using complementary techniques, like DCE-TD-NIRS or TD-NIRS and DCS, it was possible to assess brain perfusion. In particular, the accuracy of the assessment of brain perfusion with the DCE technique, although achievable with a CW-NIRS system, is greatly improved by using TD-NIRS instrumentation, notably because it helps to distinguish between the extra- and intra-cerebral tissues. Having the possibility to continuously assess the CBF together with the brain oxygenation, at the bedside, is a true benefit because it can give information about both oxygen delivery and demand, facilitating the identification of the underlying issue. The novel development in that area (see next section) will probably contribute greatly to the adoption of TD-NIRS in the clinic.

Finally, TD-NIRS proved to be able to be easily integrated with other modalities, and that even simultaneous imaging was possible with MRI, MEG or EEG. As the use of multi-modality imaging in modern medicine is growing, because it helps to get a clearer picture of a patient condition, this capability is also a great strength for TD-NIRS. In the context of brain imaging, this possibility can be used to give a complete picture of the link between anatomical and functional disorders. From a functional point of view, it has the potential to show the links between the neuronal and vascular components of various conditions.

5. Current Developments in TD-NIRS and Clinical Pertinence

We have reviewed the current clinical applications of TD-NIRS for brain monitoring. The present review shows that current generation of TD-NIRS can be used in a clinical environment on patients with different pathologies. However, this technology is still not at its final stage and current progress made in different areas of TD-NIRS might help to push even more its use in the clinic. In the next sections, we will highlight the major novelties in TD-NIRS, which we think will help to promote its growth as a clinical tool.

5.1. Toward TD-NIRS Standardisation

The main reported drawback of NIRS is the lack of standardisation. Indeed, different systems can use different wavelengths, different source or detector types, or even different theoretical models. The use of TD-NIRS makes the standardisation of the NIRS measurement easier by extracting absolute information. Recently, new protocols have emerged in order to compare the results of different TD-NIRS systems. We can cite the MEDPHOT protocol [121], consisting of a set of calibrated phantoms that can be used to test the accuracy of the measured absorption and reduced scattering coefficient; the basic instrumental performance (BIP) protocol [122], consisting of a series of tests that enable us to compare the raw characteristics of different TD-NIRS instruments; and the nEUROPt protocol [123], that addresses the characteristics of optical brain imaging to detect, localize, and quantify absorption changes in the brain. The application of those protocols enables the community to qualitatively and quantitatively compare TD-NIRS instruments, independently of their raw characteristics or independently of the model used. Thus, large-scale clinical studies, even with slightly different TD-NIRS instruments, could be designed.

5.2. Instrumental Progress

In recent years, a great deal of progress has been made in TD-NIRS instrumentation. One of the issues that has to be addressed is to reduce the footprint of the instruments, often considered as bulky. Indeed, even though current systems are useable in a clinical context, the current generation are often mounted on racks which are significantly larger than typical CW-NIRS systems [57]. Reducing the footprint of instruments is therefore crucial in a clinical context where space is limited. Figure 6 illustrates newly developed hardware that illustrates the miniaturisation of TD-NIRS hardware by showing recently developed elements like miniaturized laser sources or detectors [124,125] no larger than few millimetres. Using these newly developed components, a compact two channel, two wavelength system which fits in a box has been reported, as presented in Figure 6e [126]. This illustrates well the future possibilities of TD-NIRS with instrumental footprints comparable to CW-NIRS systems already used in the clinic. Moreover, these new components have the great advantage of being cheap, which will reduce the overall cost of a TD-NIRS system, facilitating even more its implementation.

Another line of developments of TD-NIRS systems in order to make the measurement easier to handle for medical staff is non-contact imaging. Few non-contact TD-NIRS system are reported in the literature [127]. Such systems smooth the user experience since no expertise is required in fibre positioning. However, such imaging systems would be limited to specific applications, probably to the prefrontal cortex, since hairy regions are particularly difficult to image with such systems. Nevertheless, it is an avenue that needs to be explored. Other systems based on non-fibre emission and detection, in contact with the skin of the patient, have also been reported [128]. These systems not only remove the optical fibres, but also increase the SNR by reducing light losses.

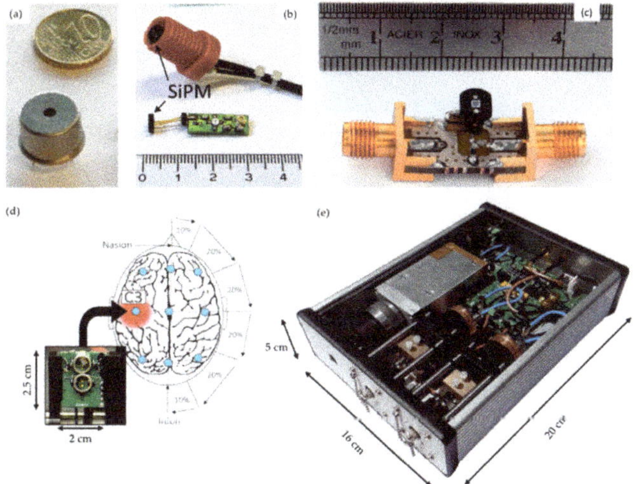

Figure 6. Examples of new TD-NIRS instrumentation. (**a**) Laser diode for compact systems (from [124]). (**b**) A compact probe holder based on a SiPM detector. No optical fibre is needed, and the detector is placed in direct contact with the skin. Extracted from [128]. (**c**) Compact SiPM detector. Extracted from [129]. (**d**) Fibreless TD-NIRS system. Extracted from [130]. (**e**) Compact all-in-one single channel TD-NIRS system. Extracted from [131].

5.3. Development of New Methodologies

In parallel with these technical developments, new methodologies have emerged that could be relevant in clinical applications, by improving the data quality, extracting more information or facilitating the operation of TD-NIRS system. For example, one of the particularities of TD measurements is the need to acquire an IRF to calibrate the measurement. This operation is time-consuming and is not always suitable in a clinical environment. A recent work by Milej and colleagues [132] reported a method based on a multi-distance TD-NIRS measurement that eliminates the need for the IRF. The removal of this extra measurement step would definitely benefit its adoption by the clinical community.

Other new methodological developments arise from hardware improvements. For example, the development of the new sources and detectors is not only beneficial in terms of miniaturisation, but also improves the quality of the raw signal. Indeed, one of the drawbacks of the TD-NIRS instrumentations over CW-NIRS systems is that it is more difficult to obtain a TPSF with a good SNR [49]. However, the newly developed sources and detectors can be used to perform a null source-detector distance measurement [133]. This approach is based on the fact that, unlike the CW technique, the source-detector distance does not influence the penetration depth in a TD measurement. Therefore, the sources and detectors can be positioned close together to increase the number of photons acquired, which greatly increases the SNR. Recently, Pifferi and colleagues reviewed the new developments in the concept and hardware of TD-NIRS [134], and showed not only that the SNR of TD-NIRS measurements would not be a problem anymore, but also that the new instrumentation and methodologies can help to improve the resolution, the contrast, or the depth sensitivity of the TD-NIRS measurement. Thus, measurements to assess structures deeper than 5–6 cm are reportedly possible [133]. Although these possibilities are still to be demonstrated practically, this could allow access to subcortical regions which would be a considerable asset in fields like TBI or stroke monitoring.

Another exciting development is the broadband or multi-wavelength TD-NIRS, which enables true spectroscopic analysis. Indeed, the systems presented in this review rarely use more than three wavelengths, which is enough to assess the haemoglobin content. However, being able to quantify other chromophores, like water or lipids, can help to increase the accuracy of the oxygen saturation, by actually quantifying contributions that were either assumed (i.e., water content) or not taken into account (i.e., fat content). Sekar and colleagues [120] have reported the development of a TD-NIRS instrument with a broadband capability, acquiring data from 600–1350 nm. This system has a slow acquisition speed (i.e., minutes per spectra) but could still be used to assess the brain status of patients at the bedside. Lange and colleagues [135] reported the development of a multiwavelength TD-NIRS system, capable of acquiring data at a rate compatible with a typical hemodynamic response, with 16 wavelengths in the range 650 to 870 nm. This system has been designed in order to target haemodynamic as well as CCO. This capability could be of interest in the clinic, as CCO has been shown to be a potential marker of brain injury in HIE [136]. In the context of HIE, some interesting work has been reported recently, using a TD-NIRS instrument to non-invasively monitor the brain temperature [137]. The method has been tested on piglet with promising results. As hypothermic therapy is now the standard treatment for neonates, but optimal protocols are yet to be established (i.e., in terms of duration and depth of treatment, for example) [138], the ability to monitor brain temperature non-invasively at the bedside would certainly be another argument in favour of the clinical use of TD-NIRS.

The last big methodological development, which was discussed in an earlier section, was the ability to measure brain perfusion. It has been shown that the use of TD-NIRS in combination with DCS could improve the data quality of the BFI retrieved by the DCS. However, the current DCS measurement suffers from the same drawback as CW-NIRS, being the contamination of the signal by extracerebral layers. Baker and colleagues in 2015 used a combination of short and long source-detector distance DCS with pressure modulation (head blood pressure cuff) to remove extracerebral contributions in the DCS signal [139]. However, a recent development called TD-diffuse correlation spectroscopy (TD-DCS) has been reported, allowing DCS measurement in the TD regime [137]. Thus, through the use of time gating, an in-depth measurement of the BFI became possible, improving the accuracy of the measurement of the cerebral blood flow [138]. Finally, it is also worth mentioning that while most of the DCS systems reported in the literature only have one channel, Delgado-Mederos and colleagues have recently demonstrated a dual channel DCS allowing bi-hemispheric measurements in stroke patients [140]. This illustrates the potential development of multi-channel DCS.

Finally, as seen in this review, imaging patients differs from imaging healthy subjects because normal tissue configuration cannot always be assumed. A recent study by Ancora and colleagues [141] showed that it might be possible to use TD-NIRS to evaluate the progression of brain atrophy in patients. Indeed, the late photons of the TPSF seems to be sensitive to the thickness of the CSF layer. However, this simulation study showed that this technique would require the acquisition of photons with very long arrival time (>2.5 ns) with a good enough SNR. In that sense, the newly developed detector might help to push this methodology forward.

On the other hand, the structural changes consecutive to some pathologies also suggests that classical models might not work in all cases. However, as it has been seen, NIRS is extremely well-suited for multimodal imaging. Therefore, combining anatomical imaging with NIRS allows for model-patient specific geometry that will increase the NIRS accuracy [142]. Here, the progress being done in terms of Monte-Carlo modelling, especially in terms of calculation speed [143], unlock new possibilities for it to be used in real-time at the bedside.

6. Conclusions

The use of NIRS technology has drawn a great deal of attention in the last four decades, due to its potential to continuously monitor the brain tissue physiology non-invasively, in real-time and at the patient bedside. However, the lack of reference values of brain tissue oxygen saturation coupled with the inherent limitations of the commonly-used CW-NIRS systems, in terms of accuracy and capability to differentiate non-cortical and cortical tissues, have hindered the popularity of NIRS. On the other hand, TD-NIRS systems have not been as popular because of their complexity and the lack of system availability commercially, with only one company selling instruments. However, in recent years, a few publications have reported successful clinical work using TD-NIRS. Indeed, this technique presents the unique advantage of measuring the distribution of the time of flight of photons. The amount of information contained in those DTOFs enables us to retrieve the optical properties (μ_a and $\mu\prime_s$) of the tissue, even with only one channel. Having access to the absolute optical properties allows us to retrieve absolute concentration of haemoglobin and tissue saturation that are more robust than the those acquired via CW-NIRS systems, which only assume a constant, non-patient-specific $\mu\prime_s$.

Moreover, it has also been shown that $\mu\prime_s$ could be used as a biomarker of TBI and that it was possible to follow the development of brain atrophy, for example in patients with neurodegenerative disease, by looking at the distribution of the late photons of the TPSF. These works reinforce the importance of the information acquired with TD-NIRS, which has the potential to not only measure brain oxygenation accurately, but also to give some insights to the brain tissue structure.

It is worth mentioning that TD-NIRS can be coupled with DCE and/or DCS in order to monitor brain perfusion. The use of the moments allows DCE-TD-NIRS to extract an in-depth measure of the brain perfusion, improving the accuracy of its estimation over DCE-CW-NIRS. Additionally, combining DCE-TD-NIRS with DCS allows us to continuously monitor the brain perfusion, reported in physiological units, which facilitates better clinical interpretation.

Finally, both the cost and the size of the TD-NIRS instruments have repeatedly been reported as the main disadvantages of this technique. However, the recent instrumental developments allow the reduction of both the cost and the size of the TD-NIRS systems. Therefore, we have no doubt that these improvements, coupled with the great possibilities offered by TD-NIRS to extract accurate brain oxygenation information, together with insights of the brain tissue structure, will enhance and further facilitate the popularity of TD-NIRS in clinical environments.

Supplementary Materials: The following are available online at http://www.mdpi.com/2076-3417/9/8/1612/s1, Video S1. Depth sensitivity of the gating method.

Author Contributions: F.L. and I.T. conceived and designed the study. F.L. conducted the literature search and drafted the first version of the paper. F.L. and I.T. commented on, revised and approved the manuscript.

Funding: This research was funded by The Wellcome Trust (grant number 104580/Z/14/Z).

Acknowledgments: The authors would like to thank Luke Dunne for providing the video presented in the supplementary material.

Conflicts of Interest: The authors declare no conflict of interest.

References

1. Smith, M. Shedding light on the adult brain: A review of the clinical applications of near-infrared spectroscopy. *Philos. Trans. A. Math. Phys. Eng. Sci.* **2011**, *369*, 4452–4469. [CrossRef] [PubMed]
2. Jacques, S.L. Optical Properties of Biological Tissues: A Review. *Phys. Med. Biol.* **2013**, *58*, R37–R61. [CrossRef]
3. Jobsis, F. Noninvasive, infrared monitoring of cerebral and myocardial oxygen sufficiency and circulatory parameters. *Science* **1977**, *198*, 1264–1267. [CrossRef] [PubMed]
4. Wolf, M.; Ferrari, M.; Quaresima, V. Progress of near-infrared spectroscopy and topography for brain and muscle clinical applications. *J. Biomed. Opt.* **2007**, *12*, 62104.

5. Grosenick, D.; Rinneberg, H.; Cubeddu, R.; Taroni, P. Review of optical breast imaging and spectroscopy. *J. Biomed. Opt.* **2016**, *21*, 091311. [CrossRef]

6. Obrig, H. NIRS in clinical neurology—A 'promising' tool? *Neuroimage* **2014**, *85*, 535–546. [CrossRef]

7. Weigl, W.; Milej, D.; Janusek, D.; Wojtkiewicz, S.; Sawosz, P.; Kacprzak, M.; Gerega, A.; Maniewski, R.; Liebert, A. Application of optical methods in the monitoring of traumatic brain injury: A review. *J. Cereb. Blood Flow Metab.* **2016**, *36*, 1825–1843. [CrossRef] [PubMed]

8. Murkin, J.M.; Arango, M. Near-infrared spectroscopy as an index of brain and tissue oxygenation. *Br. J. Anaesth.* **2009**, *103*, i3–i13. [CrossRef]

9. Nielsen, H.B. Systematic review of near-infrared spectroscopy determined cerebral oxygenation during non-cardiac surgery. *Front. Physiol.* **2014**, *5*, 1–15. [CrossRef]

10. Greisen, G.; Leung, T.; Wolf, M. Has the time come to use near-infrared spectroscopy as a routine clinical tool in preterm infants undergoing intensive care? *Philos. Trans. R. Soc. A Math. Phys. Eng. Sci.* **2011**, *369*, 4440–4451. [CrossRef]

11. Chou, P.H.; Lan, T.H. The role of near-infrared spectroscopy in Alzheimer's disease. *J. Clin. Gerontol. Geriatr.* **2013**, *4*, 33–36. [CrossRef]

12. Ehlis, A.-C.; Schneider, S.; Dresler, T.; Fallgatter, A.J. Application of functional near-infrared spectroscopy in psychiatry. *Neuroimage* **2014**, *85 Pt 1*, 478–488. [CrossRef]

13. Hillman, E.M.C. Experimental and Theoretical Investigations of Near Infrared Tomographic Imaging Methods and Clinical Applications. Ph.D. Thesis, University College London, London, UK, 2002.

14. Ehlis, A.-C.; Barth, B.; Hudak, J.; Storchak, H.; Weber, L.; Kimmig, A.-C.S.; Kreifelts, B.; Dresler, T.; Fallgatter, A.J. Near-Infrared Spectroscopy as a New Tool for Neurofeedback Training: Applications in Psychiatry and Methodological Considerations. *Jpn. Psychol. Res.* **2018**, *60*, 225–241. [CrossRef]

15. Scholkmann, F.; Kleiser, S.; Metz, A.J.; Zimmermann, R.; Mata Pavia, J.; Wolf, U.; Wolf, M. A review on continuous wave functional near-infrared spectroscopy and imaging instrumentation and methodology. *Neuroimage* **2014**, *85*, 6–27. [CrossRef]

16. SfNIRS Instrument List. Available online: https://fnirs.org/resources/instruments/ (accessed on 1 March 2019).

17. Jacques, S.L.; Pogue, B.W. Tutorial on diffuse light transport. *J. Biomed. Opt.* **2008**, *13*, 041302. [CrossRef]

18. Pifferi, A.; Farina, A.; Torricelli, A.; Quarto, G.; Cubeddu, R.; Taronia, P. Review: Time-domain broadband near infrared spectroscopy of the female breast: A focused review from basic principles to future perspectives. *J. Near Infrared Spectrosc.* **2012**, *20*, 223. [CrossRef]

19. Durduran, T.; Choe, R.; Baker, W.B.; Yodh, A.G. Diffuse optics for tissue monitoring and tomography. *Rep. Prog. Phys.* **2010**, *73*, 076701. [CrossRef]

20. Arridge, S.R.; Schotland, J.C. Optical tomography: Forward and inverse problems. *Inverse Probl.* **2009**, *25*, 123010. [CrossRef]

21. Fang, Q.; Boas, D. Monte Carlo simulation of photon migration in 3D turbid media accelerated by graphics processing units. *Opt. Express* **2009**, *17*, 20178–20190. [CrossRef]

22. Schweiger, M.; Arridge, S.R. The Toast++ software suite for forward and inverse modeling in optical tomography. *J. Biomed. Opt.* **2014**, *19*, 040801. [CrossRef]

23. Dehghani, H.; Eames, M.E.; Yalavarthy, P.K.; Davis, S.C.; Srinivasan, S.; Carpenter, C.M.; Pogue, B.W.; Paulsen, K.D. Near infrared optical tomography using NIRFAST: Algorithm for numerical model and image reconstruction. *Commun. Numer. Methods Eng.* **2008**, *25*, 711–732. [CrossRef]

24. Contini, D.; Zucchelli, L.; Spinelli, L.; Caffini, M.; Re, R.; Pifferi, A.; Cubeddu, R.; Torricelli, A. Review: Brain and muscle near infrared spectroscopy/imaging techniques. *J. Near Infrared Spectrosc.* **2012**, *20*, 15. [CrossRef]

25. Sassaroli, A.; Fantini, S. Comment on the modified Beer–Lambert law for scattering media. *Phys. Med. Biol.* **2004**, *49*, N255–N257. [CrossRef]

26. Duncan, A.; Meek, J.H.; Clemence, M.; Elwell, C.E.; Tyszczuk, L.; Cope, M.; Delpy, D. Optical pathlength measurements on adult head, calf and forearm and the head of the newborn infant using phase resolved optical spectroscopy. *Phys. Med. Biol.* **1995**, *40*, 295–304. [CrossRef] [PubMed]

27. Scholkmann, F.; Wolf, M. General equation for the differential pathlength factor of the frontal human head depending on wavelength and age. *J. Biomed. Opt.* **2013**, *18*, 105004. [CrossRef] [PubMed]

28. Matcher, S.J.; Kirkpatrick, P.J.; Nahid, K.; Cope, M.; Delpy, D.T. Absolute quantification methods in tissue near-infrared spectroscopy. In *Proceedings of the Optical Tomography, Photon Migration, and Spectroscopy of Tissue and Model Media: Theory, Human Studies, and Instrumentation*; Chance, B., Alfano, R.R., Eds.; International Society for Optics and Photonics: San Jose, CA, USA, 1995; Volume 2389, pp. 486–495.

29. Al-Rawi, P.G.; Smielewski, P.; Kirkpatrick, P.J. Evaluation of a near-infrared spectrometer (NIRO 300) for the detection of intracranial oxygenation changes in the adult head. *Stroke* **2001**, *32*, 2492–2500. [CrossRef] [PubMed]

30. Pocivalnik, M.; Pichler, G.; Zotter, H.; Tax, N.; Müller, W.; Urlesberger, B. Regional tissue oxygen saturation: Comparability and reproducibility of different devices. *J. Biomed. Opt.* **2011**, *16*, 057004. [CrossRef]

31. Diop, M.; Tichauer, K.M.; Elliott, J.T.; Migueis, M.; Lee, T.-Y.; St. Lawrence, K. Comparison of time-resolved and continuous-wave near-infrared techniques for measuring cerebral blood flow in piglets. *J. Biomed. Opt.* **2010**, *15*, 057004. [CrossRef]

32. Bale, G.; Elwell, C.E.; Tachtsidis, I. From Jöbsis to the present day: A review of clinical near-infrared spectroscopy measurements of cerebral cytochrome-c-oxidase. *J. Biomed. Opt.* **2016**, *21*, 091307. [CrossRef]

33. Ferrari, M.; Norris, H.; Sowa, M.G. Medical near infrared spectroscopy 35 years after the discovery Guest editorial. *J. Near Infrared Spectrosc.* **2012**, *20*, 7–9. [CrossRef]

34. White, B.R.; Culver, J.P. Quantitative evaluation of high-density diffuse optical tomography: In vivo resolution and mapping performance. *J. Biomed. Opt.* **2010**, *15*, 26006. [CrossRef] [PubMed]

35. Fantini, S. Frequency-domain multichannel optical detector for noninvasive tissue spectroscopy and oximetry. *Opt. Eng.* **1995**, *34*, 32. [CrossRef]

36. Gunadi, S.; Leung, T.S.; Elwell, C.E.; Tachtsidis, I. Spatial sensitivity and penetration depth of three cerebral oxygenation monitors. *Biomed. Opt. Express* **2014**, *5*, 2896–2912. [CrossRef] [PubMed]

37. Davies, D.J.; Clancy, M.; Lighter, D.; Balanos, G.M.; Lucas, S.J.E.; Dehghani, H.; Su, Z.; Forcione, M.; Belli, A. Frequency-domain vs continuous-wave near-infrared spectroscopy devices: A comparison of clinically viable monitors in controlled hypoxia. *J. Clin. Monit. Comput.* **2017**, *31*, 967–974. [CrossRef] [PubMed]

38. Patterson, M.; Chance, B.; Wilson, B.C. Time resolved reflectance and transmittance for the non-invasive measurement of tissue optical properties. *Appl. Opt.* **1989**, *28*, 2331–2336. [CrossRef]

39. Cubeddu, R.; Pifferi, A.; Taroni, P.; Torricelli, A.; Valentini, G. Experimental test of theoretical models for time-resolved reflectance. *Med. Phys.* **1996**, *23*, 1625. [CrossRef] [PubMed]

40. Liemert, A.; Kienle, A. Light diffusion in N-layered turbid media: Frequency and time domains. *J. Biomed. Opt.* **2010**, *15*, 025002. [CrossRef] [PubMed]

41. Oda, M.; Yamashita, Y.; Nakano, T.; Suzuki, A.; Shimizu, K.; Hirano, I.; Shimomura, F.; Ohmae, E.; Suzuki, T.; Tsuchiya, Y. Near-infrared time-resolved spectroscopy system for tissue oxygenation monitor. *Proc. SPIE* **2000**, *4160*, 204–210.

42. Valim, N.; Brock, J.; Leeser, M.; Niedre, M. The effect of temporal impulse response on experimental reduction of photon scatter in time-resolved diffuse optical tomography. *Phys. Med. Biol.* **2013**, *58*, 335–349. [CrossRef] [PubMed]

43. Spinelli, L.; Martelli, F.; Farina, A.; Pifferi, A.; Torricelli, A.; Cubeddu, R.; Zaccanti, G. Accuracy of the nonlinear fitting procedure for time-resolved measurements on diffusive phantoms at NIR wavelengths. In *Optical Tomography and Spectroscopy of Tissue VIII*; International Society for Optics and Photonics: San Jose, CA, USA, 2009.

44. Kohl-Bareis, M.; Nolte, C.; Heekeren, H.R.; Horst, S.; Scholz, U.; Obrig, H.; Villringer, A. Determination of the wavelength dependence of the differential pathlength factor from near-infrared pulse signals. *Phys. Med. Biol.* **1998**, *43*, 1771–1782. [CrossRef]

45. Liebert, A.; Wabnitz, H.; Grosenick, D.; Möller, M.; Macdonald, R.; Rinneberg, H. Evaluation of optical properties of highly scattering media by moments of distributions of times of flight of photons. *Appl. Opt.* **2003**, *42*, 5785. [CrossRef] [PubMed]

46. Liebert, A.; Wabnitz, H.; Steinbrink, J.; Obrig, H.; Möller, M.; Macdonald, R.; Villringer, A.; Rinneberg, H. Time-resolved multidistance near-infrared spectroscopy of the adult head: Intracerebral and extracerebral absorption changes from moments of distribution of times of flight of photons. *Appl. Opt.* **2004**, *43*, 3037. [CrossRef]

47. Zucchelli, L.; Contini, D.; Re, R.; Torricelli, A.; Spinelli, L. Method for the discrimination of superficial and deep absorption variations by time domain fNIRS. *Biomed. Opt. Express* **2013**, *4*, 2893. [CrossRef] [PubMed]

48. Dunne, L. Development of a Novel Multiwavelength, Time Resolved, Near Infrared Spectrometer Declaration of Authorship. Ph.D. Thesis, University College London, London, UK, 2016.

49. Selb, J.; Stott, J.J.; Franceschini, M.A.; Sorensen, A.G.; Boas, D. Improved sensitivity to cerebral hemodynamics during brain activation with a time-gated optical system: Analytical model and experimental validation. *J. Biomed. Opt.* **2005**, *10*, 11013. [CrossRef] [PubMed]

50. Lange, F.; Peyrin, F.; Montcel, B. Broadband time-resolved multi-channel functional near-infrared spectroscopy system to monitor in vivo physiological changes of human brain activity. *Appl. Opt.* **2018**, *57*, 6417. [CrossRef] [PubMed]

51. Alayed, M.; Deen, M. Time-Resolved Diffuse Optical Spectroscopy and Imaging Using Solid-State Detectors: Characteristics, Present Status, and Research Challenges. *Sensors* **2017**, *17*, 2115. [CrossRef] [PubMed]

52. Habermehl, C.; Schmitz, C.H.; Steinbrink, J. Contrast enhanced high-resolution diffuse optical tomography of the human brain using ICG. *Opt. Express* **2011**, *19*, 18636–18644. [CrossRef] [PubMed]

53. Boushel, R.; Langberg, H.; Olesen, J.; Nowak, M.; Simonsen, L.; Bulow, J.; Kjar, M. Regional blood flow during exercise in humans measured by near-infrared spectroscopy and indocyanine green Regional blood flow during exercise in humans measured by near-infrared spectroscopy and indocyanine green. *J. Appl. Physiol.* **2000**, *89*, 1868–1878. [CrossRef]

54. Elliott, J.T.; Milej, D.; Gerega, A.; Weigl, W.; Diop, M.; Morrison, L.B.; Lee, T.-Y.; Liebert, A.; St Lawrence, K. Variance of time-of-flight distribution is sensitive to cerebral blood flow as demonstrated by ICG bolus-tracking measurements in adult pigs. *Biomed. Opt. Express* **2013**, *4*, 206–218. [CrossRef]

55. Diop, M.; Verdecchia, K.; Lee, T.-Y.; St Lawrence, K. Calibration of diffuse correlation spectroscopy with a time-resolved near-infrared technique to yield absolute cerebral blood flow measurements: Errata. *Biomed. Opt. Express* **2012**, *3*, 1476. [CrossRef]

56. Durduran, T.; Zhou, C.; Edlow, B.L.; Yu, G.; Choe, R.; Kim, M.N.; Cucchiara, B.L.; Putt, M.E.; Shah, Q.; Kasner, S.E.; et al. Transcranial optical monitoring of cerebrovascular hemodynamics in acute stroke patients. *Opt. Express* **2009**, *17*, 3884. [CrossRef] [PubMed]

57. Torricelli, A.; Contini, D.; Pifferi, A.; Caffini, M.; Re, R.; Zucchelli, L.; Spinelli, L. Time domain functional NIRS imaging for human brain mapping. *Neuroimage* **2014**, *85*, 28–50. [CrossRef] [PubMed]

58. Weigl, W.; Milej, D.; Gerega, A.; Toczylowska, B.; Kacprzak, M.; Sawosz, P.; Botwicz, M.; Maniewski, R.; Mayzner-Zawadzka, E.; Liebert, A. Assessment of cerebral perfusion in post-traumatic brain injury patients with the use of ICG-bolus tracking method. *Neuroimage* **2014**, *85*, 555–565. [CrossRef] [PubMed]

59. Giovannella, M.; Andresen, B.; De Carli, A.; Pagliazzi, M.; Fumagalli, M.; Greisen, G.; Contini, D.; Pifferi, A.; Spinelli, L.; Durduran, T.; et al. The BabyLux device: Baseline hemodynamic and optical properties of the newborn brain and the reproducibility of the measurements. In *Proceedings of the Biophotonics Congress: Biomedical Optics Congress 2018 (Microscopy/Translational/Brain/OTS)*; Optical Society of America: Washington, DC, USA, 2018; p. OW4C.2.

60. He, L.; Baker, W.B.; Busch, D.R.; Jiang, J.Y.; Lawrence, K.S.; Kofke, W.A.; Yodh, A.G.; He, L.; Baker, W.B.; Milej, D.; et al. Noninvasive continuous optical monitoring of absolute cerebral blood flow in critically ill adults. *Neurophotonics* **2018**, *5*, 1.

61. Lee, C.W.; Cooper, R.J.; Austin, T. Diffuse optical tomography to investigate the newborn brain. *Pediatr. Res.* **2017**, *82*, 376–386. [CrossRef] [PubMed]

62. Hintz, S.R.; Benaron, D.A.; Van Houten, J.P.; Duckworth, J.L.; Liu, F.W.H.; Spilman, S.D.; Stevenson, D.K.; Cheong, W.F. Stationary Headband for Clinical Time-of-Flight Optical Imaging at the Bedside. *Photochem. Photobiol.* **1998**, *68*, 361–369. [CrossRef]

63. Hintz, S.R.; Cheong, W.-F.; Van Houten, J.P.; Stevenson, D.K.; Benaron, D.A. Bedside Imaging of Intracranial Hemorrhage in the Neonate Using Light: Comparison with Ultrasound, Computed Tomography, and Magnetic Resonance Imaging. *Pediatr. Res.* **1999**, *45*, 54. [CrossRef]

64. Hintz, S.R.; Benaron, D.A.; Siegel, A.M.; Zourabian, A.; Stevenson, D.K.; Boas, D.A. Bedside functional imaging of the premature infant brain during passive motor activation. *J. Perinat. Med.* **2001**, *29*, 335–343. [CrossRef]

65. Benaron, D.A.; Hintz, S.R.; Villringer, A.; Boas, D.; Kleinschmidt, A.; Frahm, J.; Hirth, C.; Obrig, H.; van Houten, J.C.; Kermit, E.L.; et al. Noninvasive functional imaging of human brain using light. *J. Cereb. Blood Flow Metab.* **2000**, *20*, 469–477. [CrossRef]

66. Hintz, S.R.; Benaron, D.A.; Siegel, A.M.; Stevenson, D.K.; Boas, D.A. Real-Time Functional Imaging of the Premature Infant Brain during Passive Motor Activation. *Pediatr. Res.* **1999**, *45*, 343A. [CrossRef]

67. Schmidt, F.E.W.; Fry, M.E.; Hillman, E.M.C.; Hebden, J.C.; Delpy, D.T. A 32-channel time-resolved instrument for medical optical tomography. *Rev. Sci. Instrum.* **2000**, *71*, 256. [CrossRef]

68. Cooper, R.J.; Magee, E.; Everdell, N.; Magazov, S.; Varela, M.; Airantzis, D.; Gibson, A.P.; Hebden, J.C. MONSTIR II: A 32-channel, multispectral, time-resolved optical tomography system for neonatal brain imaging. *Rev. Sci. Instrum.* **2014**, *85*, 053105. [CrossRef] [PubMed]

69. Austin, T.; Gibson, A.P.; Branco, G.; Yusof, R.M.; Arridge, S.R.; Meek, J.H.; Wyatt, J.S.; Delpy, D.T.; Hebden, J.C. Three dimensional optical imaging of blood volume and oxygenation in the neonatal brain. *Neuroimage* **2006**, *31*, 1426–1433. [CrossRef]

70. Hebden, J.C.; Gibson, A.; Yusof, R.M.; Everdell, N.; Hillman, E.M.C.; Delpy, D.T.; Arridge, S.R.; Austin, T.; Meek, J.H.; Wyatt, J.S. Three-dimensional optical tomography of the premature infant brain. *Phys. Med. Biol.* **2002**, *47*, 4155–4166. [CrossRef]

71. Hebden, J.C.; Gibson, A.P.; Austin, T.; Yusof, R.M.; Everdell, N.; Delpy, D.T.; Arridge, S.R.; Meek, J.H.; Wyatt, J.S. Imaging changes in blood volume and oxygenation in the newborn infant brain using three-dimensional optical tomography. *Phys. Med. Biol.* **2004**, *49*, 1117–1130. [CrossRef] [PubMed]

72. Gibson, A.P.; Austin, T.; Everdell, N.; Schweiger, M.; Arridge, S.R.; Meek, J.H.; Wyatt, J.S.; Delpy, D.T.; Hebden, J.C. Three-dimensional whole-head optical tomography of passive motor evoked responses in the neonate. *Neuroimage* **2006**, *30*, 521–528. [CrossRef] [PubMed]

73. Dempsey, L.A.; Cooper, R.J.; Powell, S.; Edwards, A.; Lee, C.-W.; Brigadoi, S.; Everdell, N.; Arridge, S.R.; Gibson, A.P.; Austin, T.; et al. Whole-head functional brain imaging of neonates at cot-side using time-resolved diffuse optical tomography. *Prog. Biomed. Opt. Imaging Proc. SPIE* **2015**, *9538*, 1–10.

74. Dempsey, L.A. Development and Application of Diffuse Optical Tomography Systems for Diagnosis and Assessment of Perinatal Brain Injury. Ph.D. Thesis, University College London, London, UK, 2018.

75. Koyano, K.; Kusaka, T.; Nakamura, S.; Nakamura, M.; Konishi, Y.; Miki, T.; Ueno, M.; Yasuda, S.; Okada, H.; Nishida, T.; et al. The effect of blood transfusion on cerebral hemodynamics in preterm infants. *Transfusion* **2013**, *53*, 1459–1467. [CrossRef]

76. Ishii, H.; Takami, T.; Fujioka, T.; Mizukaki, N.; Kondo, A.; Sunohara, D.; Hoshika, A.; Akutagawa, O.; Isaka, K. Comparison of changes in cerebral and systemic perfusion between appropriate- and small-for-gestational-age infants during the first three days after birth. *Brain Dev.* **2014**, *36*, 380–387. [CrossRef]

77. Fujioka, T.; Takami, T.; Ishii, H.; Kondo, A.; Sunohara, D. Difference in Cerebral and Peripheral Hemodynamics among Term and Preterm Infants during the First Three Days of Life. *Neonatology* **2014**, *106*, 181–187. [CrossRef]

78. Nakamura, S.; Koyano, K.; Jinnai, W.; Hamano, S.; Yasuda, S.; Konishi, Y.; Kuboi, T.; Kanenishi, K.; Nishida, T.; Kusaka, T. Simultaneous measurement of cerebral hemoglobin oxygen saturation and blood volume in asphyxiated neonates by near-infrared time-resolved spectroscopy. *Brain Dev.* **2015**, *37*, 925–932. [CrossRef] [PubMed]

79. Hoshino, T.; Sakatani, K.; Yokose, N.; Awano, T.; Nakamura, S.; Murata, Y.; Kano, T.; Katayama, Y. *Changes in Cerebral Blood Oxygenation and Hemodynamics After Endovascular Treatment of Vascular Malformation Measured by Time-Resolved Spectroscopy BT-Oxygen Transport to Tissue XXXI*; Takahashi, E., Bruley, D.F., Eds.; Springer: Boston, MA, USA, 2010; pp. 491–496.

80. Ijichi, S.; Kusaka, T.; Isobe, K.; Okubo, K.; Kawada, K.; Namba, M.; Okada, H.; Nishida, T.; Imai, T.; Itoh, S. Developmental changes of optical properties in neonates determined by near-infrared time-resolved spectroscopy. *Pediatr. Res.* **2005**, *58*, 568–573. [CrossRef]

81. Spinelli, L.; Zucchelli, L.; Contini, D.; Caffini, M.; Mehler, J.; Fló, A.; Ferry, A.L.; Filippin, L.; Macagno, F.; Cattarossi, L.; et al. In vivo measure of neonate brain optical properties and hemodynamic parameters by time-domain near-infrared spectroscopy. *Neurophotonics* **2017**, *4*, 1. [CrossRef] [PubMed]

82. Giacalone, G.; Zanoletti, M.; Contini, D.; Re, R.; Spinelli, L.; Roveri, L.; Torricelli, A. Cerebral time domain-NIRS: Reproducibility analysis, optical properties, hemoglobin species and tissue oxygen saturation in a cohort of adult subjects. *Biomed. Opt. Express* **2017**, *8*, 4987. [CrossRef]

83. Obrig, H.; Steinbrink, J. Non-invasive optical imaging of stroke. *Philos. Trans. R. Soc. A Math. Phys. Eng. Sci.* **2011**, *369*, 4470–4494. [CrossRef] [PubMed]

84. Liebert, A.; Wabnitz, H.; Steinbrink, J.; Möller, M.; Macdonald, R.; Rinneberg, H.; Villringer, A.; Obrig, H. Bed-side assessment of cerebral perfusion in stroke patients based on optical monitoring of a dye bolus by time-resolved diffuse reflectance. *Neuroimage* **2005**, *24*, 426–435. [CrossRef] [PubMed]

85. Liebert, A.; Milej, D.; Weigl, W.; Gerega, A.; Kacprzak, M.; Mayzner-Zawadzka, E.; Maniewski, R. Assessment of brain perfusion disorders by ICG bolus tracking with time-resolved fluorescence monitoring. In *Proceedings of the Biomedical Optics and 3-D Imaging*; OSA: Washington, DC, USA, 2012; p. BTu3A.20.

86. Steinkellner, O.; Wabnitz, H.; Jelzow, A.; MacDonald, R.; Gruber, C.; Steinbrink, J.; Obrig, H. Cerebral perfusion in acute stroke monitored by time-domain near-infrared reflectometry. *Biocybern. Biomed. Eng.* **2012**, *32*, 3–16. [CrossRef]

87. Leistner, S.; Sander-Thoemmes, T.; Wabnitz, H.; Moeller, M.; Wachs, M.; Curio, G.; Macdonald, R.; Trahms, L.; Mackert, B.-M. Non-invasive simultaneous recording of neuronal and vascular signals in subacute ischemic stroke. *Biomed. Tech. Eng.* **2011**, *56*, 85–90. [CrossRef] [PubMed]

88. Giacalone, G.; Germinario, B.; Roveri, L.; Giacalone, G.; Zanoletti, M.; Re, R.; Germinario, B.; Contini, D.; Spinelli, L. Time-domain near-infrared spectroscopy in acute ischemic stroke patients. *Neurophotonics* **2019**, *6*, 1. [CrossRef]

89. Yokose, N.; Sakatani, K.; Murata, Y.; Awano, T.; Igarashi, T.; Nakamura, S.; Hoshino, T.; Kano, T.; Yoshino, A.; Katayama, Y.; et al. *Bedside Assessment of Cerebral Vasospasms After Subarachnoid Hemorrhage by Near Infrared Time-Resolved Spectroscopy BT-Oxygen Transport to Tissue XXXI*; Takahashi, E., Bruley, D.F., Eds.; Springer: Boston, MA, USA, 2010; pp. 505–511.

90. Yokose, N.; Sakatani, K.; Murata, Y.; Awano, T.; Igarashi, T.; Nakamura, S.; Hoshino, T.; Katayama, Y. Bedside Monitoring of Cerebral Blood Oxygenation and Hemodynamics after Aneurysmal Subarachnoid Hemorrhage by Quantitative Time-Resolved Near-Infrared Spectroscopy. *World Neurosurg.* **2010**, *73*, 508–513. [CrossRef]

91. Sato, Y.; Komuro, Y.; Lin, L.; Tang, Z.; Hu, L.; Kadowaki, S.; Ugawa, Y.; Yamada, Y.; Sakatani, K. *Differences in Tissue Oxygenation, Perfusion and Optical Properties in Brain Areas Affected by Stroke: A Time-Resolved NIRS Study BT—Oxygen Transport to Tissue XL*; Thews, O., LaManna, J.C., Harrison, D.K., Eds.; Springer International Publishing: Cham, Switzerland, 2018; pp. 63–67. ISBN 978-3-319-91287-5.

92. Davies, D.J.; Su, Z.; Clancy, M.T.; Lucas, S.J.E.; Dehghani, H.; Logan, A.; Belli, A. Near-Infrared Spectroscopy in the Monitoring of Adult Traumatic Brain Injury: A Review. *J. Neurotrauma* **2015**, *32*, 933–941. [CrossRef] [PubMed]

93. Highton, D.; Tachtsidis, I.; Tucker, A.; Elwell, C.; Smith, M. *Near Infrared Light Scattering Changes Following Acute Brain Injury*; Advances in Experimental Medicine and Biology; Elwell, C.E., Leung, T.S., Harrison, D.K., Eds.; Springer: New York, NY, USA, 2016; Volume 876, ISBN 978-1-4939-3022-7.

94. Lanks, C.W.; Kim, C.B.; Fu, J.; DaCosta, D.; Chang, D.W.; Hsia, D.; Stringer, W.W.; Rossiter, H.B. A Pilot Study of Cortical Oxygenation in Septic Shock by Time-Resolved Near-Infrared Spectroscopy. In *A51. Critical Care: Risk Stratification and Prognostication—From Bedside to Big Data*; American Thoracic Society International Conference Abstracts; American Thoracic Society: Washington, DC, USA, 2017; p. A1803.

95. Lanks, C.; Kim, C.B.; Rossiter, H.B. A "NIRS" death experience: A reduction in cortical oxygenation by time-resolved near-infrared spectroscopy preceding cardiac arrest. *J. Clin. Monit. Comput.* **2018**, *32*, 683–686. [CrossRef] [PubMed]

96. Liebert, A.; Milej, D.; Weigl, W.; Toczyłowska, B.; Gerega, A.; Kacprzak, M.; Maniewski, R. Evaluation of ICG washout based on time-resolved monitoring of fluorescence in patients with severe cerebral perfusion abnormalities. In Proceedings of the Biomedical Optics 2014; Optical Society of America: Miami, FL, USA, 2014; p. BW2B.6.

97. Weigl, W.; Milej, D.; Gerega, A.; Toczyłowska, B.; Sawosz, P.; Kacprzak, M.; Janusek, D.; Wojtkiewicz, S.; Maniewski, R.; Liebert, A. Confirmation of brain death using optical methods based on tracking of an optical contrast agent: Assessment of diagnostic feasibility. *Sci. Rep.* **2018**, *8*, 7332. [CrossRef] [PubMed]

98. Shinba, T.; Nagano, M.; Kariya, N.; Ogawa, K.; Shinozaki, T.; Shimosato, S.; Hoshi, Y. Near-infrared spectroscopy analysis of frontal lobe dysfunction in schizophrenia. *Biol. Psychiatry* **2004**, *55*, 154–164. [CrossRef]

99. Hoshi, Y.; Shinba, T.; Sato, C.; Doi, N. Resting hypofrontality in schizophrenia: A study using near-infrared time-resolved spectroscopy. *Schizophr. Res.* **2006**, *84*, 411–420. [CrossRef] [PubMed]

100. Harada, H.; Nashihara, H.; Morozumi, K.; Ota, H.; Hatakeyama, E. A Comparison of Cerebral Activity in the Prefrontal Region between Young Adults and the Elderly while Driving. *J. Physiol. Anthropol.* **2007**, *26*, 409–414. [CrossRef] [PubMed]
101. Sakatani, K. Effects of Cosmetic Therapy on Cognitive Function in Elderly Women: A Near Infrared Spectroscopy Study. In Proceedings of the 29th Annual Conference of the Japanese Society for Artificial Intelligence, Hakodate, Japan, 5 June 2015; pp. 1–2.
102. Machida, A.; Shirato, M.; Tanida, M.; Kanemaru, C.; Nagai, S.; Sakatani, K. *Effects of Cosmetic Therapy on Cognitive Function in Elderly Women Evaluated by Time-Resolved Spectroscopy Study BT—Oxygen Transport to Tissue XXXVII*; Elwell, C.E., Leung, T.S., Harrison, D.K., Eds.; Springer: New York, NY, USA, 2016; pp. 289–295.
103. Murayama, Y.; Sato, Y.; Hu, L.; Brugnera, A.; Compare, A.; Sakatani, K. *Relation Between Cognitive Function and Baseline Concentrations of Hemoglobin in Prefrontal Cortex of Elderly People Measured by Time-Resolved Near-Infrared Spectroscopy BT—Oxygen Transport to Tissue XXXIX*; Halpern, H.J., LaManna, J.C., Harrison, D.K., Epel, B., Eds.; Springer International Publishing: Cham, Switzerland, 2017; pp. 269–276. ISBN 978-3-319-55231-6.
104. Oyama, K.; Hu, L.; Sakatani, K. *Prediction of MMSE Score Using Time-Resolved Near-Infrared Spectroscopy BT—Oxygen Transport to Tissue XL*; Thews, O., LaManna, J.C., Harrison, D.K., Eds.; Springer International Publishing: Cham, Switzerland, 2018; pp. 145–150. ISBN 978-3-319-91287-5.
105. Sakatani, K.; Tanida, M.; Hirao, N.; Takemura, N. *Ginkobiloba Extract Improves Working Memory Performance in Middle-Aged Women: Role of Asymmetry of Prefrontal Cortex Activity During a Working Memory Task*; Advances in Experimental Medicine and Biology; Swartz, H.M., Harrison, D.K., Bruley, D.F., Eds.; Springer: New York, NY, USA, 2014; Volume 812, ISBN 978-1-4939-0583-6.
106. Nakano, M.; Murayama, Y.; Hu, L.; Ikemoto, K.; Uetake, T.; Sakatani, K. *Effects of Antioxidant Supplements (BioPQQTM) on Cerebral Blood Flow and Oxygen Metabolism in the Prefrontal Cortex BT—Oxygen Transport to Tissue XXXVIII*; Luo, Q., Li, L.Z., Harrison, D.K., Shi, H., Bruley, D.F., Eds.; Springer International Publishing: Cham, Switzerland, 2016; pp. 215–222.
107. Abdalmalak, A.; Milej, D.; Norton, L.; Debicki, D.B.; Gofton, T.; Diop, M.; Owen, A.M.; St. Lawrence, K. Single-session communication with a locked-in patient by functional near-infrared spectroscopy. *Neurophotonics* **2017**, *4*, 1. [CrossRef]
108. Trafidło, T.; Gaszyński, T.; Gaszyński, W.; Nowakowska-Domagała, K. Intraoperative monitoring of cerebral NIRS oximetry leads to better postoperative cognitive performance: A pilot study. *Int. J. Surg.* **2015**, *16*, 23–30. [CrossRef]
109. Kacprzak, M.; Liebert, A.; Staszkiewicz, W.; Gabrusiewicz, A.; Sawosz, P.; Madycki, G.; Maniewski, R. Application of a time-resolved optical brain imager for monitoring cerebral oxygenation during carotid surgery. *J. Biomed. Opt.* **2012**, *17*, 016002. [CrossRef]
110. Ohmae, E.; Oda, M.; Suzuki, T.; Yamashita, Y.; Kakihana, Y.; Matsunaga, A.; Kanmura, Y.; Tamura, M. Clinical evaluation of time-resolved spectroscopy by measuring cerebral hemodynamics during cardiopulmonary bypass surgery. *J. Biomed. Opt.* **2007**, *12*, 062112. [CrossRef]
111. Kakihana, Y.; Okayama, N.; Matsunaga, A.; Yasuda, T.; Imabayashi, T.; Nakahara, M.; Kiyonaga, N.; Ikoma, K.; Kikuchi, T.; Kanmura, Y.; et al. *Cerebral Monitoring Using Near-Infrared Time-Resolved Spectroscopy and Postoperative Cognitive Dysfunction BT—Oxygen Transport to Tissue XXXIII*; Wolf, M., Bucher, H.U., Rudin, M., Van Huffel, S., Wolf, U., Bruley, D.F., Harrison, D.K., Eds.; Springer: New York, NY, USA, 2012; pp. 19–24.
112. Sato, C.; Yamaguchi, T.; Seida, M.; Ota, Y.; Yu, I.; Iguchi, Y.; Nemoto, M.; Hoshi, Y. Intraoperative monitoring of depth-dependent hemoglobin concentration changes during carotid endarterectomy by time-resolved spectroscopy. *Appl. Opt.* **2007**, *46*, 2785–2792. [CrossRef] [PubMed]
113. Yamazaki, K.; Suzuki, K.; Itoh, H.; Muramatsu, K.; Nagahashi, K.; Tamura, N.; Uchida, T.; Sugihara, K.; Maeda, H.; Kanayama, N. Cerebral oxygen saturation evaluated by near-infrared time-resolved spectroscopy (TRS) in pregnant women during caesarean section—A promising new method of maternal monitoring. *Clin. Physiol. Funct. Imaging* **2013**, *33*, 109–116. [CrossRef] [PubMed]
114. Suzuki, K.; Itoh, H.; Mukai, M.; Yamazaki, K.; Uchida, T.; Maeda, H.; Oda, M.; Yamaki, E.; Suzuki, H. Measurement of maternal cerebral tissue hemoglobin on near-infrared time-resolved spectroscopy in the peripartum period. *J. Obstetr. Gynaecol. Res.* **2015**, *41*, 876–883. [CrossRef] [PubMed]

115. Fujisaka, S.; Ozaki, T.; Suzuki, T.; Kamada, T.; Kitazawa, K.; Nishizawa, M.; Takahashi, A.; Suzuki, S. *A Clinical Tissue Oximeter Using NIR Time-Resolved Spectroscopy BT—Oxygen Transport to Tissue XXXVII*; Elwell, C.E., Leung, T.S., Harrison, D.K., Eds.; Springer: New York, NY, USA, 2016; pp. 427–433.

116. Visani, E.; Canafoglia, L.; Gilioli, I.; Sebastiano, D.R.; Contarino, V.E.; Duran, D.; Panzica, F.; Cubeddu, R.; Contini, D.; Zucchelli, L.; et al. Hemodynamic and EEG Time-Courses During Unilateral Hand Movement in Patients with Cortical Myoclonus. An EEG-fMRI and EEG-TD-fNIRS Study. *Brain Topogr.* **2015**, *28*, 915–925. [CrossRef] [PubMed]

117. Naseer, N.; Hong, K.-S. fNIRS-based brain-computer interfaces: A review. *Front. Hum. Neurosci.* **2015**, *9*, 1–15. [CrossRef] [PubMed]

118. Abdalmalak, A.; Milej, D.; Diop, M.; Shokouhi, M.; Naci, L.; Owen, A.M.; St. Lawrence, K. Can time-resolved NIRS provide the sensitivity to detect brain activity during motor imagery consistently? *Biomed. Opt. Express* **2017**, *8*, 2162. [CrossRef] [PubMed]

119. Selb, J.; Ogden, T.M.; Dubb, J.; Fang, Q.; Boas, D. Comparison of a layered slab and an atlas head model for Monte Carlo fitting of time-domain near-infrared spectroscopy data of the adult head. *J. Biomed. Opt.* **2014**, *19*, 16010. [CrossRef] [PubMed]

120. Konugolu Venkata Sekar, S.; Dalla Mora, A.; Bargigia, I.; Martinenghi, E.; Lindner, C.; Farzam, P.; Pagliazzi, M.; Durduran, T.; Taroni, P.; Pifferi, A.; et al. Broadband (600–1350 nm) Time-Resolved Diffuse Optical Spectrometer for Clinical Use. *IEEE J. Sel. Top. Quantum Electron.* **2016**, *22*, 406–414. [CrossRef]

121. Pifferi, A.; Torricelli, A.; Bassi, A.; Taroni, P.; Cubeddu, R.; Wabnitz, H.; Grosenick, D.; Möller, M.; Macdonald, R.; Swartling, J.; et al. Performance assessment of photon migration instruments: The MEDPHOT protocol. *Appl. Opt.* **2005**, *44*, 2104–2114. [CrossRef]

122. Wabnitz, H.; Jelzow, A.; Mazurenka, M.; Steinkellner, O.; Macdonald, R.; Milej, D.; Zolek, N.; Kacprzak, M.; Sawosz, P.; Maniewski, R.; et al. Performance assessment of time-domain optical brain imagers, part 1: Basic instrumental performance protocol. *J. Biomed. Opt.* **2014**, *19*, 86010. [CrossRef] [PubMed]

123. Wabnitz, H.; Jelzow, A.; Mazurenka, M.; Steinkellner, O.; Macdonald, R.; Milej, D.; Zolek, N.; Kacprzak, M.; Sawosz, P.; Maniewski, R.; et al. Performance assessment of time-domain optical brain imagers, part 2: nEUROPt protocol. *J. Biomed. Opt.* **2014**, *19*, 086012. [CrossRef]

124. Di Sieno, L.; Nissinen, J.; Hallman, L.; Martinenghi, E.; Contini, D.; Pifferi, A.; Kostamovaara, J.; Mora, A.D. Miniaturized pulsed laser source for time-domain diffuse optics routes to wearable devices. *J. Biomed. Opt.* **2017**, *22*, 1. [CrossRef]

125. Martinenghi, E.; Di Sieno, L.; Contini, D.; Sanzaro, M.; Pifferi, A.; Dalla Mora, A. Time-resolved single-photon detection module based on silicon photomultiplier: A novel building block for time-correlated measurement systems. *Rev. Sci. Instrum.* **2016**, *87*, 073101. [CrossRef]

126. Renna, M.; Buttafava, M.; Zappa, F.; Tosi, A.; Martinenghi, E.; Zanoletti, M.; Mora, A.D.; Pifferi, A.; Torricelli, A.; Contini, D. Compact dual-wavelength system for time-resolved diffuse optical spectroscopy. In Proceedings of the 2017 13th Conference on Ph.D. Research in Microelectronics and Electronics (PRIME), Giardini Naxos, Italy, 12–15 June 2017; pp. 293–296.

127. Wabnitz, H.; Mazurenka, M.; Di Sieno, L.; Contini, D.; Dalla Mora, A.; Farina, A.; Hoshi, Y.; Kirilina, E.; Macdonald, R.; Pifferi, A. Non-contact time-domain imaging of functional brain activation and heterogeneity of superficial signals. In *European Conference on Biomedical Optics*; Dehghani, H., Wabnitz, H., Eds.; Optical Society of America: Munich, Germany, 2017; p. 104120J.

128. Re, R.; Martinenghi, E.; Mora, A.D.; Contini, D.; Pifferi, A.; Torricelli, A. Probe-hosted silicon photomultipliers for time-domain functional near-infrared spectroscopy: Phantom and in vivo tests. *Neurophotonics* **2016**, *3*, 045004. [CrossRef]

129. Dalla Mora, A.; Martinenghi, E.; Contini, D.; Tosi, A.; Boso, G.; Durduran, T.; Arridge, S.R.; Martelli, F.; Farina, A.; Torricelli, A.; et al. Fast silicon photomultiplier improves signal harvesting and reduces complexity in time-domain diffuse optics. *Opt. Express* **2015**, *23*, 13937–13946. [CrossRef]

130. Mora, A.D.; Contini, D.; Arridge, S.R.; Martelli, F.; Tosi, A.; Boso, G.; Farina, A.; Durduran, T.; Martinenghi, E.; Torricelli, A.; et al. Towards next-generation time-domain diffuse optics for extreme depth penetration and sensitivity. *Biomed. Opt. Express* **2015**, *6*, 1749. [CrossRef]

131. Buttafava, M.; Martinenghi, E.; Tamborini, D.; Contini, D.; Mora, A.D.; Renna, M.; Torricelli, A.; Pifferi, A.; Zappa, F.; Tosi, A. A Compact Two-Wavelength Time-Domain NIRS System Based on SiPM and Pulsed Diode Lasers. *IEEE Photonics J.* **2017**, *9*, 1–14. [CrossRef]

132. Milej, D.; Abdalmalak, A.; Janusek, D.; Diop, M.; Liebert, A.; St. Lawrence, K. Time-resolved subtraction method for measuring optical properties of turbid media. *Appl. Opt.* **2016**, *55*, 1507. [CrossRef]

133. Torricelli, A.; Pifferi, A.; Spinelli, L.; Cubeddu, R.; Martelli, F.; Del Bianco, S.; Zaccanti, G. Time-Resolved Reflectance at Null Source-Detector Separation: Improving Contrast and Resolution in Diffuse Optical Imaging. *Phys. Rev. Lett.* **2005**, *95*, 078101. [CrossRef]

134. Pifferi, A.; Contini, D.; Mora, A.D.; Farina, A.; Spinelli, L.; Torricelli, A. New frontiers in time-domain diffuse optics, a review. *J. Biomed. Opt.* **2016**, *21*, 091310. [CrossRef]

135. Lange, F.; Dunne, L.; Hale, L.; Tachtsidis, I. MAESTROS: A Multiwavelength Time-Domain NIRS System to Monitor Changes in Oxygenation and Oxidation State of Cytochrome-C-Oxidase. *IEEE J. Sel. Top. Quantum Electron.* **2019**, *25*, 1–12. [CrossRef] [PubMed]

136. Bale, G.; Mitra, S.; de Roever, I.; Sokolska, M.; Price, D.; Bainbridge, A.; Gunny, R.; Uria-Avellanal, C.; Kendall, G.S.; Meek, J.; et al. Oxygen dependency of mitochondrial metabolism indicates outcome of newborn brain injury. *J. Cereb. Blood Flow Metab.* **2018**, 0271678X1877792. [CrossRef] [PubMed]

137. Bakhsheshi, M.F.; Diop, M.; St. Lawrence, K.; Lee, T.-Y. Monitoring brain temperature by time-resolved near-infrared spectroscopy: Pilot study. *J. Biomed. Opt.* **2014**, *19*, 057005. [CrossRef]

138. Davidson, J.O.; Wassink, G.; van den Heuij, L.G.; Bennet, L.; Gunn, A.J. Therapeutic hypothermia for neonatal hypoxic-ischemic encephalopathy—Where to from here? *Front. Neurol.* **2015**, *6*, 198. [CrossRef] [PubMed]

139. Kung, D.K.; Mesquita, R.C.; Ko, T.S.; Abramson, K.; Yodh, A.G.; Greenberg, J.H.; Busch, D.R.; Baker, W.B.; Durduran, T.; Tzeng, S.-Y.; et al. Pressure modulation algorithm to separate cerebral hemodynamic signals from extracerebral artifacts. *Neurophotonics* **2015**, *2*, 035004.

140. Delgado-Mederos, R.; Gregori-Pla, C.; Zirak, P.; Blanco, I.; Dinia, L.; Marín, R.; Durduran, T.; Martí-Fàbregas, J. Transcranial diffuse optical assessment of the microvascular reperfusion after thrombolysis for acute ischemic stroke. *Biomed. Opt. Express* **2018**, *9*, 1262–1271. [CrossRef]

141. Ancora, D.; Zacharakis, G.; Pifferi, A.; Torricelli, A.; Qiu, L.; Spinelli, L. Noninvasive optical estimation of CSF thickness for brain-atrophy monitoring. *Biomed. Opt. Express* **2018**, *9*, 4094. [CrossRef]

142. Chuang, C.-C.; Lee, Y.-T.; Chen, C.-M.; Hsieh, Y.-S.; Liu, T.-C.; Sun, C.-W. Patient-oriented simulation based on Monte Carlo algorithm by using MRI data. *Biomed. Eng. Online* **2012**, *11*, 21. [CrossRef]

143. Yu, L.; Nina-Paravecino, F.; Kaeli, D.; Fang, Q. Scalable and massively parallel Monte Carlo photon transport simulations for heterogeneous computing platforms. *J. Biomed. Opt.* **2018**, *23*, 010504. [CrossRef]

Article

Early Therapeutic Prediction Based on Tumor Hemodynamic Response Imaging: Clinical Studies in Breast Cancer with Time-Resolved Diffuse Optical Spectroscopy

Shigeto Ueda * and Toshiaki Saeki

Department of Breast Oncology, Saitama Medical University International Medical Center, 1397-1 Yamane, Hidaka, Saitama 350-1298, Japan; syueda2000@yahoo.co.jp
* Correspondence: syueda@saitama-med.ac.jp

Received: 22 October 2018; Accepted: 12 December 2018; Published: 20 December 2018

Abstract: This study reports data from three clinical studies using the time-resolved diffuse optical spectroscopy (TRS) system among breast cancer patients. The parameters of oxy-hemoglobin (O_2Hb), deoxy-hemoglobin (HHb), total hemoglobin (tHb), and oxygen saturation (SO_2) were evaluated using TRS, and its efficacy was tested in three trials. In trial 1, we recruited 118 patients with primary breast cancer to estimate the tumor detection rate. The cumulative detection rate was 62.7%, while that in T stage 0 was 31.3% and in T stage 1 was 44.7%. These were lower than those of T stage 2 (78.9%) and T stage 3 (100%). Next, we used TRS to monitor tumor hemodynamic response to neoadjuvant chemotherapy ($n = 100$) and found that pathological complete response (pCR) tumors had significantly lower tumor tHb than non-pCR tumors; a similar result was observed in estrogen receptor (ER)-negative tumors, but not in ER-positive tumors. The third trial monitored hemodynamic response to antiangiogenic therapy, bevacizumab ($n = 28$), and we demonstrated that sequential optical measurement of tumor SO_2 might be useful for detecting acute hypoxia 1–3 days after bevacizumab initiation. Next, response monitoring of neoadjuvant endocrine therapy ($n = 30$) suggested that changes in tumor tHb during treatment can predict and distinguish between responsive and non-responsive tumors early in letrozole therapy. In conclusion, our results show that hemodynamic monitoring of tumors by TRS could pair the unique features of tumor physiology to drug therapy and contribute to patient-tailored medicine. We recently established a platform for performing TRS in patients with breast cancer.

Keywords: breast cancer; diffuse optical spectroscopy; chemotherapy

1. Introduction

Breast cancer is one of the most common causes of death worldwide. Screening and recent advances in adjuvant therapy for primary breast cancer have reduced its mortality rate by approximately 10–16% in Japan [1]. As data suggest that one in fourteen Japanese women will develop breast cancer in their lifetimes, screening, diagnosis, and adjuvant therapy are essential strategies. X-ray mammography is the current gold standard screening modality, and ultrasonography is considered an adjunct therapy to mammography in premenopausal women [2]. Diffuse optical spectroscopic imaging (DOSI) is an emerging modality with potential applications in oncology. DOSI uses near-infrared light, in the 600–1000 nm wavelength range, to measure the concentrations of oxy-hemoglobin (O_2Hb), deoxy-hemoglobin (HHb), water (water), lipid (lipid), and oxygen saturation (SO_2) in breast tissue [3]. DOSI is a non-invasive, non-ionizing, cost-effective modality that also provides functional quantification of tumor angiogenesis, hypoxia, edema, and adipose tissue [4–6].

However, DOSI in breast cancer surveillance is not expected to provide precise anatomical information, but only to contribute towards providing evidence of tumor physiological activity. This makes DOSI an adjunct tool to conventional modalities for the early detection of breast cancer.

Neoadjuvant chemotherapy (NAC) before surgery is the standard procedure for patients with advanced breast cancer [7]. Clinical studies have shown that disease-free survival after NAC is equal to that after adjuvant chemotherapy, and the advantage of NAC is that a greater number of patients undergo breast-conserving surgery after tumor shrinkage. Patients with pathological complete response (pCR) after NAC experience longer survival compared to those with non-pCR, and prediction of pCR before NAC or early during chemotherapy can improve patient outcomes [8]. Additionally, as more effective therapies in patients with low response to chemotherapy are needed, functional imaging technologies such as positron emission tomography (PET) and magnetic resonance imaging (MRI) play a definite role in determining whether to continue, change, or abandon treatment [9]. DOSI has substantial advantages during longitudinal monitoring as it captures physiological changes in breast tumors, especially in the vasculature and the microenvironment.

Many optical imaging methods for diagnosis and treatment monitoring have been reported [10–13], and we have established a two-dimensional optical imaging system with time-resolved spectroscopy (TRS) for breast cancer surveillance, as previously reported [14]. TRS is a technique that measures the time of flight, in addition to the light intensity, at the boundary of a medium. This method uses a short light pulse as a light source, and measures the time point spread function of the light passing through the tissue. Although TRS has the disadvantages of being complicated, expensive and having a long measuring time, the short pulsed light used in the time domain contains all the frequency components, and the measurement is stabilized compared to the measurement using the limited frequency component of the frequency domain. In this report, we present findings obtained from three clinical studies and discuss the usefulness and limitations of TRS in breast cancer treatment.

2. Materials and Methods

2.1. Patients, Settings, and Study Design

We conducted three clinical studies in breast cancer patients between July 2012 and March 2018 using the TRS system, and recruited 436 patients for all three studies. Using this system, we acquired 1474 measurements from these subjects. All three were prospective clinical studies: (1) clinical study of in vivo optical imaging of breast cancer using diffuse optical spectroscopy (UMIN000011888); (2) value of usefulness of diffuse optical spectroscopic imaging for monitoring the efficacy of bevacizumab followed by paclitaxel in breast cancer patients (UMIN000015837); (3) early prediction of tumor response using imaging and molecular biomarkers of hormone sensitive breast cancer in a neoadjuvant hormonal therapy setting (UMIN000013815), and were registered at the UMIN Clinical trials registry.

2.2. Establishment of the TRS Breast Imaging System

A dual-channel three-wavelength time-resolved spectroscopy system (TRS-20, Hamamatsu Photonics K.K., Hamamatsu, Japan) was used to measure optical breast tissue parameters. The TRS-20 consists of two pulsed light sources, photomultiplier tubes (GaAs PMT, Hamamatsu Photonics K.K., Hamamatsu, Japan), a single photon counting (SPC) unit, and optical fibers. Each pulsed light source, called Picosecond Light Pulses (PLPs, Hamamatsu Photonics K.K., Hamamatsu, Japan), is composed of three laser diodes (760 nm, 800 nm, and 834 nm) which generate a light pulse with full-width at half-maximum (FWHM) of 70–100 ps at repetition frequency of 5 MHz, and has an average output power at the irradiation optical fiber end of ~100 μW. The SPC unit (custom-designed, Hamamatsu Photonics K.K., Hamamatsu, Japan), based on the time-correlated single-photon counting method (TCSPC) for measuring the temporal point spread functions (TPSFs) of tissue, is composed of a constant fraction discriminator (CFD), a time-to-amplitude converter (TAC), an analog-to-digital (A/D) converter, and a histogram memory. The irradiation fiber is a single fiber with a numerical aperture (NA) of 0.25 (GC.200/250L, FUJIKURA, Tokyo, Japan), and a fiber bundle 3 mm in diameter with NA of 0.29 (Sumita Opcical Glass, Inc., Saitama, Japan) is applied to the optical detection fiber. To estimate the absorption and reduced scattering coefficients, TPSF is first derived from analytical solution of the diffusion equation, on the assumption that the breast is a semi-infinite homogeneous medium. Then, the TPSF, which is convoluted with the instrumental response function, is fitted to the observed temporal profile. In the fitting procedure, the non-linear least-squares method is used [15,16]. Oxy- and deoxy-hemoglobin concentrations are determined using an expression in which the absorption coefficient consists of a linear combination of the extinction coefficients and concentrations of hemoglobin. The TRS imaging system is shown in Figure 1.

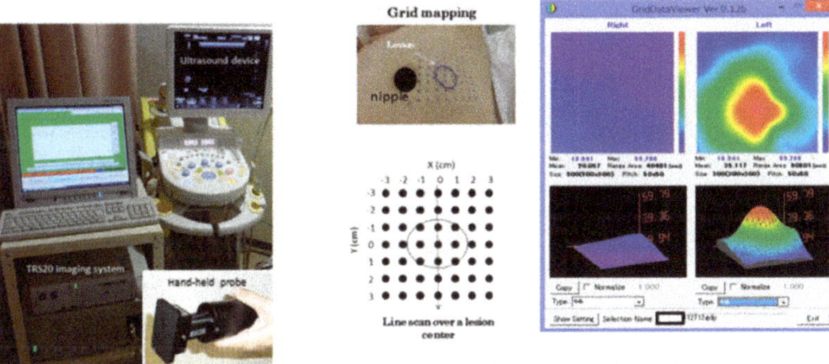

Figure 1. Time-resolved spectroscopy (TRS) breast imaging system.

2.3. Procedure of TRS Measurements

All patients were histologically diagnosed with mammary carcinoma from a core biopsy sample before optical measurement. The average optical measurement time for each patient was about 20 min. A handheld TRS probe with a 2.8 cm source-detection distance was used to measure absorption and scattering of the breasts with patients in the supine position. After ultrasonography (US)-based tumor detection, an optical probe was placed on the skin surface corresponding to tumor location and its surrounding breast tissues, and point sensing was initiated. The grid maps of the tissue with the tumors comprised 7×7 points with a 10-mm interval between two points in the x–y dimension, resulting in a minimum of 49 measurement points in each map [14]. Acquisitions were made such that the tumor was located at the center of the grid map. A lesion region of interest (ROI) was used for 2D image reconstruction for Hb distribution. In the contralateral, normal breast, a grid map with 5×5 points was obtained and used as a mirror image. The average concentration of Hb and percentage of SO_2 were recorded as representative parameters.

2.4. Monitoring Tumor Response to Neoadjuvant Chemotherapy

Functional imaging can predict tumor response to chemotherapy. Further, as changes in tumor metabolism precede tumor shrinkage, fluorodeoxyglucose-positron emission tomography (FDG-PET) imaging and dynamic-enhanced MRI are widely accepted as noninvasive modalities for monitoring tumor response during NAC [17,18]. However, these modalities are expensive, and their cost precludes integration into routine clinical practice. Many clinical studies have reported that DOSI has great potential for early assessment of tumor hemoglobin response to NAC, which serves as a predictor of pathological outcome [19–21]. Therefore, between September 2013 and February 2015, we prospectively enrolled 100 patients who were eligible for NAC at two centers (Saitama Medical University International Medical Center and Hamamatsu University Medical School) [22]; of these, 84 patients completed the full course of NAC and underwent definitive surgery (Figure 2A). The average tumor size was 37.5 mm (SD, 16.8 mm). Hormonal receptor-positive breast cancer was present in 47 patients (55.9%), and HER2-positive breast cancer was present in 19 patients (21.4%). The pathological outcome after surgery was determined to be pCR in 16 (19%) and non-pCR in 68 (81%) patients, where pCR was defined as the absence of invasive cancer cells in the breast irrespective of axillary status (ypT0/is). The chemotherapy regimen varied among patients and was at the treating physician's discretion; however, the majority (70.2%, $n = 59$) received anthracycline- and taxane-based regimens, 17.9% ($n = 15$) received carboplatin- and docetaxel-based regimens, and 11.9% ($n = 10$) of the patients were administered bevacizumab- and paclitaxel-based regimens. All patients with HER2-positive breast cancer received chemotherapy combined with trastuzumab. TRS measurement was performed before initiation of chemotherapy (at day 2 to day 1 before initial infusion of the drug), and at day 2 to day 1 before the second or third infusion. The ROI of the tumor, corresponding to peak tumor tHb, was monitored during the early courses of NAC, and percentage change in average tHb between the baseline and after chemotherapy was calculated using the formula: (interim tHb − baseline tHb)/baseline tHb × 100 (%).

A) Study protocol of hemodynamic monitoring of tumor response to neoadjuvant chemotherapy in patients with primary breast cancer

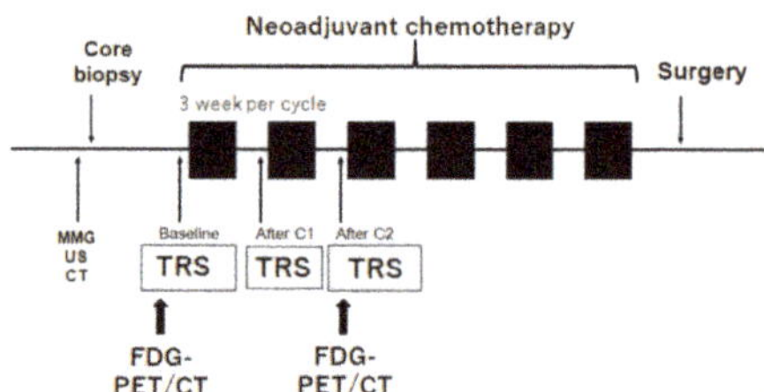

B) Study protocol of hemodynamic monitoring of tumor response to single-agent bevacizumab in patients with advanced breast cancer

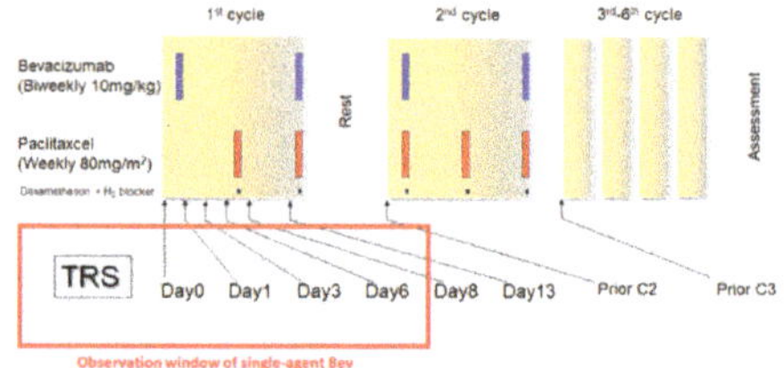

C) Study protocol of FDG-PET guided neoadjuvant endocrine therapy and monitoring of tumor hemodynamic response

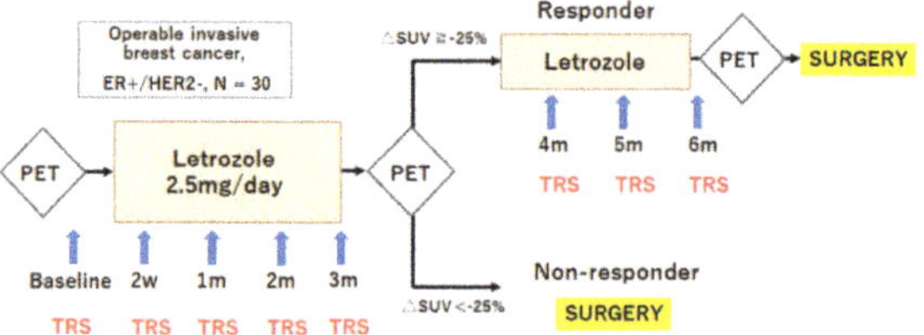

Figure 2. (**A**) Study protocol of hemodynamic monitoring of tumor response during neoadjuvant chemotherapy. (**B**) Study protocol of hemodynamic monitoring of tumor response to single-agent bevacizumab in patients with advanced breast cancer. (**C**) Study protocol for fluorodeoxyglucose-positron emission tomography-guided (FDG-PET-guided) neoadjuvant endocrine therapy and hemodynamic monitoring of tumor response.

2.5. Optical Visualization of Cancer Vascular Remodeling after Antiangiogenic Therapy

Angiogenesis is a key driver of growth and metastatic cancer spread [23]. However, the anti-angiogenic concept of cancer therapy posits that anti-angiogenic drugs should restore oxygenation in the presence of proper vascular remodeling [24]. Visualization of vascular remodeling and oxygenation is, therefore, an intriguing approach for exploring normalization in the cancer microenvironment after initiation of antiangiogenic drugs [25]. TRS can be a powerful tool to monitor vascular remodeling, since visualization of breast cancer hemodynamics can help elucidate the mechanisms underlying changes in vascularity and tissue oxygenation [26]. Bevacizumab, an antibody against endothelial growth factor, is well-known as a key drug in patients with advanced and/or metastatic breast cancer [27,28]. Thus, we conducted a clinical study in 28 patients with advanced stage 3/4 breast cancer using the standard regimen of bevacizumab (10 mg/kg body weight), administered intravenously on days 0 and 14, in combination with paclitaxel (90 mg/m^2 body surface area), administered on days 0, 7, and 14 in every cycle (Figure 2B) [29]. To evaluate the physiological effect of bevacizumab alone, paclitaxel was omitted on the first day of the first cycle. Patients continued the regimen for six cycles unless disease progression or unacceptable toxicity precluded chemotherapy continuation. Patients underwent serial FDG-PET/CT scans at baseline and after two cycles of chemotherapy. Then, the change in tumor maximal standard uptake value (SUV_{max}) was evaluated. The tumor metabolic response was classified based on the change in SUV_{max}, with a cutoff value of -20% used to categorize patients as either responders (change in $SUV_{max} > 20\%$; $n = 18$) or non-responders (change in $SUV_{max} < 20\%$; $n = 10$). Patients also underwent repeat TRS measurements every day during single-agent bevacizumab, and we observed changes in Hb and SO_2 in the tumor at baseline, and on days 1, 3, 6, 8, and 13. For the clinical study, we hypothesized that if vascular normalization occurs after successful remodeling, tumor tHb should decrease with a simultaneous increase in SO_2 levels.

2.6. Monitoring Tumor Response to Neoadjuvant Endocrine Therapy

Although neoadjuvant endocrine therapy is not the standard treatment for breast cancer, it has been widely used among postmenopausal women with hormone receptor (HR)-positive breast cancer, because neoadjuvant endocrine therapy has reportedly contributed to tumor shrinkage in 30–75% of the patients and because it can increase the conversion rate from mastectomy to breast-conserving surgery [30–32]. Therefore, early identification of responding and non-responding tumors allows patients with resistant tumors to receive alternative treatments such as surgery, molecular-targeting agents, or chemotherapy.

We have previously assessed whether the early PET response, evaluated by FDG PET/CT, could predict morphological and pathological responses to neoadjuvant endocrine therapy in patients with hormone receptor-positive breast cancer [33]. The pilot study enrolled 12 patients who received a daily dose of 2.5 mg of letrozole for 12 weeks, followed by surgery. When serial FDG-PET scans were acquired at baseline and after 4 weeks of letrozole therapy, metabolic responders showed a 40% or greater reduction in FDG uptake, as evidenced by SUV_{max} changes, and a significant decrease in the Ki67 proliferative index; these were not observed in non-responders. A large randomized clinical trial (Intermediate Marker Project: anastrozole, combination or tamoxifen; IMPACT) also reported that a low Ki67 index following endocrine therapy was significantly correlated with relapse-free survival among patients after surgery [31]. Similarly, our findings reveal that an FDG-PET/CT-guided therapeutic strategy could be promising during neoadjuvant endocrine therapy [33]. Recently, we conducted a feasibility study using TRS to monitor hemodynamic response during neoadjuvant endocrine therapy (Figure 2C). We enrolled 30 patients with HR-positive primary breast cancer, and categorized PET responders ($n = 20$) and PET non-responders ($n = 8$) based on a 25% reduction in tumor SUV_{max} between serial scans of FDG-PET/CT, obtained before therapy and at 3 months after initiation letrozole therapy.

3. Results

In the initial study, we recruited 118 patients with an established histological diagnosis of breast carcinoma, and TRS-based values such as HbO_2, HHb, tHb, and SO_2 of breast tumors were found to be significantly higher than those of the contralateral normal tissue (Figure 3A) [34]. There were no significant differences in average rtHb levels among the various histological types such as ductal carcinoma in situ, invasive ductal carcinoma, invasive lobular carcinoma, or mucinous carcinoma. On the other hand, higher tumor rtHb was significantly and positively correlated with tumor size, nuclear grade, and lymphatic vascular invasion. The overall detection rate for breast tumors using TRS was 62.7%, while its sensitivity according to tumors stratified by T stage was: T0, 31.3%; T1, 44.7%; T2, 78.9%; and T3, 100% (Figure 3B). These results imply that our approach using TRS is not suitable for early detection of breast cancer; however, Hb parameters measured by TRS may be useful for assessing tumor aggressiveness.

A) Tissue concentration of hemoglobin: Tumor vs. Normal breast

	Breast tumor		Normal breast		t-test
	mean	SD	mean	SD	
HbO_2	22.4	14	14.9	11.2	<0.0001
HHb	9.4	4.8	6.5	3.8	<0.0001
tHb	31.8	18.7	21.5	14.8	<0.0001
SO_2	68.8	5.2	66.4	8.2	0.01

B) Detection rate of a tumor using TRS breast imaging system

T stage	Number of pts	Number of positive case	Sensitivity (%)
Tis	16	5	31.3
T1	38	17	44.7
T2	57	45	78.9
T3	7	7	100
Total	118	74	62.7

Figure 3. (**A**) Tissue concentration of hemoglobin: tumor vs. normal breast. (**B**) Tumor detection rate using the TRS breast imaging system. Tis: in situ carcinoma; T1: tumor size of 2 cm or less; T2: tumor size of 2–5 cm; T3: tumor size of 5 cm or more, or tumor with skin invasion, or tumor with muscle invasion. SD: standard deviation.

Regarding neoadjuvant chemotherapy, we found that pCR tumors showed a significantly greater reduction in percentage tHb change after the first course of chemotherapy compared to non-pCR tumors (average, $-23.4\% \pm 4.3$ SE vs. $-14.1\% \pm 1.7$ SE; $p = 0.02$), which increased after the second course of chemotherapy (average, $-33.9\% \pm 3.8$ SE vs. $-20.2\% \pm 1.7$ SE; $p = 0.001$, Figure 4A-1,A-2). Receiver-operating-characteristic (ROC) curve analysis yielded an Area under the curve (AUC) of 0.69 for the first chemotherapy infusion, and of 0.75 for the second chemotherapy infusion. Thus, we considered that tHb measurement alone may be insufficient for monitoring tumor response to NAC and for precisely predicting histological outcome, as many neoadjuvant studies using FDG-PET/CT have reported a diagnostic performance of 75–90% when serial scans of FDG-PET/CT were administered at baseline and during the initial stages of 1–3 courses of chemotherapy. Next, patients were categorized into a hormonal-receptor-dependent sub-group and a hormonal-independent sub-group based on ER status. A total of 47 patients had ER-positive breast cancer, including eight patients with pCR (Figure 4B-1), while 36 patients had ER-negative breast cancer, including seven patients with pCR (Figure 4B-2). When ER-positive tumors and ER-negative tumors were separately evaluated, the ROC curve analysis yielded AUC values of 0.6 (0.15 SE) and 0.5 (0.13 SE) for ER-positive breast cancer, and 0.81 (0.08 SE) and 0.82 (0.07 SE) for ER-negative breast cancer after the first and the second courses of chemotherapy, respectively. When stratified by intrinsic subtypes, 40 patients had luminal breast cancer, including three with pCR (Figure 4C-1). Importantly, tHb in pCR tumors did not decrease after the first chemotherapy infusion, and showed no difference after the second infusion compared to non-pCR. Among 27 patients with triple-negative breast cancer (TNBC), three patients had pCR. The decrease in tHb in pCR tumors was higher compared to that of non-pCR tumors (Figure 4C-2), and among 19 HER2-positive patients, nine achieved pCR (Figure 4C-3). Thus, it appears that pCR tumors show a trend of greater reduction in tHb than non-pCR tumors, but these differences were smaller than those seen in TNBC patients.

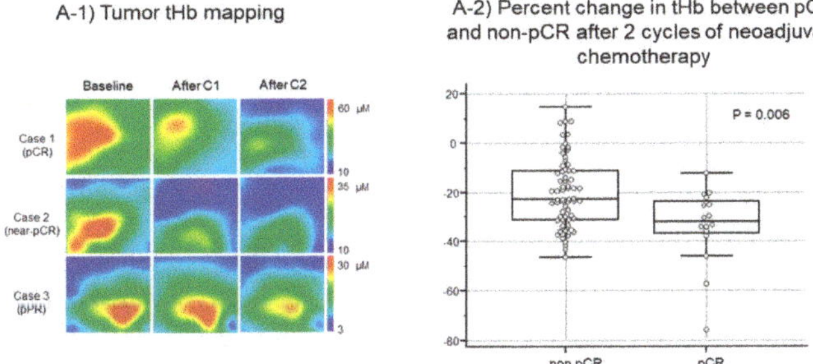

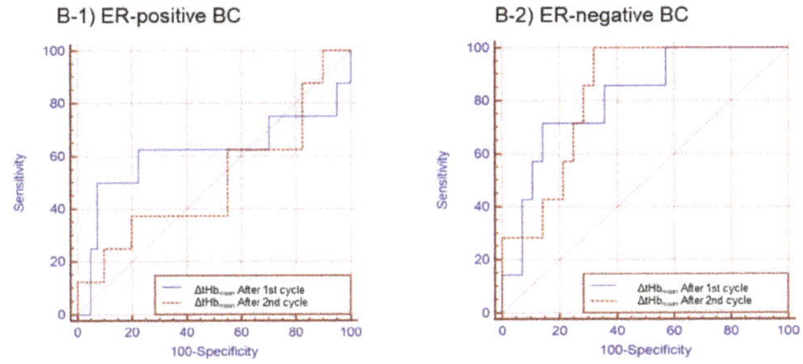

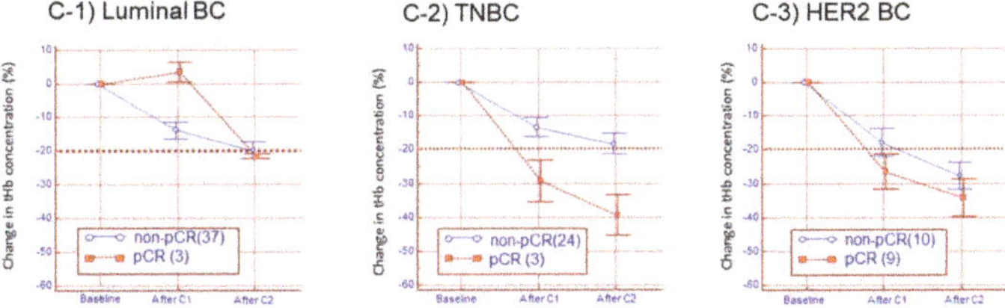

Figure 4. (**A**) Hemodynamic monitoring of breast tHb during neoadjuvant chemotherapy. C1, the first cycle of chemotherapy; C2, the second cycle of chemotherapy; pCR, pathological complete response. (**B**) Diagnostic performance of pCR stratified by ER status. ER, estrogen receptor; BC, breast cancer; (**C**) Percent change in tHb among pCR and non-pCR patients stratified by breast cancer subtypes. TNBC: triple-negative breast cancer.

Regarding antiangiogenic therapy, non-responders to bevacizumab, determined based on FDG-PET/CT, showed both lower tumor shrinkage (mean ± SE; −11.9 ± 10.5%) than responders (−38.7 ± 8.8%; $p = 0.07$) and poorer survival than the responders ($p = 0.1$, Kaplan–Meier method, data not shown). Representative case studies of a patient with a tumor responsive to bevacizumab and a patient with a non-responsive tumor to bevacizumab are shown in Figure 5A-1,A-2. Non-responders also showed markedly lower tumor SO_2 immediately after bevacizumab infusion compared to responders, and the SO_2 level in non-responders was significantly lower than that of the responders from days 1 to 3 (Figure 5B). Tumor Hb (tHb) of non-responders transiently decreased on day 1 after bevacizumab infusion, but recovered to baseline between days 3 and 6. In contrast, tHb among responders showed a sustained decrease during the observation period. Specifically, tHb levels on days 3 and 6 were significantly higher in non-responders compared to responders. The findings imply that tumors that are non-responsive to bevacizumab display acute hypoxia and further angiogenesis after drug infusion.

A) Hemodynamic monitoring of single-agent bevacizumab in breast cancer

A-1) A case study of a patient with responding tumor to bevacizumab

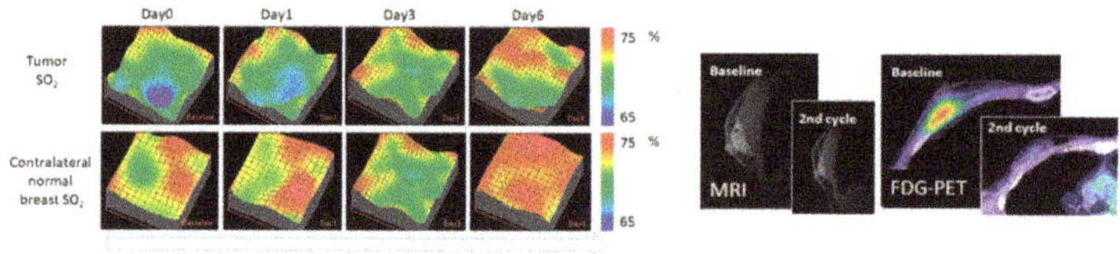

A-2) A case study of a patient with non-responding tumor to bevacizumab

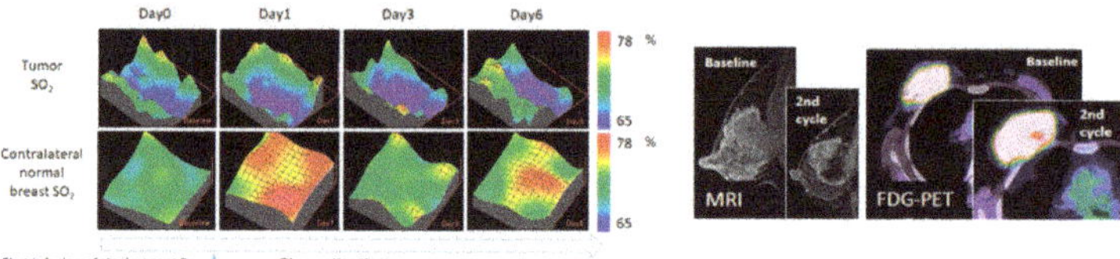

B) Tumor tHb and SO₂ change during Bevacizumab followed by Paclitaxel

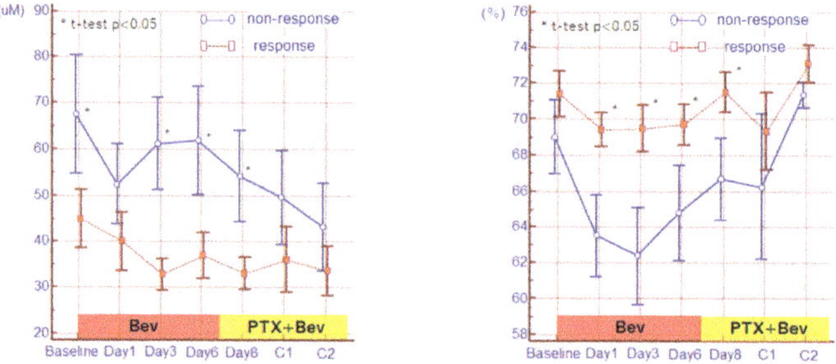

Figure 5. (**A**) Hemodynamic monitoring of single-agent bevacizumab in breast cancer. (**B**) Tumor tHb and SO_2 change during bevacizumab followed by paclitaxel. Bev: bevacizumab; PTX: paclitaxel; MRI: magnetic resonance imaging.

Regarding neoadjuvant endocrine therapy, PET non-responders represented PD ($n = 0$), SD ($n = 1$), PR ($n = 18$), and CR ($n = 1$), while PET non-responders represented PD ($n = 2$), SD ($n = 3$), PR ($n = 2$), and CR ($n = 0$). There were significant differences between the two responder groups regarding clinical response ($p = 0.003$, data not shown). Although PET responders and non-responders did not differ regarding invasive size at post-surgical histological evaluation, the Ki67 proliferative index among PET responders was significantly lower than that of non-responders. TRS was performed before and after 2 weeks, 1 month, 2 months, and 3 months of letrozole therapy initiation. Representative case studies of a patient with PET response and a patient with PET non-response are shown in Figure 6A-1,A-2. Among some patients with PET-responsive tumors, peak tHb disappeared immediately after letrozole therapy initiation, while tHb values increased in some patients with non-responsive tumors despite continued letrozole therapy. The decline in tHb value at 1 and 3 months after letrozole therapy among PET responders was significantly larger than that seen among PET non-responders ($p = 0.01$) (Figure 6B). Our findings reveal a close relationship between glycolysis and the hemodynamics of malignant tumors, implying that hemodynamic response monitoring using TRS is a feasible approach even in patients who receive endocrine therapy.

A) Tumor tHb change during neoadjuvant endocrine therapy

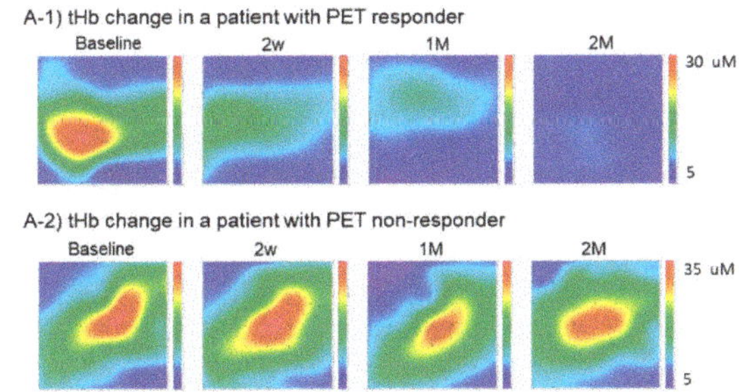

B) Differential response of tumor tHb change between PET responders
and non-responders

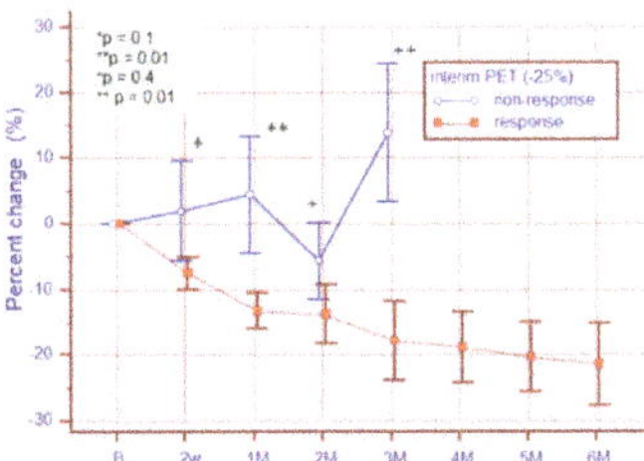

Figure 6. (**A**) Tumor tHb change during neoadjuvant endocrine therapy; (**B**) Differential response of tumor tHb change between PET responders and non-responders.

4. Discussion

Our initial clinical studies conveyed important information on breast cancer imaging using TRS. The first was that the TRS imaging system is not useful for screening during early breast cancer as the overall detection rate of 62.7% is unacceptably low, with even lower rates for T stages 0 and 1. Nonetheless, it was useful for monitoring neoadjuvant therapy in patients with advanced breast cancer, because detection rates were acceptable among patients with T-stages 2 and 3 tumors. During neoadjuvant chemotherapy monitoring, pCR tumors showed a significantly greater decrease in tHb compared to non-pCR tumors. However, when we compared the diagnostic performance of TRS to FDG-PET/CT for predicting pCR during neoadjuvant chemotherapy, DOSI showed a lower diagnostic performance than FDG-PET/CT. Interestingly, we found that the responsiveness of ER-positive tumors and ER-negative tumors was different in patients who received cytotoxic chemotherapy. Specifically, in patients with ER-negative tumors, TNBC was an excellent predictor of pCR based on the AUC; however, it was completely unacceptable among patients with ER-positive tumors. These findings imply that tumor hemodynamics depend on breast cancer's hormonal status. Since ER-negative tumors are characterized by higher metabolic and angiogenic activities, chemotherapy monitoring using TRS could be useful in patients with ER-negative tumors, especially among TNBC patients. Thus, attention to differential response of tHb stratified by breast cancer subtypes may be essential.

TRS is feasible among patients with HR-positive tumors who receive endocrine therapy. Further, while letrozole therapy (neoadjuvant endocrine therapy using aromatase inhibitor) can distinguish between metabolic responders and non-responders based on serial FDG-PET, metabolic responders had significantly lower tHb than non-responders.

A recent clinical trial (JFMC34-0601) among 107 patients treated with 25 mg/day exemestane for 16 weeks, followed by a further 8 weeks depending on the clinical response, reported that eight patients with progressive disease had markedly poorer disease-free survival and overall survival compared to patients with partial response or stable disease [34].

The discrepancy in clinical results between cytotoxic chemotherapy and endocrine therapy indicates that hemodynamic response is dependent on drug pharmacology. In our trial using the antiangiogenic drug bevacizumab, tumor SO_2 among non-responders, but not responders, decreased dramatically after bevacizumab administration. Thus, TRS could be useful for monitoring the dynamics of the hypoxic tumor microenvironment in breast tissue. These observations may help explain the negative impact of bevacizumab, wherein the drug fails to promote vascular remodeling and destroys the tumor microenvironment in some patients. In fact, published data demonstrate that circulating angiogenic biomarker levels, such as vascular endothelial growth factor (VEGF), basic fibroblast growth factor (FGF), and Transforming Growth Factor Beta (TGFβ), markedly increase on days 3 and 4 after infusion of bevacizumab in non-responders, but not in responders [30]. Thus, our TRS study with bevacizumab has provided physiological insight into drug-induced hypoxia and cancer progression, and may be useful for in vivo biomarker imaging in the future for assessing the effects of antiangiogenic therapy.

There are several limitations to the TRS technology. First, TRS can only be used for primary tumors and the surrounding normal tissue, not on axillary nodes. Further, we cannot precisely evaluate pCR status, including axillary node status. Optical absorption is limited to the reach of the photons, which corresponds to a few centimeters of depth from the skin. In addition, as chest wall thickness varies at different angles, it can compromise measurements when the distance between the skin and the muscle is too low. The relatively low sensitivity and poor spatial resolution of TRS can affect the reproducibility of the results, although MRI or US-guided measurement combined with TRS can improve measurement quality. Third, larger datasets on the water and lipid content of tissue were required to precisely assess tHb values, as tissue water and lipid concentrations are known to change during drug therapy drastically.

Despite these limitations, TRS, which does not use either ionizing radiation or extra contrast agents, has the advantage of being safe and painless even if multiple measurements need to be acquired from the same patient. Vascular and hypoxia imaging measured by TRS is an intriguing approach, with demonstrable implications for functional diagnosis in breast cancer. Although TRS is not suitable for early detection of primary breast cancer, hemodynamic stratification can identify unique alterations in tumor angiogenesis and hypoxia during treatment. Further studies will be required to explore optical phenotypes associated with molecular profiles of breast cancer, and to develop patient-tailored medicine.

Author Contributions: S.U. and T.S. conducted the clinical studies, and S.U. wrote the manuscript.

Acknowledgments: The authors acknowledge funding from KAKEN 16K10293 and 17H03591 [Grant-in-Aid for Scientific Research C and B, JSPS]. We are grateful to all patients who participated in the clinical studies and thank Yukio Ueda, Yutaka Yamashita, Hiroyuki Suzuki, and Kenji Yoshimoto for their technical support with TRS20 (Hamamatsu K.K.); Noriko Wakui and Midori Nakajima for their help with optical measurement and all the medical doctors, Akihiko Osaki, Hideki Takeuchi, Takashi Shigekawa, Eiko Hirokawa, Ikuko Sugitani, Hiroko Shimada, and Aya Asano, who recruited eligible patients for the studies.

Conflicts of Interest: The authors declare no conflict of interest.

References

1. Machida, Y.; Tozaki, M.; Shimauchi, A.; Yoshida, T. Breast density: The trend in breast cancer screening. *Breast Cancer* **2015**, *22*, 253–261. [CrossRef] [PubMed]

2. Suzuki, A.; Ishida, T.; Ohuchi, N. Controversies in breast cancer screening for women aged 40–49 years. *Jpn. J. Clin. Oncol.* **2014**, *44*, 613–618. [CrossRef] [PubMed]

3. Tromberg, B.J.; Cerussi, A.E. Imaging breast cancer chemotherapy response with light. Commentary on soliman et al., p. 2605. *Clin. Cancer Res.* **2010**, *16*, 2486–2488. [CrossRef] [PubMed]

4. Tromberg, B.J.; Cerussi, A.; Shah, N.; Compton, M.; Durkin, A.; Hsiang, D.; Butler, J.; Mehta, R. Imaging in breast cancer: Diffuse optics in breast cancer: Detecting tumors in pre-menopausal women and monitoring neoadjuvant chemotherapy. *Breast Cancer Res.* **2005**, *7*, 279–285. [CrossRef] [PubMed]

5. Chung, S.H.; Cerussi, A.E.; Klifa, C.; Baek, H.M.; Birgul, O.; Gulsen, G.; Merritt, S.I.; Hsiang, D.; Tromberg, B.J. In vivo water state measurements in breast cancer using broadband diffuse optical spectroscopy. *Phys. Med. Biol.* **2008**, *53*, 6713–6727. [CrossRef]

6. Taroni, P.; Quarto, G.; Pifferi, A.; Abbate, F.; Balestreri, N.; Menna, S.; Cassano, E.; Cubeddu, R. Breast tissue composition and its dependence on demographic risk factors for breast cancer: Non-invasive assessment by time domain diffuse optical spectroscopy. *PLoS ONE* **2015**, *10*, e0128941. [CrossRef] [PubMed]

7. Kaufmann, M.; Morrow, M.; von Minckwitz, G.; Harris, J.R.; Biedenkopf Expert Panel, M. Locoregional treatment of primary breast cancer: Consensus recommendations from an international expert panel. *Cancer* **2010**, *116*, 1184–1191. [CrossRef]

8. Von Minckwitz, G.; Untch, M.; Blohmer, J.U.; Costa, S.D.; Eidtmann, H.; Fasching, P.A.; Gerber, B.; Eiermann, W.; Hilfrich, J.; Huober, J.; et al. Definition and impact of pathologic complete response on prognosis after neoadjuvant chemotherapy in various intrinsic breast cancer subtypes. *J. Clin. Oncol.* **2012**, *30*, 1796–1804. [CrossRef]

9. Li, H.; Yao, L.; Jin, P.; Hu, L.; Li, X.; Guo, T.; Yang, K. Mri and pet/ct for evaluation of the pathological response to neoadjuvant chemotherapy in breast cancer: A systematic review and meta-analysis. *Breast* **2018**, *40*, 106–115. [CrossRef]

10. Zhu, Q.; Ricci, A., Jr.; Hegde, P.; Kane, M.; Cronin, E.; Merkulov, A.; Xu, Y.; Tavakoli, B.; Tannenbaum, S. Assessment of functional differences in malignant and benign breast lesions and improvement of diagnostic accuracy by using us-guided diffuse optical tomography in conjunction with conventional us. *Radiology* **2016**, *280*, 387–397. [CrossRef]

11. Zhao, Y.; Mastanduno, M.A.; Jiang, S.; Ei-Ghussein, F.; Gui, J.; Pogue, B.W.; Paulsen, K.D. Optimization of image reconstruction for magnetic resonance imaging-guided near-infrared diffuse optical spectroscopy in breast. *J. Biomed. Opt.* **2015**, *20*, 56009. [CrossRef] [PubMed]

12. Leproux, A.; O'Sullivan, T.D.; Cerussi, A.; Durkin, A.; Hill, B.; Hylton, N.; Yodh, A.G.; Carp, S.A.; Boas, D.; Jiang, S.; et al. Performance assessment of diffuse optical spectroscopic imaging instruments in a 2-year multicenter breast cancer trial. *J. Biomed. Opt.* **2017**, *22*, 121604. [CrossRef] [PubMed]

13. Lee, S.; Kim, J.G. Breast tumor hemodynamic response during a breath-hold as a biomarker to predict chemotherapeutic efficacy: Preclinical study. *J. Biomed. Opt.* **2018**, *23*. [CrossRef] [PubMed]

14. Ueda, S.; Nakamiya, N.; Matsuura, K.; Shigekawa, T.; Sano, H.; Hirokawa, E.; Shimada, H.; Suzuki, H.; Oda, M.; Yamashita, Y.; et al. Optical imaging of tumor vascularity associated with proliferation and glucose metabolism in early breast cancer: Clinical application of total hemoglobin measurements in the breast. *BMC Cancer* **2013**, *13*, 514. [CrossRef] [PubMed]

15. Patterson, M.S.; Chance, B.; Wilson, B.C. Time resolved reflectance and transmittance for the non-invasive measurement of tissue optical properties. *Appl. Opt.* **1989**, *28*, 2331–2336. [CrossRef] [PubMed]

16. Ijichi, S.; Kusaka, T.; Isobe, K.; Okubo, K.; Kawada, K.; Namba, M.; Okada, H.; Nishida, T.; Imai, T.; Itoh, S. Developmental changes of optical properties in neonates determined by near-infrared time-resolved spectroscopy. *Pediatr. Res.* **2005**, *58*, 568–573. [CrossRef] [PubMed]

17. Wahl, R.L.; Jacene, H.; Kasamon, Y.; Lodge, M.A. From recist to percist: Evolving considerations for pet response criteria in solid tumors. *J. Nucl. Med.* **2009**, *50* (Suppl. 1), S122–S150. [CrossRef] [PubMed]

18. Drisis, S.; Metens, T.; Ignatiadis, M.; Stathopoulos, K.; Chao, S.L.; Lemort, M. Quantitative dce-mri for prediction of pathological complete response following neoadjuvant treatment for locally advanced breast cancer: The impact of breast cancer subtypes on the diagnostic accuracy. *Eur. Radiol.* **2016**, *26*, 1474–1484. [CrossRef]

19. Xu, C.; Vavadi, H.; Merkulov, A.; Li, H.; Erfanzadeh, M.; Mostafa, A.; Gong, Y.; Salehi, H.; Tannenbaum, S.; Zhu, Q. Ultrasound-guided diffuse optical tomography for predicting and monitoring neoadjuvant chemotherapy of breast cancers: Recent progress. *Ultrason. Imaging* **2016**, *38*, 5–18. [CrossRef] [PubMed]

20. Tromberg, B.J.; Zhang, Z.; Leproux, A.; O'Sullivan, T.D.; Cerussi, A.E.; Carpenter, P.M.; Mehta, R.S.; Roblyer, D.; Yang, W.; Paulsen, K.D.; et al. Predicting responses to neoadjuvant chemotherapy in breast cancer: Acrin 6691 trial of diffuse optical spectroscopic imaging. *Cancer Res.* **2016**, *76*, 5933–5944. [CrossRef]

21. Roblyer, D.; Ueda, S.; Cerussi, A.; Tanamai, W.; Durkin, A.; Mehta, R.; Hsiang, D.; Butler, J.A.; McLaren, C.; Chen, W.P.; et al. Optical imaging of breast cancer oxyhemoglobin flare correlates with neoadjuvant chemotherapy response one day after starting treatment. *Proc. Natl. Acad. Sci. USA* **2011**, *108*, 14626–14631. [CrossRef] [PubMed]

22. Ueda, S.; Saeki, T.; Takeuchi, H.; Shigekawa, T.; Yamane, T.; Kuji, I.; Osaki, A. In vivo imaging of eribulin-induced reoxygenation in advanced breast cancer patients: A comparison to bevacizumab. *Br. J. Cancer* **2016**, *114*, 1212–1218. [CrossRef] [PubMed]

23. Folkman, J. Tumor angiogenesis: Therapeutic implications. *N. Engl. J. Med.* **1971**, *285*, 1182–1186. [PubMed]

24. Jain, R.K. Normalization of tumor vasculature: An emerging concept in antiangiogenic therapy. *Science* **2005**, *307*, 58–62. [CrossRef]

25. Jain, R.K. Antiangiogenesis strategies revisited: From starving tumors to alleviating hypoxia. *Cancer Cell* **2014**, *26*, 605–622. [CrossRef] [PubMed]

26. Ueda, S.; Kuji, I.; Shigekawa, T.; Takeuchi, H.; Sano, H.; Hirokawa, E.; Shimada, H.; Suzuki, H.; Oda, M.; Osaki, A.; et al. Optical imaging for monitoring tumor oxygenation response after initiation of single-agent bevacizumab followed by cytotoxic chemotherapy in breast cancer patients. *PLoS ONE* **2014**, *9*, e98715. [CrossRef]

27. Miller, K.; Wang, M.; Gralow, J.; Dickler, M.; Cobleigh, M.; Perez, E.A.; Shenkier, T.; Cella, D.; Davidson, N.E. Paclitaxel plus bevacizumab versus paclitaxel alone for metastatic breast cancer. *N. Engl. J. Med.* **2007**, *357*, 2666–2676. [CrossRef]

28. Robert, N.J.; Dieras, V.; Glaspy, J.; Brufsky, A.M.; Bondarenko, I.; Lipatov, O.N.; Perez, E.A.; Yardley, D.A.; Chan, S.Y.; Zhou, X.; et al. Ribbon-1: Randomized, double-blind, placebo-controlled, phase iii trial of chemotherapy with or without bevacizumab for first-line treatment of human epidermal growth factor receptor 2-negative, locally recurrent or metastatic breast cancer. *J. Clin. Oncol.* **2011**, *29*, 1252–1260. [CrossRef]

29. Ueda, S.; Saeki, T.; Osaki, A.; Yamane, T.; Kuji, I. Bevacizumab induces acute hypoxia and cancer progression in patients with refractory breast cancer: Multimodal functional imaging and multiplex cytokine analysis. *Clin. Cancer Res.* **2017**. [CrossRef]

30. Cataliotti, L.; Buzdar, A.U.; Noguchi, S.; Bines, J.; Takatsuka, Y.; Petrakova, K.; Dube, P.; de Oliveira, C.T. Comparison of anastrozole versus tamoxifen as preoperative therapy in postmenopausal women with hormone receptor-positive breast cancer: The pre-operative "arimidex" compared to tamoxifen (proact) trial. *Cancer* **2006**, *106*, 2095–2103. [CrossRef]

31. Smith, I.E.; Dowsett, M.; Ebbs, S.R.; Dixon, J.M.; Skene, A.; Blohmer, J.U.; Ashley, S.E.; Francis, S.; Boeddinghaus, I.; Walsh, G.; et al. Neoadjuvant treatment of postmenopausal breast cancer with anastrozole, tamoxifen, or both in combination: The immediate preoperative anastrozole, tamoxifen, or combined with tamoxifen (impact) multicenter double-blind randomized trial. *J. Clin. Oncol.* **2005**, *23*, 5108–5116. [CrossRef] [PubMed]

32. Chia, Y.H.; Ellis, M.J.; Ma, C.X. Neoadjuvant endocrine therapy in primary breast cancer: Indications and use as a research tool. *Br. J. Cancer* **2010**, *103*, 759–764. [CrossRef] [PubMed]

33. Ueda, S.; Tsuda, H.; Saeki, T.; Omata, J.; Osaki, A.; Shigekawa, T.; Ishida, J.; Tamura, K.; Abe, Y.; Moriya, T.; et al. Early metabolic response to neoadjuvant letrozole, measured by fdg pet/ct, is correlated with a decrease in the ki67 labeling index in patients with hormone receptor-positive primary breast cancer: A pilot study. *Breast Cancer* **2011**, *18*, 299–308. [CrossRef] [PubMed]
34. Nakamiya, N.; Ueda, S.; Shigekawa, T.; Takeuchi, H.; Sano, H.; Hirokawa, E.; Shimada, H.; Suzuki, H.; Oda, M.; Osaki, A.; et al. Clinicopathological and prognostic impact of imaging of breast cancer angiogenesis and hypoxia using diffuse optical spectroscopy. *Cancer Sci.* **2014**, *105*, 833–839. [CrossRef] [PubMed]

 *applied sciences*

Article

Compression Stockings Suppressed Reduced Muscle Blood Volume and Oxygenation Levels Induced by Persistent Sitting

Misato Kinoshita [1], Yuko Kurosawa [1], Sayuri Fuse [1], Riki Tanaka [1], Nobuko Tano [1], Ryota Kobayashi [2], Ryotaro Kime [1] and Takafumi Hamaoka [1,*]

[1] Department of Sports Medicine for Health Promotion, Tokyo Medical University, 6-1-1 Shinjuku, Shinjuku-ku, Tokyo 160-8402, Japan; kino3885@aol.com (M.K.); kurosawa@tokyo-med.ac.jp (Y.K.); fuse@tokyo-med.ac.jp (S.F.); s118042@tokyo-med.ac.jp (R.T.); kobutainlondon@gmail.com (N.T.); kime@tokyo-med.ac.jp (R.K.)

[2] Center for Fundamental Education, Teikyo University of Science, Tokyo 120-0045, Japan; rkobayashi@ntu.ac.jp

* Correspondence: hamaoka@tokyo-med.ac.jp; Tel.: +81-3-3351-6141 (ext. 420); Fax: +81-3-3226-5277

Received: 4 March 2019; Accepted: 28 April 2019; Published: 30 April 2019

Abstract: This study quantitatively analyzed the effects of 3 h of constant sitting on skeletal muscle oxygenation in the lower extremities, using near-infrared time-resolved spectroscopy (NIR_{TRS}). The effects of compression stockings were also evaluated. Eleven healthy men (age, 30.0 ± 6.7 years) maintained their knee joints at 90° flexion during 3 h of constant sitting and wore a compression stocking on either the right or left leg. The side the stocking was worn was chosen randomly. Subsequently, leg circumference and extracellular water were measured. After 3 h of sitting, both factors increased significantly in uncompressed limbs. Furthermore, intracellular water and muscle oxygenation had significantly decreased. In contrast, extracellular water had not increased in the limbs wearing compression stockings. Furthermore, the increased circumference of compressed limbs was significantly smaller than that of uncompressed limbs. Decreases in oxygenated hemoglobin and total hemoglobin were significantly smaller in compressed limbs than in uncompressed limbs (oxy-Hb; $p = 0.021$, total-Hb; $p = 0.013$). Three hours of sitting resulted in decreased intracellular water and increased extracellular water in the lower extremities, leading to reduced blood volume and oxygenation levels in skeletal muscle. Compression stockings successfully suppressed these negative effects.

Keywords: 3-hour sitting; near infrared time-resolved spectroscopy; compression stocking; tissue oxygenation; extracellular water; intracellular water; circumference; gastrocnemius

---

1. Introduction

Sitting has gained attention because sitting for a long period of time can cause deep vein thrombosis (DVT). Long-distance travel has become common; therefore, DVT can occur in healthy individuals as well as in those at greater risk for DVT due to spending long periods in a sitting position [1]. It is reported that thrombi formed in the deep veins of the lower extremities can be carried to the lungs through the bloodstream, leading to pulmonary thromboembolism and death [2]. The time spent in a sitting position has gained attention over the past 10 years because it is a factor independent of exercise levels [3,4]. The longer an individual (even a healthy individual) remains in a sitting position during the course of daily activities, the higher the presence of biomarkers that are indicative of the development of diabetes and cardiovascular disease [5,6], resulting in increased prevalence of these diseases [7] and increased all-cause mortality [8]. Current data regarding acute changes in the body as a result of extended sitting are not sufficient. Several problems have been reported as a result of sitting for long periods, such as increased calf circumference, decreased blood flow velocity in the lower extremities, and worsening vascular endothelial function [9,10]. However, there are very few reports focusing on the influence of 3 h of sitting on lower limb skeletal muscle tissue, which is the final oxygen utilization site.

Near-infrared spectroscopy (NIRS) is a technique used to measure tissue oxygenation levels. It involves irradiating tissues with two or three types of near-infrared light of different wavelengths and then detecting and analyzing the light passing through the tissue, thereby allowing for quantitative measurements of oxygen saturation and evaluations of changes in oxygenated hemoglobin (oxy-Hb) and deoxygenated Hb (deoxy-Hb) levels. Because this technique is both convenient and non-invasive, its use for assessing oxygen dynamics in various tissues, including skeletal muscle and brain tissue, has become widespread. A broadly adopted form of NIRS is near-infrared continuous-wave spectroscopy (NIR_{CWS}), in which irradiation is performed using a continuous beam of light [11,12]. Hb concentrations assessed using NIR_{CWS} are calculated based on Lambert-Beer's law. However, NIR_{CWS} is unable to determine the optical path length. Therefore, this value can be calculated based on relative changes in concentration even though quantitative measurement is impossible [11,12]. Near-infrared time-resolved spectroscopy (NIR_{TRS}) was developed as a measurement method that could compensate for this defect. This technique involves irradiating the target tissue with a short pulse of light, thereby allowing for independent determination of the scatter and absorption coefficients, which indicate the optical characteristics of the object, based on the time responsiveness. Therefore, when measurements were performed using NIR_{TRS}, it was possible to determine the optical path lengths that could not be measured with NIR_{CWS}, due to which it was possible to evaluate the oxygen kinetics in tissues with absolute values [13,14]. It is now possible to measure changes over time in the same subject after maintaining a sitting position and to compare measured values among different individuals.

Several methods are recommended to prevent DVT and lower extremity varices, such as walking and active exercise, wearing compression stockings, intermittent pneumatic compression, and administration of heparin and warfarin [15–17]. The advantages of compression stocking usage are not only high cost performance but also no side effects of hemorrhage, which often occurs when heparin and warfarin are administered. Compression stockings can also be used by patients who cannot move and frail elderlies. The recommended compression stockings have a variable pressure design that exerts the highest pressure at the ankle joint and gradually decreases the pressure proximally toward the limbs; the ankle joint pressure is less than 20 mmHg (26.7 hPa). However, compression stockings have been criticized as being difficult to wear because of low flexibility, and patients with weak or damaged skin are at risk for developing dermatitis [18]. Although many people commonly maintain sitting positions for periods of 3 h or longer in daily life during activities such as office work, many previous studies investigated only the short-term impact of wearing compression stockings while sitting for periods of 1 h.

Furthermore, few studies have been performed to verify the effect of wearing compression stockings on oxygen kinetics in the lower leg skeletal muscles. To our knowledge, no study discussing the relationship between the level of lower leg tissue oxygenation, lower extremity extracellular water, and limb circumference has been published previously. Therefore, this study investigated the relationship between the oxygen dynamics of the lower leg skeletal muscles and changes in lower limb extracellular water and circumference based on quantitative measurements using near-infrared time-resolved spectroscopy (NIR_{TRS}) in 11 healthy adult male subjects before and after completion of a 3-hour sitting task. Additionally, we evaluated the impact of wearing a compression stocking on physiological changes in individual subjects during the 3-hour sitting task by applying the stocking to only a single leg.

2. Materials and Methods

2.1. Study Design

This study was approved by the Medical Ethics Committee at Tokyo Medical University (approval number: 2018-045) and was performed in accordance with the Declaration of Helsinki. Prior to study initiation, informed written consent was obtained from all participants. All subjects were briefed regarding the details and purpose of the study, related experimental procedures, and potential risks. The study was performed from May to July 2018.

The subjects were 11 healthy men (age: 30.0 ± 6.7 years, height: 171.5 ± 5.1 cm, weight: 73.0 ± 10.4 kg, BMI: 24.8 ± 3.3, body fat percentage: 20.7 ± 5.9%). Ten of the 11 subjects worked in professions that required them to stand for the majority of the day. All statistics are expressed as mean ± standard deviation. Subjects who were using medication were excluded from participation. In addition, subjects with a history of or a current diagnosis of hematological disorders, such as thrombosis, hypertension, and venous/arterial diseases, were excluded.

To eliminate the influence of diurnal variations, all baseline measurements were performed between 8 a.m. and 10 a.m. After measuring each item at baseline, subjects were instructed to maintain a sitting posture for 3 h with their knee joints in 90° flexion. During this task, subjects were instructed to wear a compression stocking (Elaction Pro, Asahi Kasei Corp, Tokyo, Japan) from the inguinal region to the ankle on a single leg selected randomly; no stocking was worn on the other leg. Subjects were surveyed regarding the subjective degrees of fatigue and stress using a self-reported visual analog scale (VAS), as well as the heart rate, blood pressure, tympanic membrane temperature, compression stocking pressure fit, lower leg circumference, body water measurements using bioelectrical impedance, and skeletal muscle oxygen dynamics, using NIR$_{TRS}$. The chair used for the sitting task had a width of 46 cm and a backrest angle of 110°; these standard dimensions are the same as those of an economy class seat on a passenger aircraft. During the 3-hour sitting task, spontaneous muscle activity in the lower limbs was completely prohibited. The room temperature of the laboratory was controlled at 22 ± 2 °C, and dehydration was prevented by allowing 100 mL of drinking water per hour during the 3-hour sitting task. No subjects withdrew before completion of the task.

2.2. Subjective Fatigue and Stress Survey

Using the self-reported VAS, the subjective degrees of physical fatigue, subjective degrees of mental fatigue degree, subjective physical stress levels, and subjective mental stress levels were evaluated. When no fatigue or stress was felt, the percentage was recorded as 0%; the maximum level was 100%.

2.3. Heart Rate, Systolic and Diastolic Blood Pressures, and Tympanic Membrane Temperature

Heart rate, systolic blood pressure, and diastolic blood pressure were measured using an automatic sphygmomanometer (HEM-1025; Omron KK, Kyoto, Japan). The tympanic membrane temperature was measured using a near-infrared thermometer (Omron Corporation, Kyoto, Japan).

2.4. Compression Stocking Pressure

In this study, the pressure of the ankle joint part was set at 15.7 ± 2.8 hPa, which is approximately 10 hPa lower than the recommended pressure, and the study investigation was performed using the compression stockings offering the highest pressure for the calf region, which exhibits the muscle pump action (compression stocking pressure ratio of the ankle joint: abdominal gastrocnemius muscle: thigh = 8:10:8). To measure compression pressure, the pressure sensor was attached at the ankle joint (1 cm above the lateral malleolus), calf (1 cm lateral to the location of the NIR probe placed over the lateral head of the gastrocnemius muscle), and thigh (15 cm above the upper end of the patella over the vastus lateralis). Compression stocking pressure was measured with a contact pressure gauge before and after the 3-hour sitting task (AMI 3037-SB with display; AMI Techno, Tokyo, Japan). Measurements were performed three times at each measurement point, and the average value of the three measurements was used as the compression stocking pressure.

2.5. Lower Limb Circumference

The circumference of the calf was measured by the same examiner using the same method with a tape measure. The measurement site was one-third (or the largest circumference) the distance from the midpoint of the straight line connecting the head of the fibula to the lateral malleolus and the knee joint was maintained at 90° of flexion. After determining the measurement site, the same site was repeatedly measured by marking it with flexible tape. Measurements were performed three times at each measurement site, and the average value was used as the limb circumference.

2.6. Body Water Volume

The body water volume (intracellular water/extracellular water) was evaluated using bioelectrical impedance (InBody 720; BioSpace Corporation, Tokyo, Japan). This technique uses a multifrequency bioelectrical impedance device that displays a high correlation with body water content calculated by using the deuterium oxide (D_2O) dilution technique (r = 0.974) [19]. The ratio of extracellular water to total body water (extracellular water + intracellular water) was obtained as a relevant index of edema. The ratio of intracellular water to total body water was also evaluated.

2.7. Skeletal Muscle Oxygen Kinetics

The oxygen kinetics of the lateral head of the gastrocnemius was measured using NIR_{TRS} (TRS-20; Hamamatsu Photonics K.K., Hamamatsu, Japan) [13,14]. The distance between the components irradiating and receiving the near-infrared light was set to 3.0 cm, and detection of the light in a region of skeletal muscle tissue with a depth of 1–2 cm was attempted. The measurement was performed for 3 min before and after the 3-hour sitting task was started, and the average values during these 3-minute periods were used as the measurement value.

The methods for calculating oxy-Hb, deoxy-Hb, total Hb, and oxygen saturation SO_2 were as follows: The target tissue was repeatedly irradiated using semiconductor pulse laser lights of three different wavelengths (760, 800, 830 nm) under the conditions of half-width 100 ps, pulse rate of 5 MHz, and average power level of 100 μW. The pulsed light scattered and absorbed inside the living tissue was detected by a photomultiplier tube capable of single-photon detection. Time-resolved measurements were conducted using the time-correlated single photon counting method. The absorption coefficient and reduced scattering coefficient for the obtained tissue were determined by fitting the photon diffusion equation R (ρ,t) (Equation (1)) that convolved the instrument response function to the time-response properties of the samples.

Where (t) is the response time, (ρ) is the distance between light source and detection, μ_a and μ_s' are the absorption coefficient and equivalent scattering coefficient, D = $1/3\mu_s'$] is the photon diffusion coefficient, c is the light velocity inside the light scattering medium (20 cm ns^{-1}), Z_0 (= $1/\mu_s'$) is the transport mean free path, and average path length (L) (= $\int[R(\rho, t)\, t\, dt]\, c/\int [R (\rho, t)\, dt]$. [Deoxy-Hb] and [oxy-Hb] were obtained using simultaneous equations with μ_a and μ_s' obtained from Equation (1) using the least-squares fitting method [20], and [total-Hb] and SO_2 were calculated from the following calculation formula.

$$R\,(\rho, t) = (4\pi Dc)^{-3/2}{\cdot}Z_0 t^{-5/2} \exp(-\mu_a\, ct)\, \exp[-(\rho^2 + Z_0^2)/4Dct] \tag{1}$$

$$[\text{total-Hb}] = [\text{deoxy-Hb}] + [\text{oxy-Hb}] \tag{2}$$

$$SO_2\,(\%) = [\text{oxy-Hb}] \times 100/[\text{total-Hb}] \tag{3}$$

2.8. Statistical Analysis

The measured values are expressed as mean ± standard deviation. All statistical analyses were performed using SPSS statistics version 25 (IBM SPSS, Tokyo, Japan). Data were analyzed in three stages. First, Wilcoxon's rank-sum test was used for statistical analysis of subjective fatigue before and after the 3-hour sitting task, and the subjective stress level, heart rate, systolic and diastolic blood pressures, tympanic membrane temperature, and compression stocking pressure were also analyzed. Similarly, the rates of change in calf circumference, extracellular water, intracellular water, and NIR_{TRS} indices after the 3-hour sitting task were also analyzed using the Wilcoxon rank-sum test, and the limbs with and without compression stockings were compared. Second, the effects of the 3-hour sitting task on lower limb circumference, lower limb extracellular water, and lateral head of gastrocnemius muscle oxygen kinetics were compared regarding compression stocking wear status and wear time (at baseline vs. after the 3-hour sitting task), after which a repeated two-way analysis of variance was conducted, followed by Bonferroni's multiple comparisons tests as necessary. Third, Spearman's correlation analysis was performed to determine the relationship between the rate of change in limb circumference, extracellular water, and each index measured with NIR_{TRS} before and after the 3-hour sitting task. The level of statistical significance was set at less than 5% for all tests (two-tailed test).

3. Results

3.1. Subjective Degree of Fatigue and Mental Fatigue Levels

After maintaining sitting for 3 h, the subjects' degree of physical fatigue showed an increasing trend, and mental fatigue increased significantly (physical fatigue: 32.4% ± 19.7% to 54.1 ± 20.4%, $p = 0.075$; mental fatigue: 34.7% ± 27.8% to 59.2 ± 26.6%, $p < 0.05$). Furthermore, the subjective levels of physical and mental fatigue increased significantly (physical fatigue: 19.6 ± 15.2% to 51.1 ± 23.7%; mental fatigue: 25.4% ± 28.2% to 48.6 ± 29.3%; $p < 0.05$).

3.2. Heart Rate, Systolic and Diastolic Blood Pressures, and Tympanic Membrane Temperature

Heart rate, systolic and diastolic blood pressures in the right upper arm, and tympanic membrane temperature did not show any significant changes after the 3-hour sitting task (heart rate: 58.5 ± 10.6 to 58.5 ± 10.6 beats/min; systolic blood pressure: 127.2 ± 11.2 to 127.7 ± 11.9 mmHg; diastolic blood pressure: 81.0 ± 8.0 to 80.7 ± 6.9 mmHg; tympanic membrane temperature: 35.2 ± 0.5 to 35.2 ± 0.3 °C).

3.3. Compression Stocking Pressure

Changes in compression stocking pressure at each of the three measurement sites after the 3-hour sitting task were as follows: ankle joint, 15.7 ± 2.8 to 16.4 ± 2.8 hPa; gastrocnemius muscle, 20.4 ± 1.1 to 21.1 ± 1.3 hPa; and thigh, 15.6 ± 2.2 to 16.5 ± 2.5 hPa. Significant increases were observed for pressure at the gastrocnemius muscle and thigh (all $p < 0.05$).

3.4. Calf Circumference

Figure 1 illustrates the changes in the lower limb circumference after the 3-hour sitting task. There was a significant interaction of the lower limb circumference between the compression stocking wear status and wear time (baseline value and after the 3-hour sitting task, $p < 0.05$). For limbs without compression stockings, the circumference of the gastrocnemius muscle significantly increased (+2.0 ± 0.6%; minimum value: 1.0%; maximum value: 3.1%; $p < 0.05$, Figure 1) after the 3-hour sitting task. For limbs with compression stockings, this increase was significantly lower (0.4 ± 0.3%); however, the circumference still slightly increased after the 3-hour sitting task ($p < 0.05$, Figure 1).

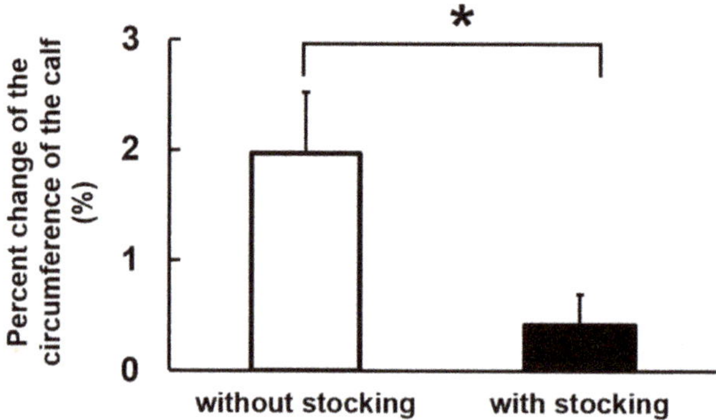

Figure 1. Percent changes of the circumference of the calf before and after 3-hour constant sitting with- or without compression stocking. * $p < 0.05$ between groups.

3.5. Extracellular and Intracellular Water in the Lower Limbs

Figure 2A shows the change in the extracellular water/total body water ratio in the lower limbs with and without the compression stocking before and after the 3-hour sitting task. There was a significant interaction of the extracellular water/total body water ratio in the lower limbs between the compression stocking wear status and wear time (baseline value and after the 3-hour sitting task, $p < 0.05$). The main effect of compression stocking use and wear time were also found ($p < 0.05$, respectively). For limbs without compression stockings, the extracellular water/total body water ratio significantly increased after the 3-hour sitting task (0.371 ± 0.008 to 0.375 ± 0.009 $p < 0.05$, Figure 2A), and edema was observed. In contrast, for limbs with the compression stocking, extracellular water did not increase (0.370 ± 0.009 to 0.371 ± 0.009, $p = 0.062$, Figure 2A).

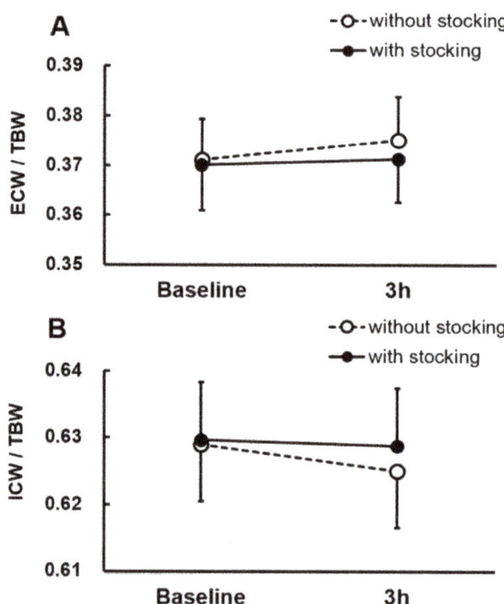

Figure 2. (**A**) Extracellular water/total water ratio (ECW/TBW) and (**B**) intracellular water/total water ratio (ICW/TBW) in the lower limbs before and after 3-hour constant sitting with or without compression stocking.

Changes in the intracellular water/total body water ratio (an indicator of intracellular water in the lower extremities) after the 3-hour sitting task are shown in Figure 2B. There was a significant interaction of intracellular water/total body water ratio in the lower limbs between the compression stocking wear status and wear time ($p < 0.05$, Figure 2B). The main effect of wear time was also found ($p < 0.05$, Figure 2B). However, the main effect of the presence or absence of wearing compression stockings remained insignificant ($p = 0.07$, Figure 2B). Intracellular water in the lower limbs without compression stockings decreased (from 0.629 ± 0.008 to 0.625 ± 0.008, $p < 0.05$, Figure 2B) whereas intracellular water in the limbs with compression stockings did not change (0.630 ± 0.009 to 0.629 ± 0.009, $p < 0.05$, Figure 2B).

3.6. NIR$_{TRS}$ Data

Table 1 displays the values of the NIR$_{TRS}$ indicators before and after 3 h of sitting with and without compression stockings. Figure 3 displays the rates of change in these values after 3 h of sitting with and without compression stockings.

Table 1. Near-infrared time-resolved spectroscopy parameters before and after 3-hour constant sitting with or without compression stocking.

	without Stocking		with Stocking	
	Bassline	**3 h**	**Bassline**	**3 h**
μ_a (cm^{-1})	$0.24 + 0.08$	$0.22 + 0.07$ *	0.25 ± 0.08	0.24 ± 0.08 *
μ_s' (cm^{-1})	14.8 ± 6.5	15.4 ± 6.7 *	17.7 ± 12.1	17.9 ± 12.1
Path Length (cm)	15.6 ± 2.9	16.5 ± 3.3 *	15.1 ± 2.7	15.5 ± 2.6 *
[oxy-Hb] (μM)	103.7 ± 37.4	91.1 ± 33.3 *	108.1 ± 32.3	100.9 ± 35.2 *
[deoxy-Hb] (μM)	48.4 ± 13.0	48.6 ± 12.0	45.6 ± 12.1	47.2 ± 11.2
[total-Hb] (μM)	152.1 ± 50.2	139.9 ± 44.4 *	153.8 ± 43.8	148.1 ± 44.8 *
SO$_2$ (%)	67.7 ± 2.4	64.4 ± 4.1 *	70.2 ± 1.8	67.5 ± 3.6 *

* $p < 0.05$ vs. Baseline.

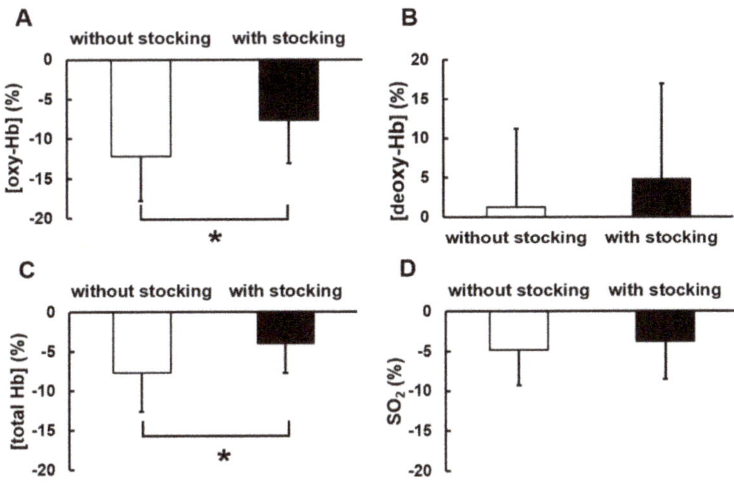

Figure 3. Percent changes in the near-infrared time-resolved spectroscopy parameters before and after 3-hour constant sitting with or without compression stocking; (**A**) oxy-Hb, (**B**) deoxy-Hb, (**C**) total Hb, (**D**) SO$_2$. * $p < 0.05$ between groups.

There was a significant decrease in the absorption coefficient of both limbs with and without compression stockings after the 3-hour sitting task (each $p < 0.05$) (Table 1). However, the reduced scattering coefficient showed a significant increase in limbs without the compression stocking, but it did not change in limbs with the compression stocking after the 3-hour sitting task (Table 1). The average optical path length increased significantly in both limbs (each $p < 0.05$) (Table 1). The oxy-Hb levels decreased significantly in both limbs after the 3-hour sitting task (Table 1), and the rate of decrease was significantly greater in the limbs without compression stockings (limbs without compression stockings: $-12.2 \pm 5.6\%$; limbs with compression stockings: $-7.7 \pm 5.3\%$; $p < 0.05$) (Figure 3A). Total Hb decreased significantly in both limbs after the 3-hour sitting task (Table 1), but the rate of decrease was lower in the limbs with compression stockings (limbs without compression stockings: $-7.7 \pm 4.9\%$; limbs with compression stockings: $-4.0 \pm 3.7\%$; $p < 0.05$) (Figure 3C). The deoxy-Hb levels showed no significant changes before and after the 3-hour sitting task (Table 1, Figure 3B). In addition, although SO_2 decreased in both limbs after the 3-hour sitting task, there was no significant difference between the rates of decrease (Figure 3D).

3.7. Relationship Between Lower Limb Circumference, Lower Limb Extracellular Water, and NIR$_{TRS}$ Indicators

Figure 4 shows the relationship between the rate of change in lower limb extracellular water/total body water ratio and the rate of change in the lower limb circumference with and without compression stockings.

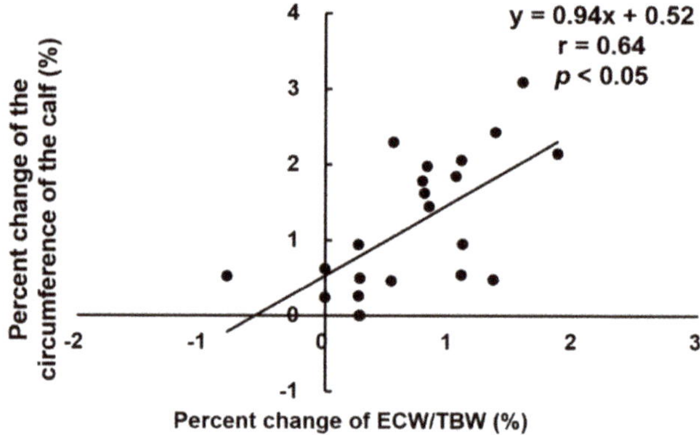

Figure 4. Relationship between percent change of ECW/TBW ratio and percent change of calf circumference before and after 3-hour constant sitting with or without compression stocking.

There was a significant positive correlation between the rates of change in extracellular water and lower limb circumference before and after 3 h of constant sitting (r = 0.64; $p < 0.05$), and we believe this was caused by an increase in extracellular water.

Figure 5 shows the relationship between the rate of change in the extracellular water/total body water ratio of the lower limbs and the rate of change in each NIR$_{TRS}$ index in lower limbs with and without compression stockings due to 3-hour sitting task. A significant negative correlation was observed between the changes in extracellular water/total body water ratio and oxy-Hb (r = −0.40, $p < 0.05$) and SO_2 percentages (r = −0.49, $p < 0.05$).

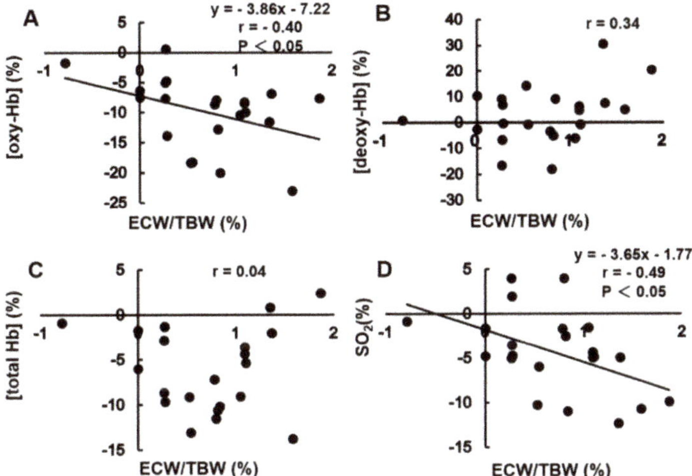

Figure 5. Relationship between percent change of ECW/TBW ratio and percent change of near-infrared time-resolved spectroscopy parameters (**A**) oxy-Hb, (**B**) deoxy-Hb, (**C**) total Hb, (**D**) SO_2, before and after 3-hour constant sitting with or without compression stocking.

4. Discussion

Time spent in a sitting position is believed to have increased dramatically in recent years, particularly in industrialized countries, and situations that require people to sit for long periods are increasing. Long periods of sitting during not only long-distance travel but also deskwork have been found to increase the risks of lifestyle-related diseases such as diabetes and cardiovascular disease [5,6]. They also present the possibility of increasing the all-cause mortality risk [4,8]. In this study, we investigated the effects of sitting for 3 h on the body; we placed particular focus on indices related to the lower extremities. The circumference of the calf region was found to have significantly increased after the 3-hour sitting task. The extracellular water/total body water ratio also increased, indicating a significant positive correlation between these two factors. Moreover, oxy-Hb and total Hb levels at the lateral head of the lower leg gastrocnemius muscle measured using NIR_{TRS} decreased after the 3-hour sitting task, as did the degree of oxygen saturation. In contrast, for limbs with compression stockings, extracellular water did not change after the 3-hour sitting task and, although the lateral head of the gastrocnemius muscle circumference increased, the rate of increase was significantly lower compared to limbs without compression stockings. In addition, although oxy-Hb, total Hb, and SO_2 in the lateral head of the gastrocnemius muscle decreased in the limbs with compression stockings after the 3-hour sitting task, the rates of decrease for the oxy-Hb and total Hb levels were significantly lower compared to those of limbs without compression stockings. Decreases in the level of oxygen saturation in the lower leg skeletal muscle tissue were also suppressed.

The results of the self-reported survey conducted during this study revealed that the physical and mental fatigue experienced by the subjects increased significantly as a result of the 3-hour sitting task. Ten of the 11 subjects had professions that required them to stand for the majority of the day. Therefore, these subjects were not accustomed to sitting for long periods. Additionally, during this study, we restricted subjects from moving their lower limbs during the 3-hour sitting task. This may have led to increased subjective mental fatigue, which could have also contributed to elevated stress levels.

Typically, the rate of water filtration in the capillary is determined by the hydrostatic pressure gradient and the osmotic pressure gradient between the capillary and the interstitial fluid. Sitting for a long period is strongly influenced by gravity, particularly in the lower extremities, and can lead to increased hydrostatic pressure in capillary blood vessels, eventually disrupting the equilibrium of the hydrostatic pressure gradient. In typical cases, venous pressure decreases due to the muscle pump action in the vasculature of the lower extremities, and outflow of intravascular fluid to the interstitium is suppressed by decreasing the hydrostatic pressure in capillaries. However, in this study, we believed that this muscle pump action was ineffective because voluntary contractions of the lower limb skeletal muscles were strictly restricted during the 3-hour sitting task. Therefore, extracellular water/total body water ratio increased by 1.1% (from 0.371 to 0.375; $p < 0.05$). As a result, the circumference of the gastrocnemius muscle was found to have increased significantly by 1.9% after the 3-hour sitting task (from 39.6 ± cm to 40.4 cm; $p < 0.05$).

In contrast, for limbs with compression stockings, the rate of change in the gastrocnemius muscle circumference increased only slightly after the 3-hour sitting task (+1.9 ± 0.6% for limbs without compression stocking vs. +0.4 ± 0.3% limbs with compression stocking; $p < 0.05$), and the lower limb extracellular water/total body water ratio did not exhibit any significant change. This is believed to be due to the following effects of wearing compression stockings: Tightening of the veins due to increased skin surface pressure, reduction of blood stagnation, and increased venous reflux while preventing physiological counterflow due to the physiological venous valve function. Due to these effects, compression stockings were successful for optimizing the equilibrium of the hydrostatic pressure gradient between the intravascular and interstitial spaces, thereby suppressing the onset of edema.

The elastic stockings recommended by the Lower Leg Ulcer/Varices Diagnosis and Treatment Guidelines published by the Japanese Dermatological Association [16] are those that place the greatest pressure on the ankle and gradually lower the pressure toward the proximal aspect of the leg; the pressure applied to the ankle joint is less than 20 mmHg (26.7 hPa). In clinical practice, it is generally recommended that compression stockings should be applied to the ankle joint (ankle joint:calf:thigh pressure ratio of 10:7:4 or 26.7:18.7:10.6 hPa). Variable pressure is recommended to increase blood reflux from the periphery to the center. However, the robustness of evidence concerning the prevention of DVT and varices of the lower extremities in Japan is still extremely poor, and few details regarding the mechanism of the effects of wearing compression stockings on edema have been reported. Compression stockings used in this study had a pressure ratio of 8:10:8, exerting 15.7 ± 2.8 hPa at the ankle joint, 20.1 ± 1.1 hPa at the lateral head of the gastrocnemius muscle, and 15.6 ± 2.2 hPa at the thigh; it was designed to place the greatest pressure at the calf region to promote muscle pump action. The pressure exerted by the compression stocking used in this study (15.7 hPa) was lower than the clinically recommended ankle pressure of 20 mmHg (26.7 hPa). Nevertheless, we observed the suppression of increases in limb circumference, of increases in extracellular water, and of decreases in skeletal muscle tissue oxygenation. Because of the low elasticity characteristic of conventional compression stockings, the potential difficulty of wearing such stockings can become an issue, resulting in decreased use [18] and an increased risk of dermatitis. The results of this study suggested the effectiveness of wearing compression stockings with lower ankle joint pressure than what is currently recommended. We investigated the effects of the compression stockings only on men in this study. However, the prevalence rate of edema among healthy subjects is 2.8 times higher in women than in men [21] The incidence of edema among aged dialysis patients is much higher in women than in men as well [22]. With this variable pressure design, we believe the utilization of compression stockings will increase in the future, in women as well as men.

The absorption coefficient decreased in both limbs after the 3-hour sitting task. This was believed to be the result of decreased absorbing substance in the NIR_{TRS} measurement region. In addition, in limbs without compression stockings, increases in the reduced scattering coefficient after the 3-hour sitting task were attributed to increased photon migration distances due to the decrease in the absorbing substance (mainly hemoglobin) in the measurement region after sitting (Table 1). The average optical path length actually increased in both limbs after the 3-hour sitting task with and without the compression stockings (Table 1). The oxy-Hb, total Hb, and SO_2 levels decreased in both limbs after the 3-hour sitting task as a result of gravity effects and decreased blood flow in the lower extremities due to reduced muscle pump action, leading to the possibility of stagnation and edema [23]. Among previous studies, one report stated that the average blood flow velocity in the popliteal artery decreased significantly compared to the task value before sitting [10]. Therefore, it can be inferred that decreased blood flow in the lower extremities can lead to decreases in the central circulation volume and that the level of oxygenation in peripheral skeletal muscle tissue decreased as a result of the decreased supply of arterial blood to the periphery. Additionally, limbs without compression stockings were determined to exhibit edema (increase in interstitial water volume) in the measurement region based on the observation of increased extracellular water and lower limb circumference compared to limbs with compression stockings, which was thought to have resulted in decreased (dilution) oxy-Hb, deoxy-Hb, and total Hb levels. The deoxy-Hb level exhibited no significant changes in both limbs after the 3-hour sitting task. This result is believed to be attributable to a decrease in concentration caused by an increase in interstitial fluid in the NIR_{TRS} measurement region as well as the accumulation of deoxy-Hb caused by decreased peripheral blood circulation volume.

5. Conclusions

The use of compression stockings with pressure lower than that currently recommended was found to help prevent increases in extracellular water and decreases in intracellular water, as well as to suppress decreases in oxy-Hb, total Hb, and SO_2 levels in the gastrocnemius. However, this study had a small sample size and included only men; therefore, the results cannot be generalized to a normal population. Wearing this type of compression stockings might have a higher positive impact on the elderly and/or women, who often have higher subjective symptoms of edema in the lower extremities. In the future, we plan to conduct further investigations to clarify the full effects of sitting for extended periods on the body and to further verify the effects of wearing devices such as compression stockings to prevent DVT and other conditions during the perioperative period, to generally improve quality of life, and to establish measures to prevent increases in the prevalence of lifestyle-related diseases and the all-cause mortality rate for healthy individuals.

Author Contributions: M.K.: Writing, data analysis; Y.K.: Review and editing, data analysis; S.F.: Acquisition of data; R.T.: Acquisition of data; N.T.: Interpretation of data; R.K. (Ryota Kobayashi): Interpretation of data; R.K. (Ryota Kime): Revising the manuscript; T.H.: Review and editing, funding acquisition.

Funding: This work was supported by research funds in Tokyo Medical University.

Acknowledgments: The authors wish to thank Haruka Maesawa (Nihon Igaku Jusei Shinkyu College) for her assistance.

Conflicts of Interest: The authors declare that no conflicts of interest exist.

References

1. Giangrande, P.L. Air travel and thrombosis. *Int. J. Clin. Pract.* **2001**, *55*, 690–693. [PubMed]
2. Di Nisio, M.; van Es, N.; Büller, H.R. Deep vein thrombosis and pulmonary embolism. *Lancet* **2016**, *388*, 3060–3073. [CrossRef]
3. Ekelund, U.; Steene-Johannessen, J.; Brown, W.J.; Fagerland, M.W.; Owen, N.; Powell, K.E.; Bauman, A.; Lee, I.M. Lancet Physical Activity Series 2 Executive Committe; Lancet Sedentary Behaviour Working Group. Does physical activity attenuate, or even eliminate, the detrimental association of sitting time with mortality? A harmonised meta-analysis of data from more than 1 million men and women. *Lancet* **2016**, *388*, 1302–1310.
4. van der Ploeg, H.P.; Chey, T.; Korda, R.J.; Banks, E.; Bauman, A. Sitting time and all-cause mortality risk in 222 497 Australian adults. *Arch. Intern. Med.* **2012**, *172*, 494–500. [CrossRef]
5. Grace, M.S.; Dempsey, P.C.; Sethi, P.; Mundra, P.A.; Mellett, N.A.; Weir, J.M.; Owen, N.; Dunstan, D.W.; Meikle, P.J.; Kingwell, B.A. Breaking Up Prolonged Sitting Alters the Postprandial Plasma Lipidomic Profile of Adults With Type 2 Diabetes. *J. Clin. Endocrinol. Metab.* **2017**, *102*, 1991–1999. [CrossRef] [PubMed]
6. Bellettiere, J.; Winkler, E.A.H.; Chastin, S.F.M.; Kerr, J.; Owen, N.; Dunstan, D.W.; Healy, G.N. Associations of sitting accumulation patterns with cardio-metabolic risk biomarkers in Australian adults. *PLoS ONE* **2017**, *12*, e0180119. [CrossRef]

7. Åsvold, B.O.; Midthjell, K.; Krokstad, S.; Rangul, V.; Bauman, A. Prolonged sitting may increase diabetes risk in physically inactive individuals: An 11 year follow-up of the HUNT Study, Norway. *Diabetologia* **2017**, *60*, 830–835. [CrossRef] [PubMed]

8. Katzmarzyk, P.T.; Church, T.S.; Craig, C.L.; Bouchard, C. Sitting time and mortality from all causes, cardiovascular disease, and cancer. *Med. Sci. Sports Exerc.* **2009**, *41*, 998–1005. [CrossRef]

9. Vranish, J.R.; Young, B.E.; Kaur, J.; Patik, J.C.; Padilla, J.; Fadel, P.J. Influence of sex on microvascular and macrovascular responses to prolonged sitting. *Am. J. Physiol. Heart Circ. Physiol.* **2017**, *312*, 800–805. [CrossRef]

10. Restaino, R.M.; Holwerda, S.W.; Credeur, D.P.; Fadel, P.J.; Padilla, J. Impact of prolonged sitting on lower and upper limb micro- and macrovascular dilator function. *Exp. Physiol.* **2015**, *100*, 829–838. [CrossRef]

11. Hamaoka, T.; Iwane, H.; Shimomitsu, T.; Katsumura, T.; Murase, N.; Nishio, S.; Osada, T.; Kurosawa, Y.; Chance, B. Noninvasive measures of oxidative metabolism on working human muscles by near-infrared spectroscopy. *J. Appl. Physiol.* **1996**, *81*, 1410–1417. [CrossRef]

12. Endo, K.; Matsukawa, K.; Liang, N.; Nakatsuka, C.; Tsuchimochi, H.; Okamura, H.; Hamaoka, T. Dynamic exercise improves cognitive function in association with increased prefrontal oxygenation. *J. Physiol. Sci.* **2013**, *63*, 287–298. [CrossRef]

13. Oda, M.; Ohmae-Yamaki, E.; Suzuki, H.; Suzuki, T.; Yamashita, Y. Tissue oxygenation measurements using near-infrared time-resolved spectroscopy. *J. Jpn. Coll. Angiol.* **2009**, *49*, 131–137.

14. Hamaoka, T.; Katsumura, T.; Murase, N.; Nishio, S.; Osada, T.; Sako, T.; Higuchi, H.; Kurosawa, Y.; Shimomitsu, T.; Miwa, M.; et al. Quantification of ischemic muscle deoxygenation by near infrared time-resolved spectroscopy. *J. Biomed. Opt.* **2000**, *5*, 102–105. [CrossRef]

15. Nicolaides, A.N.; Fareed, J.; Kakkar, A.K.; Comerota, A.J.; Goldhaber, S.Z.; Hull, R.; Myers, K.; Samama, M.; Fletcher, J.; Kalodiki, E.; et al. Prevention and treatment of venous thromboembolism–International Consensus Statement. *Int. Angiol.* **2013**, *32*, 111–260. [PubMed]

16. The Japanese Journal of Dermatology. Guild Line for Lower Leg Ulcer/Lower Extremity Varix Diagnosis URL. Available online: https://www.dermatol.or.jp/uploads/uploads/files/guideline/1380006822_2.pdf (accessed on 15 February 2019).

17. JCS Joint Working Group. Guidelines for the diagnosis, treatment and prevention of pulmonary thromboembolism and deep vein thrombosis (JCS 2009). *Circ. J.* **2011**, *75*, 1258–1281. [CrossRef]

18. Raju, S.; Hollis, K.; Neglen, P. Use of compression stockings in chronic venous disease: Patient compliance and efficacy. *Ann. Vasc. Surg.* **2007**, *21*, 790–795. [CrossRef] [PubMed]

19. Cha, K.; Chertow, G.M.; Gonzalez, J.; Lazarus, J.M.; Wilmore, D.W. Multifrequency bioelectrical impedance estimates the distribution of body water. *J. Appl. Physiol.* **1995**, *79*, 1316–1319. [CrossRef] [PubMed]

20. Ohmae, E.; Oda, M.; Suzuki, T.; Yamashita, Y. Clinical evaluation of time-resolved spectroscopy by measuring cerebral hemodynamics during cardiopulmonary bypass surgery. *J. Biomed. Opt.* **2007**, *12*, 062112. [CrossRef] [PubMed]

21. Comprehensive Survey of Living Conditions in Japan 2009. Ministry of Health, Labor and Welfare. Available online: https://www.mhlw.go.jp/toukei/saikin/hw/k-tyosa/k-tyosa10/ (accessed on 4 February 2019).

22. Nakagawa, F.; Terashima, Y.; Kamiyoshi, K.; Takenaka, S.; Yoshida, K.; Seto, N.; Koyama, T.; Ootsuki, Y.; Hatanaka, N.; Hirata, K.; et al. Relationship between nutritional status, age and sex and edema by bioelectrical impedance analysis. *J. Osaka Soc. Dial. Ther.* **2012**, *30*, 29–34.

23. Vena, D.; Rubianto, J.; Popovic, M.; Yadollahi, A. Leg fluid accumulation during prolonged sitting. *Conf. Proc. IEEE Eng. Med. Biol. Soc.* **2016**, 4284–4287. [CrossRef]

Article

Measurement of the Absolute Value of Cerebral Blood Volume and Optical Properties in Term Neonates Immediately after Birth Using Near-Infrared Time-Resolved Spectroscopy: A Preliminary Observation Study

Aya Morimoto [1,2], Shinji Nakamura [1,*], Masashiro Sugino [3], Kosuke Koyano [4], Yinmon Htun [1,2], Makoto Arioka [1], Noriko Fuke [1], Ami Mizuo [1], Takayuki Yokota [1], Ikuko Kato [1], Yukihiko Konishi [1], Sonoko Kondo [1], Takashi Iwase [1], Saneyuki Yasuda [4] and Takashi Kusaka [1]

[1] Department of Pediatrics, Faculty of Medicine, Kagawa University, 1750-1, Mikicho, Kitagun, Kagawa 761-0793, Japan; ayamoto@med.kagawa-u.ac.jp (A.M.); ymhtun0612@gmail.com (Y.H.); marioka@med.kagawa-u.ac.jp (M.A.); noriko-f@med.kagawa-u.ac.jp (N.F.); inoue26@med.kagawa-u.ac.jp (A.M.); yokotakayuki@med.kagawa-u.ac.jp (T.Y.); i-kato@med.kagawa-u.ac.jp (I.K.); lilwest@med.kagawa-u.ac.jp (Y.K.); ijichi@med.kagawa-u.ac.jp (S.K.); t.iwase@med.kagawa-u.ac.jp (T.I.); kusaka@med.kagawa-u.ac.jp (T.K.)

[2] Graduate School of Medicine, Faculty of Medicine, Kagawa University, 1750-1, Mikicho, Kitagun, Kagawa 761-0793, Japan

[3] Division of Neonatology, Shikoku Medical Center for Children and Adults, 2-1-1, Senyucho, Zentsuji, Kagawa 765-8507, Japan; masashiro@med.kagawa-u.ac.jp

[4] Maternal Perinatal Center, Faculty of Medicine, Kagawa University, 1750-1, Mikicho, Kitagun, Kagawa 761-0793, Japan; kosuke@med.kagawa-u.ac.jp (K.K.); yas@med.kagawa-u.ac.jp (S.Y.)

* Correspondence: shinji98@med.kagawa-u.ac.jp; Tel.: +81-87-891-2171; Fax: +81-87-891-2172

Received: 5 April 2019; Accepted: 21 May 2019; Published: 27 May 2019

Abstract: The aim of this study was to use near-infrared time-resolved spectroscopy (TRS) to determine the absolute values of cerebral blood volume (CBV) and cerebral hemoglobin oxygen saturation (ScO_2) during the immediate transition period in term neonates and the changes in optical properties such as the differential pathlength factor (DPF) and reduced scattering coefficient (μ_s'). CBV and ScO_2 were measured using TRS during the first 15 min after birth by vaginal delivery in term neonates who did not need resuscitation. Within 2–3 min after birth, CBV showed various changes such as increases or decreases, followed by a gradual decrease until 15 min and then stability (mean (SD) mL/100 g brain: 2 min, 3.09 (0.74); 3 min, 3.01 (0.77); 5 min, 2.69 (0.77); 10 min, 2.40 (0.61), 15 min, 2.08 (0.47)). ScO_2 showed a gradual increase, then kept increasing or became a stable reading. The DPF and μ_s' values (mean (SD) at 762, 800, and 836 nm) were stable during the first 15 min after birth (DPF: 4.47 (0.38), 4.41 (0.32), and 4.06 (0.28)/cm; μ_s': 6.54 (0.67), 5.82 (0.84), and 5.43 (0.95)/cm). Accordingly, we proved that TRS can stably measure cerebral hemodynamics, despite the dramatic physiological changes occurring at this time in the labor room.

Keywords: neonate; vaginal delivery; cerebral blood volume; cerebral hemoglobin oxygen saturation; near-infrared time-resolved spectroscopy

---

1. Introduction

Evaluation of the immediate postnatal cerebral oxygen metabolism and hemodynamics is essential for understanding extrauterine adaptation. In particular, the hemodynamic changes occurring immediately after birth are explosive and can induce fetal distress and asphyxia. The physiological processes after birth remain unclear and include cerebral vasodilation, vasoconstriction, and oxygenation. Because the brain is the most sensitive organ system of the infant, it is important to assess its autoregulation after birth.

Near-infrared spectroscopy (NIRS) using near-infrared light (700–900 nm) enables detection of changes in the oxygenation state of hemoglobin (Hb) and water content in biological tissue. Near-infrared light is safe and penetrates deeply in the body, and NIRS has recently been applied for functional evaluation of cerebral circulation and oxygenation state in neonates [1]. Several studies have reported on the measurement of cerebral oxygenation via the tissue oxygenation index or regional saturation of oxygen (rSO_2) in neonates during the immediate transition after birth using NIRS [2–8]. Furthermore, NIRS can measure changes in venous Hb concentration (tHb) and cerebral blood volume (CBV), with some work showing a decrease in CBV in term neonates in the first 15 min after birth [9]. However, commonly used NIRS modalities, such as continuous wave spectroscopy, which only measures changes in the Hb concentration, and spatially resolved spectroscopy does not provide CBV but tissue oxygen saturation [1]. The absolute value of CBV is required for its use as a clinical parameter to manage circulation and determine oxygen use. Near-infrared time-resolved spectroscopy (TRS) is a unique method for calculating quantitative CBV and ScO_2 using a light absorption coefficient (μ_a) without inducing changes in light-absorbing materials, such as Hb, because the reduced scattering coefficient (μ_s') and μ_a can be determined by resolving photon diffusion equation (PDE). Our group has already reported on the measurement of the absolute value of CBV in preterm and term neonates, determining that it sheds light on their cerebral hemodynamic development [10]. However, there have been no reports on how the absolute value of CBV is altered and whether optical properties such as the differential pathlength factor (DPF) and μ_s' are affected during this adaptation period by certain factors such as the vernix, amniotic fluid, blood, and systemic hemodynamic changes in the immediate transition period.

In this study, we used TRS to examine the absolute value of CBV and ScO_2 during the immediate transition period in term neonates and the changes in optical properties such as DPF and μ_s'.

2. Materials and Methods

2.1. Study Design

This prospective observational study was performed at Kagawa University Hospital. The Regional Committee on Biomedical Research Ethics approved all of the included studies. Between October 2012 and April 2018, term neonates (gestational age >37 weeks, birth weight >2,300 g) were born by vaginal delivery. The study was conducted in accordance with the Declaration of Helsinki, and the protocol was approved by the local ethics committee (ethics number: H29-042). The parents of all neonates enrolled in this study provided written informed consent after receiving a full explanation of the study prior to birth. After cord clamping, routinely performed after 30 s, neonates were placed on the resuscitation table under an overhead heater. The newborn infants were dried and stimulated by using warm cotton diapers to induce effective breathing. TRS calibration was performed before each test by measuring the instrumental response function with the source and detector fibers facing each other through a neutral-density filter in a black tube [11].

Before measurement, we checked that the light intensity was adjusted within the dynamic range of the photon counter using the adult forearm. As soon as possible, another neonatologist attached the TRS probe to the newborn's right forehead (Figure 1).

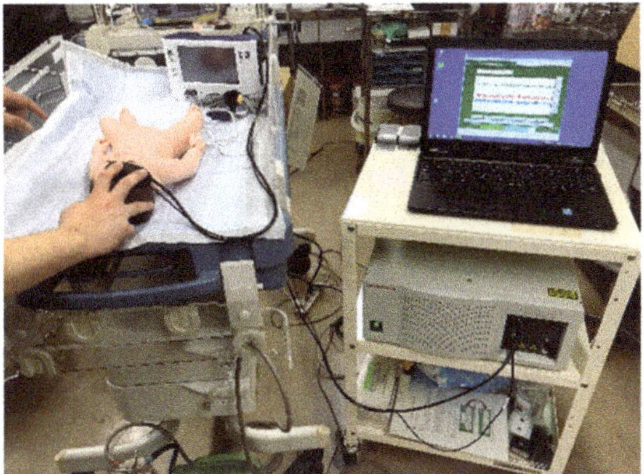

Figure 1. Photograph showing the actual near-infrared time-resolved spectroscopy (TRS) setting.

Then, we held the TRS transducer against the neonates' head for 15 min. Concurrently, a transcutaneous pulse oximeter was applied to the right hand (Nellcor™, COVIDIEN, Tokyo, Japan). The neonates were placed in a supine position and breathed room air. A neonatologist observed the transition of the newborn infant and recorded Apgar scores at 1 and 5 min. Resuscitation was performed following Neonatal Cardiac Pulmonary Resuscitation 2015 guidelines. We excluded the following neonates: (1) those that needed any respiratory support such as oxygen, continuous positive airway pressure, and artificial ventilation; (2) those who were hospitalized due to hypoglycemia or infection; (3) those with abnormalities such as congenital heart disease; and (4) those with an abnormal value of μ_a, μ_s', and DPF retrospectively.

2.2. Near-Infrared Time-Resolved Spectroscopy

We used a portable three-wavelength near-infrared TRS system (TRS-21, Hamamatsu Photonics K.K., Hamamatsu, Japan). This system uses a time-correlated single-photon counting technique for detection. The system was controlled by a computer through a digital I/O interface that consisted of a three-wavelength (762, 800, and 836 nm) picosecond light pulser (PLP) as the pulse light source, a photon-counting head for single-photon detection, and signal-processing circuits for time-resolved measurement. The PLP emitted near-infrared light with a pulse duration of approximately 100 ps, a mean power of at least 200 µW at each wavelength, and pulse repetitions at a frequency of 5 MHz. The input light power to the patient was approximately 300 µW. The light from the PLP was sent to a patient from a source fiber with a length of 3 m, and the photon re-emitted from the patient was collected simultaneously by a detector fiber bundle with a length of 3 m. We obtained a set of histograms that displayed the photon flight time or re-emission profile. In this study, the emerging light was collected over a period of 1 s to exceed a photon count of at least 1000 in the peak channel of the re-emission profiles. The re-emission profiles observed at each measurement point were fitted by the time-resolved reflectance derived from the analytical solution of the PDE proposed by Patterson et al. [12,13], which is convoluted with the instrumental response function, to calculate the µa and µs' values of the head at wavelengths of 762, 800, and 836 nm.

In each iterative calculation, the analytical solution of the PDE was calculated in reflectance mode; it was then fitted to the observed re-emission profile. After determination of the μ_a and μ_s' values at three wavelengths, the oxyHb and deoxyHb concentrations were calculated from the extinction coefficients of oxyHb and deoxyHb with the following equations, based on the assumption that the background absorption was due to 85% (by volume) water.

$$\mu_{a\,762\,nm} = \varepsilon_{762\,nm}^{oxyHb}[oxyHb] + \varepsilon_{762\,nm}^{deoxyHb}[deoxyHb] + \varepsilon_{762\,nm}^{water}[water\ volume\ fraction] \tag{1}$$

$$\mu_{a\ 800\ nm} = \varepsilon^{oxyHb}_{800\ nm}[oxyHb] + \varepsilon^{deoxyHb}_{800\ nm}[deoxyHb] + \varepsilon^{water}_{800\ nm}[water\ volume\ fraction] \tag{2}$$

$$\mu_{a\ 836\ nm} = \varepsilon^{oxyHb}_{836\ nm}[oxyHb] + \varepsilon^{deoxyHb}_{836\ nm}[deoxyHb] + \varepsilon^{water}_{836\ nm}[water\ volume\ fraction] \tag{3}$$

where $\varepsilon_{\lambda\ nm}$ is the extinction coefficient at the wavelength of λ nm and [oxyHb] and [deoxyHb] are the concentrations of oxyHb and deoxyHb, respectively. First, water absorption was subtracted from μ_a at each of the three wavelengths and then the concentrations of oxyHb and deoxyHb were estimated by applying the least-squares fitting method. The extinction coefficients for oxyHb, deoxyHb, and water shown in Table 1 were used.

Table 1. Extinction coefficients for oxyHb, deoxyHb, and water.

	oxyHb $(mM^{-1}cm^{-1})$	deoxyHb $(mM^{-1}cm^{-1})$	Water (cm^{-1})
762 nm	1.4320	3.8145	0.0272
800 nm	1.9924	1.9339	0.0204
836 nm	2.4985	1.7974	0.0363

The ratio of the optical pathlength to the interoptode distance was defined as the DPF. We used the prism-type probe. The source and detector optodes were positioned on the frontal region at a 30 mm interoptode distance. The total cerebral Hb (totalHb) concentration, ScO_2, and CBV were calculated as follows:

$$[totalHb] = [oxyHb] + [deoxyHb] \tag{4}$$

$$ScO_2\ (\%) = \{[oxyHb]/([oxyHb] + [deoxyHb])\} \times 100 \tag{5}$$

$$CBV\ (ml/100\ g) = [totalHb] \times MW_{Hb} \times 10^{-6}/(tHb \times 10^{-2} \times Dt \times 10) \tag{6}$$

where [ ] indicates Hb concentration (μM), MW_{Hb} is the molecular weight of Hb (64,500), tHb is the venous Hb concentration (g/dL) and Dt is brain tissue density (1.05 g/mL).

All neonates underwent blood gas analysis, and the CBV was calculated via the venous Hb concentration at 2 h after birth. The mean values of the DPF, μ_a, μ_s', CBV, and ScO_2 were calculated every 10 s for 15 min after birth.

3. Results

The study participants were seven healthy term neonates. One of the seven neonates was born via forceps delivery but the others were born via a normal vaginal delivery. Two neonates were excluded from the analysis because the μ_s' was abnormal according to previous work [10]. However, none of the neonates needed any respiratory support, had abnormalities, or required hospitalization. The gestational ages of the neonates were 38–40 weeks, and their Apgar scores at 1, 5min were 8 or more (Table 2). Five neonates (V1–V5) did not need resuscitation until 15 min after birth.

Table 2. Summary of neonates' clinical data in this study.

Neonate No.	Gestational Age	Body Weight (g)	Apgar: 1 min	Apgar: 5 min	pH Umbilical Artery	Venous Hemoglobin at 2 h (g/dL)
V1	37 wk 3 d	3,212	8	9	7.309	15.8
V2	39 wk 6 d	3,170	8	9	7.322	13.2
V3	38 wk 5 d	2,830	8	9	7.371	20.3
V4	39 wk 4 d	3,406	8	8	7.345	18.9
V5	39 wk 3 d	2,894	8	9	7.256	19.7

Within 2–3 min after birth, CBV showed various changes such as increases or decreases, and was then followed by a gradual decrease until 15 min, and was then stable (mean (SD) mL/100 g brain: 2 min, 3.09 (0.74); 3 min, 3.01 (0.77); 5 min, 2.69 (0.77); 10 min, 2.40 (0.61); 15 min, 2.08 (0.47)) (Figure 2).

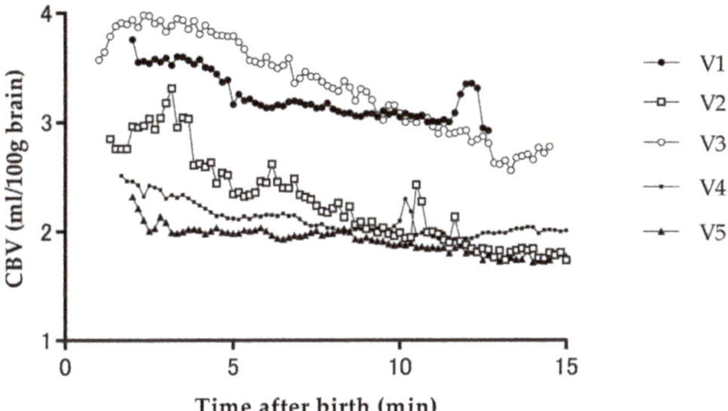

Figure 2. Cerebral blood volume (CBV) in five neonates during the first 15 min of life. Values are means. Compared with the reference value at 15 min, a significant decrease in CBV was observed at each time point.

ScO_2 showed a gradual increase, then kept increasing or became a stable reading (mean (SD)%: 2 min, 48.0 (12.3); 3 min, 53.9 (14.2); 5 min, 62.5 (13.3); 10 min, 67.2 (10.2); and 15 min, 64.3 (4.7)) (Figure 3).

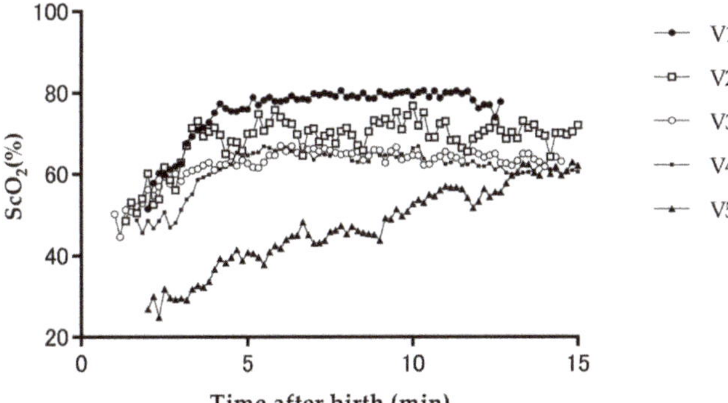

Figure 3. Cerebral hemoglobin oxygen saturation (ScO_2) in five neonates during the first 15 min of life. Values are means. ScO_2 showed the same pattern as arterial Hb oxygen saturation with a gradual increase, peak at 5–10 min, and then stable values.

The values of DPF, μ_a, and μ_s' are shown in Table 3 and correspond to those of a previous report [10]. The optical properties were stable for the first 15 min after birth: DPF (mean (SD) at 762, 800, and 836 nm), 4.47 (0.38), 4.41 (0.32), and 4.06 (0.28)/cm; μ_s': 6.54 (0.67), 5.82 (0.84), and 5.43 (0.95)/cm.

Table 3. The characteristics of the mean value of DPF, μa, and μs' in all neonates.

Neonate No.	Time after Birth (min)	DPF			μa (/cm)			μs' (/cm)		
		762 nm	800 nm	836 nm	762 nm	800 nm	836 nm	762 nm	800 nm	836 nm
	2	4.06	4.21	4.08	0.22	0.17	0.19	6.42	5.95	5.97
	3	4.36	4.38	4.16	0.19	0.16	0.19	6.64	6.10	6.18
V1	5	4.70	4.47	4.15	0.15	0.15	0.17	6.35	5.89	5.84
	10	4.71	4.39	4.05	0.14	0.14	0.17	6.09	5.54	5.56
	15	NA	NA	NA	NA	NA	NA	NA	NA	NA
	2	4.14	4.34	4.07	0.26	0.22	0.23	7.08	6.91	6.49
	3	4.23	4.46	4.13	0.26	0.23	0.26	7.81	7.57	7.50
V2	5	4.96	4.94	4.56	0.18	0.17	0.20	7.82	7.32	7.05
	10	5.21	5.13	4.63	0.15	0.15	0.17	7.36	6.97	6.59
	15	5.25	5.22	4.76	0.14	0.14	0.16	7.14	6.81	6.41
	2	3.97	4.17	3.79	0.23	0.19	0.21	6.11	5.83	5.18
	3	4.05	4.14	3.74	0.22	0.19	0.21	6.14	5.70	5.01
V3	5	4.13	4.18	3.77	0.21	0.19	0.21	6.13	5.74	5.06
	10	4.44	4.48	3.99	0.17	0.16	0.18	6.04	5.62	4.97
	15	4.68	4.69	4.19	0.16	0.14	0.16	6.21	5.80	5.08
	2	4.28	4.22	3.93	0.24	0.18	0.20	7.08	5.82	5.45
	3	4.46	4.35	4.03	0.22	0.17	0.19	7.07	5.87	5.58
V4	5	4.55	4.27	3.93	0.19	0.16	0.19	6.58	5.45	5.10
	10	4.81	4.54	4.12	0.17	0.16	0.18	7.02	5.88	5.42
	15	4.74	4.52	4.11	0.18	0.15	0.18	6.95	5.70	5.37
	2	3.82	4.13	4.14	0.25	0.16	0.17	6.30	5.07	4.51
	3	4.14	3.95	3.66	0.22	0.15	0.15	5.30	4.35	3.85
V5	5	4.36	4.12	3.74	0.20	0.14	0.15	5.63	4.55	4.03
	10	4.36	4.18	3.76	0.17	0.14	0.15	5.63	4.56	4.03
	15	4.75	4.36	3.89	0.15	0.13	0.15	6.00	4.75	4.20

4. Discussion

This is the first report to use TRS to measure the absolute value of CBV and optical properties in term neonates immediately after birth. We obtained the following main results: (1) CBV shows a tendency to decrease after birth, but we found large variability in the absolute value of CBV across neonates; and (2) DPF and μ_s' were stable immediately after birth. These findings indicate that TRS can be a useful, simple, and noninvasive tool for the stable measurement of the absolute values of CBV and ScO_2 when we need to assess the immediate postnatal cerebral oxygen metabolism and hemodynamics in neonates.

Schwaberger et al. [14] monitored the changes in CBV during the immediate postnatal transition period after elective cesarean delivery using NIRS. The mean decrease in total Hb from 2 to 15 min after birth represented a CBV decrease of 1.0 mL/100 g brain. They hypothesized that the CBV decrease after birth was mainly caused by postnatal increases in cerebral PaO_2 levels. This likely reflects a physiological process involving changes in the autoregulatory capacity of cerebral vessels in reaction to increasing pO_2 and decreasing pCO_2 levels after elective cesarean delivery. Our results following natural vaginal delivery in this study are consistent with those of previous reports. We speculate that this CBV decrease can be explained in two possible ways. Firstly, in the transition period, neonates are exposed to the stress of natural labor, which would cause greater hypoxia and hypercapnia. Secondly, as this would cause cerebral vasodilation, their reactions would improve within 1–2 min after birth and CBV would decrease by 15 min after birth.

However, our findings indicate that neonates do not always show a clear decrease in CBV and that it has considerable variability. Thus, we consider that evaluation of the absolute value of CBV would more accurately reflect each characteristic cerebral hemodynamic pattern compared with the relative change in CBV alone.

Both ScO_2 and SpO_2 gradually increased until 15 min after birth, as in previous reports (Figures 3 and 4). Although there have been many reports on whether this cerebral oxyHb/totalHb ratio would become a useful parameter for monitoring oxygen during the transition period, its value remains unknown. In this study, the ScO_2 pattern of V3 and V4 neonates was observed to increase similarly, whereas the CBV of V3 neonates was clearly decreased and that of V4 neonates showed little decrease. These results indicated that CBV may shows a different pattern, despite the similar pattern of ScO_2. Therefore, it is critical to monitor both CBV and ScO_2 to detect dramatic changes in cerebral hemodynamics and oxygen metabolism in the neonatal transition period. For neonates with hypoxic ischemic encephalopathy, our group reported that combined CBV plus ScO_2 at 24 h after birth using TRS had the best predictive ability for neurological outcomes [15]. In addition, we reported the ScO_2 and CBV before and after transfusion in very low birth weight infants with anemia of prematurity and suggested that CBV and ScO_2 may be useful markers for determining the need for transfusion [16]. Hence, monitoring of both CBV and ScO_2 using TRS may help to determine oxygen use in the neonatal transition period.

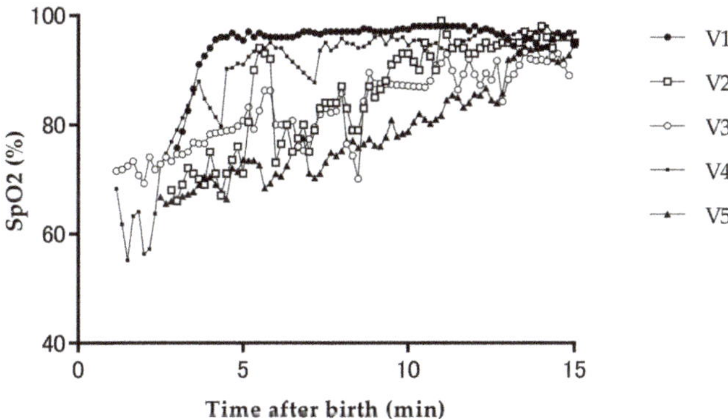

Figure 4. Arterial Hb oxygen saturation (SpO_2) in five neonates during the first 15 min of life. Values are means.

The DPF and μ_s' values found here during the immediate transition period correspond to those of the postnatal period in a previous report [10] and μ_s' value was stable from 2 to 15 min after birth. Furthermore, μ_s' is thought to depend on the conditions of tissue microstructure, such as neuron number, myelination, and edema, but is scarcely influenced by the oxygenation state and Hb concentration [10]. On the other hand, the DPF value slightly increased after birth in this study. When CBV decreased after birth, optical path length increased. This is because the contribution of photons through the deeper tissue to the temporal point spread function (TPSF) increases and the gravity of the TPSF shifts to the right due to less Hb concentration. From these results, we proved that TRS would be able to stably measure the cerebral hemodynamics, despite the dramatic physiological changes occurring in the labor room. In the next step, we will increase the size of the study population and clarify the absolute value of CBV within the standard range in term neonates, capture CBV changes in preterm neonates or neonates with asphyxia, and establish a method for resuscitation based on cerebral hemodynamics.

5. Limitations

First, we could not assess systemic hemodynamics and cannot provide immediate blood gas data, with data only from 2 h after birth. Second, the number of neonates in this study was too small. Thus, we will increase the number of neonates and additionally evaluate not only neonates delivered by normal vaginal delivery, but also by elective cesarean delivery, because we have to consider the possibility that the CBV might be influenced by subcutaneous congestion due to compression of the forehead during vaginal delivery. Third, our TRS system is highly portable and can measure the absolute value of CBV via the transit time of each photon through the tissue of interest. However, the system needs more than 20 min for warmup and calibration. During the immediate transition after birth, we have to adjust the attenuation level of light emission to keep the values stable. Due to neonatal body motion, another team member is required for system fitting.

6. Conclusions

We observed that CBV showed a tendency to decrease after birth, but had large variability in its absolute value across individual neonates, and the DPF and μ_s' values immediately after birth were stable and similar to those in previous studies. By measuring the absolute value of CBV, TRS has the potential to more accurately evaluate the cerebral hemodynamic pattern than the use of relative changes in CBV, not only ScO_2, during the immediate transition period, despite dramatic physiological changes occurring immediately after birth. However, the number of neonates in this study was too small to reach any definite conclusions and further work is required.

Author Contributions: A.M., S.N., and T.K.: substantial contributions to the conception or design of the work; Y.H, S.N., S.K., and T.K.: the necessary financial support for this project and provided study materials; A.M., S.N., M.S., K.K., M.A., T.Y., and N.F.: acquisition, analysis, or interpretation of data for the work; I.K., Y.K., S.K., S.Y., and T.I.: agreement to be accountable for all aspects of the work in ensuring that questions related to the accuracy or integrity of any part of the work are appropriately investigated and resolved.

Funding: This study was financially supported by JSPS KAKENHI Grant Numbers 17K10178, 16K19685, 16K10092, 17K10179, and RIKEN Healthcare and Medical Data Platform Project.

Acknowledgments: We thank Hamamatsu Photonics K.K. and the staff at the Faculty of Medicine Kagawa University, Kagawa, for their cooperation.

Conflicts of Interest: The authors have no conflicts of interest relevant to this manuscript.

References

1. Kusaka, T.; Isobe, K.; Yasuda, S.; Koyano, K.; Nakamura, S.; Nakamura, M.; Ueno, M.; Miki, T.; Itoh, S. Evaluation of cerebral circulation and oxygen metabolism in infants using near-infrared light. *Brain Dev.* **2014**, *36*, 277–283. [CrossRef]

2. Pichler, G.; Schmolzer, G.M.; Urlesberger, B. Cerebral tissue oxygenation during immediate neonatal transition and resuscitation. *Front. Pediatr.* **2017**, *5*, 29. [CrossRef] [PubMed]

3. Isobe, K.; Kusaka, T.; Fujikawa, Y.; Okubo, K.; Nagano, K.; Yasuda, S.; Kondo, M.; Itoh, S.; Hirao, K.; Onishi, S. Measurement of cerebral oxygenation in neonates after vaginal delivery and cesarean section using full-spectrum near infrared spectroscopy. *Comp. Biochem. Physiol. A Mol. Integr. Physiol.* **2002**, *132*, 133–138. [CrossRef]

4. Ziehenberger, E.; Urlesberger, B.; Binder-Heschl, C.; Schwaberger, B.; Baik-Schneditz, N.; Pichler, G. Near-infrared spectroscopy monitoring during immediate transition after birth: Time to obtain cerebral tissue oxygenation. *J. Clin. Monit. Comput.* **2018**, *32*, 465–469. [CrossRef] [PubMed]

5. Baik, N.; Urlesberger, B.; Schwaberger, B.; Schmolzer, G.M.; Mileder, L.; Avian, A.; Pichler, G. Reference ranges for cerebral tissue oxygen saturation index in term neonates during immediate neonatal transition after birth. *Neonatology* **2015**, *108*, 283–286. [CrossRef] [PubMed]

6. Pichler, G.; Binder, C.; Avian, A.; Beckenbach, E.; Schmolzer, G.M.; Urlesberger, B. Reference ranges for regional cerebral tissue oxygen saturation and fractional oxygen extraction in neonates during immediate transition after birth. *J. Pediatr.* **2013**, *163*, 1558–1563. [CrossRef] [PubMed]

7. Urlesberger, B.; Kratky, E.; Rehak, T.; Pocivalnik, M.; Avian, A.; Czihak, J.; Muller, W.; Pichler, G. Regional oxygen saturation of the brain during birth transition of term infants: Comparison between elective cesarean and vaginal deliveries. *J. Pediatr.* **2011**, *159*, 404–408. [CrossRef] [PubMed]

8. Urlesberger, B.; Grossauer, K.; Pocivalnik, M.; Avian, A.; Müller, W.; Pichler, G. Regional oxygen saturation of the brain and peripheral tissue during birth transition of term infants. *J. Pediatr.* **2010**, *157*. [CrossRef] [PubMed]

9. Schwaberger, B.; Pichler, G.; Binder-Heschl, C.; Baik, N.; Avian, A.; Urlesberger, B. Transitional changes in cerebral blood volume at birth. *Neonatology* **2015**, *108*, 253–258. [CrossRef] [PubMed]

10. Ijichi, S.; Kusaka, T.; Isobe, K.; Okubo, K.; Kawada, K.; Namba, M.; Okada, H.; Nishida, T.; Imai, T.; Itoh, S. Developmental changes of optical properties in neonates determined by near-infrared time-resolved spectroscopy. *Pediatr. Res.* **2005**, *58*, 568–573. [CrossRef] [PubMed]

11. Koga, S.; Poole, D.C.; Kondo, N.; Oue, A.; Ohmae, E.; Barstow, T.J. Effects of increased skin blood flow on muscle oxygenation/deoxygenation: Comparison of time-resolved and continuous-wave near-infrared spectroscopy signals. *Eur. J. Appl. Physiol.* **2015**, *115*, 335–343. [CrossRef] [PubMed]

12. Patterson, M.S.; Chance, B.; Wilson, B.C. Time resolved reflectance and transmittance for the non-invasive measurement of tissue optical properties. *Appl. Opt.* **1989**, *28*, 2331–2336. [CrossRef] [PubMed]

13. Ijichi, S.; Kusaka, T.; Isobe, K.; Islam, F.; Okubo, K.; Okada, H.; Namba, M.; Kawada, K.; Imai, T.; Itoh, S. Quantification of cerebral hemoglobin as a function of oxygenation using near-infrared time-resolved spectroscopy in a piglet model of hypoxia. *J. Biomed. Opt.* **2005**, *10*, 024026. [CrossRef] [PubMed]

14. Schwaberger, B.; Pichler, G.; Binder-Heschl, C.; Baik-Schneditz, N.; Avian, A.; Urlesberger, B. Cerebral blood volume during neonatal transition in term and preterm infants with and without respiratory support. *Front. Pediatr.* **2018**, *6*, 132. [CrossRef] [PubMed]

15. Nakamura, S.; Koyano, K.; Jinnai, W.; Hamano, S.; Yasuda, S.; Konishi, Y.; Kuboi, T.; Kanenishi, K.; Nishida, T.; Kusaka, T. Simultaneous measurement of cerebral hemoglobin oxygen saturation and blood volume in asphyxiated neonates by near-infrared time-resolved spectroscopy. *Brain Dev.* **2015**, *37*, 925–932. [CrossRef] [PubMed]

16. Koyano, K.; Kusaka, T.; Nakamura, S.; Nakamura, M.; Konishi, Y.; Miki, T.; Ueno, M.; Yasuda, S.; Okada, H.; Nishida, T.; et al. The effect of blood transfusion on cerebral hemodynamics in preterm infants. *Transfusion* **2013**, *53*, 1459–1467. [CrossRef] [PubMed]

Article

Effects of Aging, Cognitive Dysfunction, Brain Atrophy on Hemoglobin Concentrations and Optical Pathlength at Rest in the Prefrontal Cortex: A Time-Resolved Spectroscopy Study

Kaoru Sakatani [1,2,*], Lizhen Hu [1], Katsunori Oyama [3] and Yukio Yamada [4]

[1]   Laboratory of Universal Sport Health Sciences, Department of Human and Engineered Environmental Studies, Graduate School of Frontier Sciences, The University of Tokyo, Chiba 277-8561, Japan; lizhen_hu@hotmail.com

[2]   Institute for Healthcare Robotics, Future Robotics Organization, Waseda University, Tokyo 169-8050, Japan

[3]   Department of Computer Science, College of Engineering, Nihon University, Fukushima 963-8642, Japan; oyama@cs.ce.nihon-u.ac.jp

[4]   Center for Neuroscience and Biomedical Engineering, The University of Electro-Communications, Tokyo 182-8585, Japan; yukioyamada@uec.ac.jp

*   Correspondence: k.sakatani@edu.k.u-tokyo.ac.jp; Tel.: +81-4-7136-4613

Received: 9 April 2019; Accepted: 20 May 2019; Published: 29 May 2019

Featured Application: Time-resolved spectroscopy may be a useful tool for screening test of cognitive dysfunction in the elderly, particularly who cannot respond to cognitive tasks.

Abstract: Background: In order to evaluate usefulness of a time-resolved spectroscopy (TRS) in screening test of cognitive dysfunction, we studied the effects of aging, cognitive dysfunction, brain atrophy on hemoglobin (Hb) concentrations and optical pathlengths (OPLs) in the prefrontal cortex (PFC) at rest, using TRS. Methods: Employing TRS, we measured Hb concentrations and OPLs at rest in the PFC, and evaluated the relationship between the TRS parameters and cognitive function assessed by Mini-Mental State Examination (MMSE). In addition, we evaluated the relationship between the TRS parameters and the brain atrophy assessed by MRI. Results: We found positive correlations between MMSE scores and oxygen saturation (SO_2), oxy-Hb in the PFC, suggesting that the greater the degree of PFC activity, the higher the cognitive function. In addition, we found the negative correlation between the subject's age and SO_2 and oxy-Hb in the PFC, suggesting that the older the subject, the lower the PFC activity at rest. Moreover, the OPLs in the right PFC negatively correlated with degree of brain atrophy evaluated by MRI, indicating that the shorter the OPL, greater degree of brain atrophy. Conclusions: TRS allowed us to evaluate the relation between the cerebral blood oxygenation (CBO) in the PFC at rest and cognitive function.

Keywords: near infrared spectroscopy; aging; prefrontal cortex; TRS; magnetic resonance imaging; brain atrophy; VSRAD; optical pathlength; hemoglobin; cognitive function

---

1. Background

As the world's population is rapidly aging, dementia becomes a major global health problem. Currently, it is difficult to cure patients with progressive dementia, so emphasis is placed on early diagnosis and early intervention to prevent the onset of dementia [1]. The screening test of cognitive dysfunction, therefore, is important for early diagnosis of dementia. Currently, the Mini Mental State Examination (MMSE) is the most commonly used scale in cognitive function evaluation [2,3]. The MMSE is sensitive and cost-effective screening test; however, it is a subjective examination. Positron emission tomography (PET) and functional magnetic resonance imaging (fMRI) have been used for the diagnosis of dementia [4]; however, these techniques require large facilities and have high costs for examination and maintenance. A simple and less costly method to assess cognitive functions is still required for an objective screening test of dementia.

Near-infrared spectroscopy (NIRS), a non-invasive optical technique, appears to be an attractive alternative method since NIRS is compact and less expensive than fMRI or PET. NIRS evaluates cognitive functions by measuring evoked cerebral blood oxygenation (CBO) changes during cognitive tasks; NIRS provides concentration changes of oxyhemoglobin (oxy-Hb) and deoxyhemoglobin (deoxy-Hb) in cerebral vessels based on measurements of the absorption spectra of hemoglobin in the near-infrared wavelength range [5]. However, it should be noted that, in general, commercially available (conventional) NIRS systems employ continuous wave (CW) light allowing only qualitative measurements of relative changes in hemoglobin (Hb) concentrations during tasks [6]. Therefore, it is difficult to apply conventional CW-NIRS systems to evaluation of cognitive functions of aged people, particularly who cannot perform cognitive tasks due to dementia.

In contrast, time-resolved near infrared spectroscopy (TRS), which employs light sources of laser diodes emitting picosecond light pulses and a time-resolved detector with a picosecond time-resolution, can provide quantitative measurements of Hb concentrations as well as relevant optical parameters such as the absorption coefficient (μ_a), reduced scattering coefficient (μ_s') and optical pathlength (OPL) in the tissues interrogated by the light pulses [7,8]. μ_a and μ_s' are the optical properties averaged over the interrogated tissues and change mainly with the change in the HB concentration associated with the brain activation. OPL is the mean total optical pathlength of the light pulses travelling from the source to detector positions on the head surface, and it will change mainly with the change in the thicknesses of the various layers in the head tissues such as the scalp, cerebrospinal fluid (CSF), gray matter and white matter layers. One measurement procedure of TRS ends within a few minutes, and portable TRS systems cost much less than fMRI or PET. Because a TRS system used in this study is at an initial stage with an assumption of an optically homogeneous medium for multi-layered tissues of the human head, it can provide less spatial information than fMRI. Although the performance of the TRS system used in this study is limited, the measured HB concentrations and relevant optical parameters are very valuable for objective evaluation of dementia. Actually, TRS can measure hemodynamic conditions at rest due to its capability of acquiring baseline Hb concentrations quantitatively, and measurements at rest help evaluate the absolute changes in CBO from the rest to task states. By employing TRS in a previous study, we measured not only the Hb concentrations but also the OPLs in various regions of the brain of healthy adults [9]. In addition, for patients with subarachnoid hemorrhage (SAH), we measured baseline Hb concentrations at rest using TRS for detection of cerebral ischemia induced by vasospasms [10].

In the present study, by employing TRS, we focused on measuring the Hb concentrations and OPLs at rest in the prefrontal cortex (PFC) of patients under rehabilitation and investigated the relationship between the TRS parameters and cognitive functions assessed by the MMSE. In addition, we investigated the relationship between the TRS parameters and the brain atrophy assessed by MRI.

2. Methods

2.1. Subjects

We studied 202 subjects (87 males, 115 females; age 73.4 ± 13.0 years (mean ± SD) who admitted to Southern Touhoku Kasuga Rehabilitation Hospital (Sukagawa city, Japan) for rehabilitation; 68.8% of the subjects suffered from cerebrovascular diseases including 79 cases of cerebral infarction, 41 cases of cerebral haemorrhage, 21 cases of subarachnoid haemorrhage. In addition, 94.6% of the subjects suffered from at least one life-style diseases. Tables 1 and 2 show the clinical profiles of patients and age distribution, respectively.

The subjects provided written informed consents as required by the Human Subjects Committee of the Rehabilitation Hospital. When the subject had a difficulty to understand the informed consent due to cognitive dysfunction, their family provided it.

Table 1. Clinical profiles of patients.

	HT	DM	HL	HT DM	HT HL	HT G	HL G	HT HL DM	HT HL DM G	HT DM G	HT HL G	none	Total
CH	18	1	0	6	6	0	0	6	1	0	0	3	41
SAH	9	0	4	1	3	0	0	3	0	0	0	1	21
CI	16	3	6	12	10	2	2	14	1	1	5	7	79
HI	2	0	0	0	0	0	0	0	0	0	0	1	3
BF	16	1	0	4	4	0	0	2	0	0	0	12	39
others	8	0	0	4	0	0	0	1	0	1	1	4	19
Total	69	5	10	27	23	2	2	26	2	2	6	28	202

HT = Hypertension, DM = Diabetes Mellitus, HL = Hyperlipidemia, G = Gout CH = Cerebral hemorrhage, SAH = Subarachnoid hemorrhage, CI = Cerebral infarction, HI = Head injury, BF = Born fracture.

Table 2. Distribution of patient's age.

Age	Male	Female	Total
$\leq$50	11	4	15
51–60	15	8	23
61–70	20	14	34
71–80	19	38	57
$\geq$80	22	51	73
total	87	115	202

2.2. TRS Measurement

We tried to measure Hb concentrations at rest in the bilateral PFC with a two-channel TRS system (TRS-20, Hamamatsu Photonics K.K., Hamamatsu, Japan). Details of this system have been described [11]. Briefly, it consists of three pulsed laser diodes with different wavelengths (761 nm, 791 nm, and 836 nm) having a pulse duration of 100 ps at a repetition frequency of 5 MHz, a photomultiplier tube (PMT; H6279-MOD, Hamamatsu Photonics K.K., Japan), and a circuit for time-resolved measurement based on the time-correlated single photon counting technique.

Two optical probes having a pair of source and detector optical fibers each of the two-channel TRS system were attached onto the forehead with a bilateral symmetry using a flexible fixation pad, so that the midpoints between the source and detector positions were 30 mm above the centers of the upper edges of the bilateral orbital sockets. The distance between the source and detector of each probe was set at 40 mm. These positionings of the midpoints are similar to those of the midpoints between electrode positions Fp1/F3 (left) and Fp2/F4 (right) of the international electroencephalographic 10–20 system. MRI images confirmed that the optical probes were located over the dorsolateral and frontopolar areas of the PFC. Based on a simulation study of photon migration in the adult head [12], we believe that the TRS measurements in this study provided the CBO changes in the surface region of the PFC under the two midpoints between the source and detector positions.

The TRS system acquired time-resolved reflectances using the two probes. From the acquired time-resolved reflectances, various parameters were obtained assuming that the interrogated forehead tissue was an optically homogeneous semi-infinite medium with the absorption and reduced scattering coefficients of $\mu_a(\lambda)$ and $\mu_s'(\lambda)$ for the wavelength of λ, respectively. The time-resolved reflectance derived from the analytical solution of the photon diffusion equation for a homogeneous semi-infinite medium is given by Equation (1) under the zero-boundary condition [8],

$$R(t, \rho; \lambda) = \frac{z_0}{(4\pi Dc)^{3/2} t^{5/2}} \exp[-\mu_a(\lambda)ct] \exp(-\frac{\rho^2 + z_0^2}{4Dct}),$$

(1)

where $R(t, \rho; \lambda)$ is the time-resolved reflectance at time of t with the distance between the source and detector positions of ρ (= 40 mm) for the wavelength of λ, c is the speed of light in the tissue, D is the diffusion coefficient given as $1/[3\mu_s'(\lambda)]$, and $z_0 = 1/\mu_s'(\lambda)$. Here the wavelength dependences of D and z_0 are omitted for simplicity. Equation (1), convoluted by the instrumental response function of TRS-20, was fitted to the measured time-resolved reflectance to estimate $\mu_a(\lambda)$ and $\mu_s'(\lambda)$ of the medium using a non-linear least-squared technique. The absorption coefficient of tissue was assumed to be the sum of the absorption coefficients of oxy-Hb, deoxy-Hb and background tissue as Equation (2),

$$\mu_a(\lambda) = \varepsilon_{oxy\text{-}Hb}(\lambda)C_{oxy\text{-}Hb} + \varepsilon_{deoxy\text{-}Hb}(\lambda)C_{deoxy\text{-}Hb} + \mu_{a,BG}(\lambda)$$

(2)

where $\varepsilon(\lambda)$ and C are the extinction (or molar absorption) coefficient and the molar concentration, respectively, with the subscript indicating oxy-Hb, deoxy-Hb or background tissue. Solving the simultaneous equations of Equation (2) for the three wavelengths (761 nm, 791 nm, and 836 nm) gave the concentrations of oxy-Hb ($C_{oxy\text{-}Hb}$), deoxy-Hb ($C_{deoxy\text{-}Hb}$) and total-Hb ($C_{t\text{-}Hb} = C_{oxy\text{-}Hb} + C_{deoxy\text{-}Hb}$). Then the oxygen saturation, $SO_2 = C_{oxy\text{-}Hb}/C_{t\text{-}Hb}$, was calculated. The unit of the concentrations of oxy-Hb and deoxy-Hb are μM. The mean total optical pathlength of the detected light for the wavelength of λ, $OPL(\lambda)$, was calculated by Equation (3),

$$OPL(\lambda) = \frac{c\int_0^\infty tR(t, \rho; \lambda)dt}{\int_0^\infty R(t, \rho; \lambda)dt}$$

(3)

It should be noted that the Hb concentrations and SO_2 obtained by Equations (2) and (3) are the averages over the whole regions interrogated by the light pulses, and for separating those of the individual layers, i.e., the scalp, CSF, gray and white matter layers, it is necessary to know the partial optical pathlengths of the individual layers. Although it is very difficult to know the partial optical pathlengths of the individual layers, from the reference [12] the partial optical pathlength of the cortex, mainly the gray matter, is estimated to be about 5 to 10 % of the OPL. Therefore, the Hb concetrations and SO_2 of the cortex occupy only 5 to 10% of those obtained by this study using Equations (2) and (3).

2.3. Assessment of Cognitive Function

We evaluated cognitive functions of the subjects using the MMSE, which is effective as a screening tool that can be used to systematically assess mental status [2]. It was reported that sociocultural variables, ages and education could affect individual MMSE scores [13,14]; however, traditionally, a 23/24 cut off has been used to select patients with suspected cognitive impairment or dementia [15]. In the present study, the mean MMSE scores of all subjects were 24.8 ± 4.6; 108 cases for suspected normal (MMSE ≥ 24), 94 cases for suspected cognitive impairment or dementia (MMSE ≤ 23).

2.4. MRI

55 subjects underwent an MRI study on a 1.5T Vision Plus imager (Siemens, Erlangen, Germany). One hundred forty 3D sections of a T1-weighted magnetization-prepared rapid acquisition of gradient echo sequence were obtained in a sagittal orientation as 1.2-mm thick sections (FOV _ 23, TR _ 9.7 ms, TE _ 4 ms, flip angle _ 12°, and TI _ 300 ms, with no intersection gaps).

We analyzed the morphological changes of the brain using the voxel-based specific regional analysis system for Alzheimer's disease (VSRAD), a diagnosis-aiding program, which runs on Windows, for voxel-based morphometry based on statistical parametric mapping (SPM8) and diffeomorphic anatomical registration using the exponentiated lie (DARTEL) [16]. VSRAD is widely used in current clinical practice in the treatment of AD [17].

VSRAD generates the following scores [16]: (1) Severity; the severity of atrophy obtained from the averaged positive z score in the target volume of interest (VOI) (i.e., hippocampus and its surroundings); (2) Extent of VOI atrophy (%); the extent of a region showing significant atrophy in the target VOI—that is, the percentage rate of the coordinates with a z value exceeding the threshold value of 2 in the target VOI; (3) Ratio; the extent of a region showing significant atrophy in the whole brain—that is, the percentage rate of the coordinates with a z value exceeding the threshold value of 2 in the whole brain; (4) Whole Brain Extent (%); the ratio of the extent of a region showing significant atrophy in the target VOI to the extent of a region showing significant atrophy in the whole brain.

2.5. Data Analysis

We evaluated correlations between ages, the MMSE scores, VSRAD scores, Hb concentrations and OPLs measured by TRS, employing Pearson's correlation analysis.

3. Results

3.1. Correlations between MMSE Scores and Subject's Age

The patients exhibited a variety of cognitive functions between normal and dementia; the mean MMSE scores were 25.3 ± 4.0. There was a significant negative correlation between the MMSE score and patient's age ($r = -0.48$, $p < 0.01$), indicating that the MMSE score decreases with age.

3.2. Correlations Between TRS Parameters and MMSE Scores, Subject's Age

TRS measurements revealed significant positive correlations between the MMSE score and SO_2 in the bilateral PFC ($r = 0.40$, $p < 0.01$). In addition, $C_{oxy\text{-}Hb}$ ($r = 0.23$, $p < 0.01$) and $C_{t\text{-}Hb}$ ($r = 0.14$, $p < 0.05$) in the right PFC exhibited significant positive correlations with the MMSE scores. In contrast, $C_{deoxy\text{-}Hb}$ in the left PFC exhibited a significant negative correlation with the MMSE score ($r = -0.19$, $p < 0.01$). Moreover, SO_2 and $C_{oxy\text{-}Hb}$ in the bilateral PFC exhibited negative correlations with age.

However, OPL in the PFC did not exhibit significant correlations with the MMSE score and patient's age ($p > 0.05$). Table 3 summarizes the correlations between the MMSE scores, subject's ages, Hb concentrations and SO_2 at rest in the PFC.

Table 3. Correlations between the MMSE score, subject's ages, Hb concentrations and SO_2 at rest in the right and left PFC ($n = 202$).

		Age	MMSE
	$C_{oxy\text{-}Hb}$ [μM]	−0.196 **	0.230 **
Right PFC	$C_{deoxy\text{-}Hb}$ [μM]	0.029	−0.047
	$C_{t\text{-}Hb}$ [μM]	−0.127	0.142 *
	SO_2 [%]	−0.270 **	0.396 **
	$C_{oxy\text{-}Hb}$ [μM]	−0.189 **	0.135
Left PFC	$C_{deoxy\text{-}Hb}$ [μM]	0.084	−0.191 **
	$C_{t\text{-}Hb}$ [μM]	−0.107	0.022
	SO_2 [%]	−0.302 **	0.398 **

* indicates $p < 0.05$. ** indicates $p < 0.01$.

3.3. Correlations between VSRAD Parameters and MMSE Scores, Subject's Age

There were negative correlations between the MMSE scores and the VSRAD scores including Severity ($r = -0.453$, $p < 0.01$), Extent ($r = -0.484$, $p < 0.01$), Ratio ($r = -0.402$. $p < 0.01$), and Whole Brain Extent ($r = -0.409$, $p < 0.01$). In contrast, the subject's age positively correlated with the VSRAD scores including Severity ($r = 0.406$, $p < 0.01$), Extent ($r = 0.476$, $p < 0.01$), Ratio ($r = 0.400$ $p < 0.01$), and Whole Brain Extent ($r = 0.404$, $p < 0.01$) (Table 4).

Table 4. Correlations between VSRAD parameters and MMSE scores, subject's age.

	VSRAD			
	Severity	**Brain Extent (%)**	**Extent (%)**	**Ratio**
Age	0.406 **	0.404 **	0.476 **	0.400 **
MMSE	−0.453 **	−0.409 **	−0.484 **	−0.402 **

* indicates $p < 0.05$. ** indicates $p < 0.01$.

3.4. Correlations between OPL and VSRAD Parameters

The degree of brain atrophy affected the OPLs. Figure 1 compares MRI images (fluid attenuated IR) of subjects with no brain atrophy (A) and severe brain atrophy (B). Table 5A compares the MMSE scores and VRSAD scores between no brain atrophy (Case A) and severe brain atrophy (Case B). The subarachnoid space in case B is larger than that in case A due to brain atrophy. Interestingly, the OPLs in case B were shorter than those in case A (Table 5B).

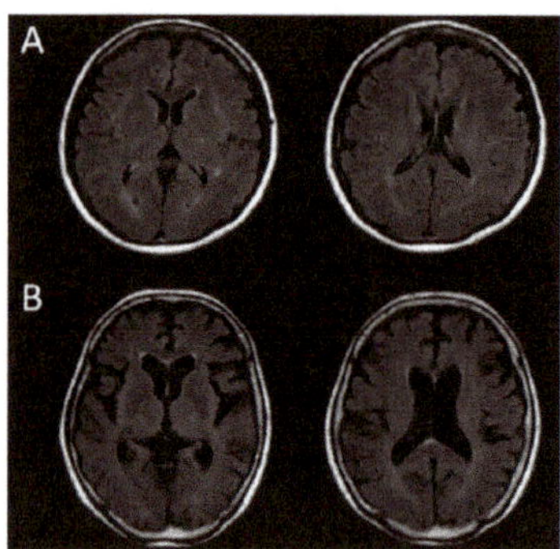

Figure 1. MRI images (fluid attenuated IR) of subjects with (**A**) no brain atrophy and (**B**) severe brain atrophy.

Table 5A. Comparison of VRSAD scores between no brain atrophy (Case A) and severe brain atrophy (Case B).

Case	Age/sex	MMSE	Severity	Whole Brain Extent (%)	Extent of VOI atrophy (%)	Ratio
A	50/F	30	0.27	1.40	0.00	0.00
B	77/F	9	3.48	6.48	86.17	13.29

VOI indicates Volume of interest.

Table 5B. Comparison of OPLs (in mm) between no brain atrophy (Case A) and severe brain atrophy (Case B).

Case	Right			Left			Average
	OPL1 (761 nm)	OPL2 (791 nm)	OPL3 (836 nm)	OPL1 (761 nm)	OPL2 (791 nm)	OPL3 (836 nm)	
A	217.9	220.8	208.4	226.2	227.2	211.4	218.7 (1.00)
B	177.1	179.0	169.6	171.8	174.0	162.9	172.4 (0.79)

On average, the OPLs (761 nm, 791 nm, and 836 nm) in the right PFC negatively correlated with VSRAD scores including Severity, Extent, and Ratio, but not Whole Brain Extent. In contrast, the correlations between the OPLs in the left PFC and VSRAD scores were limited. Table 6 summarizes the correlations between the OPLs and VSRAD scores.

Table 6. Correlations between OPLs and VSRAD parameters.

		VSRAD			
		Severity	Whole Brain Extent (%)	Extent of VOI Atrophy (%)	Ratio
Right PFC	OPL1	−0.305 *	0.073	−0.312 *	−0.396 **
	OPL2	−0.284 *	0.095	−0.292 *	−0.386 **
	OPL3	−0.306 *	0.052	−0.312 *	−0.395 **
Left PFC	OPL1	−0.211	−0.038	−0.242	−0.248
	OPL2	−0.228	−0.023	−0.262	−0.283 *
	OPL3	−0.240	−0.038	−0.276 *	−0.294 *

* indicates $p < 0.05$. ** indicates $p < 0.01$.

4. Discussion

In the present study using TRS, we evaluated the relationship between the cognitive functions (i.e., the MMSE scores) and optical parameters of the head regions including the PFC in elderly subjects with systemic disorders. It is difficult to measure Hb concentrations in the cortex selectively by TRS; however, the following findings suggest that the Hb concentrations measured by TRS reflected CBO in the PFC. First, simultaneous measurements of TRS and PET demonstrated that $C_{t\text{-}Hb}$ and SO_2 measured by TRS correlated with an increase of regional cerebral blood flow and volume induced by acetazolamide [18]. Second, the result of TRS functional study was consistent with the result obtained by fMRI; TRS demonstrated an increase in $C_{deoxy\text{-}Hb}$ in the PFC during driving simulation while fMRI demonstrated a decrease in the Blood oxgenation level dependent (BOLD) signal (i.e., deactivation) in the PFC [19]. Third, TRS could detect cerebral ischemia caused by vasospasms after SAH by demonstrating a decrease in $C_{oxy\text{-}Hb}$ and SO_2 [10]. These findings suggest that $C_{oxy\text{-}Hb}$ and SO_2 measured by TRS at rest reflected the PFC activity at rest.

The present study revealed positive correlations between the MMSE score and SO_2, $C_{oxy\text{-}Hb}$ in the PFC, suggesting that the greater the degree of PFC activity at rest, the higher the cognitive function. These observations are consistent with our recent TRS study on elderly women, which demonstrated that mild cognitive impairment exhibited higher baseline $C_{oxy\text{-}Hb}$ in the PFC than those in severe cognitive impairment [20]. Moreover, the negative correlation between the subject's age and SO_2 and $C_{oxy\text{-}Hb}$ in the PFC suggest that the older the subject, the lower the PFC activity at rest, which is consistent with the studies on the effect of aging on regional cerebral blood flow in the PFC [21].

The close correlations between the MMSE score and the TRS parameters suggest that machine learning may allow prediction of cognitive function based on the TRS parameters. For the prediction, deep learning, a subset of machine learning, may be useful since it allows analyzing regularity and relevance from a large amount of data, make judgments and predictions [22]. Indeed, the deep learning has been applied to imaging diagnosis [23], including computer- aided-diagnosis of Alzheimer's Disease (AD) based on MRI images [24]. In our preliminary study, we evaluated the variable importance for the prediction, and found that the subject's age showed the highest rank (1.0) while the right and left SO2 showed the second (0.77) and third (0.73) highest rank, respectively [25]. It should be noted that these parameters with high variable importance showed high correlation coefficients. Combination of MRI image and TRS measurement may be useful to predict cognitive dysfunction. In order to develop a deep learning based-diagnostic method of cognitive dysfunction, further study is necessary based on the present study.

The OPLs in the right PFC negatively correlated with VSRAD scores (Severity, Extent, and Ratio), indicating that the shorter the OPL, the greater the VSRAD scores which means greater degree of brain atrophy [16,17]. This might be caused by an increase of subarachnoid space due to brain atrophy; an increase of cerebrospinal fluid (CSF) layer in subarachnoid space caused a shortening of OPL since light scattering of CSF is much less than that of brain tissue. However, these findings are inconsistent with our previous TRS study on chronic stroke patients; we observed that the OPLs on the affected side, where subarachnoid space increased due to cortical atrophy in chronic stroke, was longer than that on the normal side [26]. It should be noted, however, that VSRAD scores does not indicate the degree of the PFC atrophy selectively. Therefore, further study, such as selective measurements of the degree of atrophy of the PFC, is necessary to clarify the relation between OPLs and brain atrophy.

Interestingly, there was a difference in correlation of OPLs and VSRAD between left and right PFC. Some evidence suggests that neurodegeneration related to aging and disease may preferentially affect the left-usually language- and motor-dominant-hemisphere; however, a recent meta-analysis provided no evidence for increased left-hemisphere vulnerability [27]. Further work is needed to provide a better understanding of the role of gray matter asymmetries.

We discuss the relation between the brain atrophy and the OPL from the view point of light propagation in the head. According to the numerical study of light propagation inside a human head model by Koyama et al. [28], the OPL between the source and detector depends on the reduced scattering coefficient of the CSF layer ($\mu'_{s\text{-}CSF}$), which corresponds to the subarachnoid space. The head model consists of four layers, i.e., the superficial layer including the scalp and skull (thickness of 10 mm), the CSF layer (2 mm), the gray matter (4 mm) and the white matter (6 mm). Light injected from the source propagates through the four layers to reach the detector with the partial optical pathlengths of l_{sup}, l_{CSF}, l_{gray} and l_{white} in the four layers, respectively, and the OPL is the sum of all the partial optical pathlengths, OPL $= l_{sup} + l_{CSF} + l_{gray} + l_{white}$. As the results of their numerical calculation, for the case of the source-detector distance of $\rho = 40$ mm, the OPL decreases with the decrease in $\mu'_{s\text{-}CSF}$ as shown in Table 7.

Table 7. Changes in the total and partial optical pathlengths with the change in $\mu'_{s\text{-}CSF}$ of the CSF layer for the case of the source-detector distance of $\rho = 40$ mm. Data from [28].

$\mu'_{s\text{-}CSF}$ (mm^{-1})	OPL (mm)	l_{sup} (mm)	l_{CSF} (mm)	l_{gray} (mm)	l_{white} (mm)
1.0 (soft tissue)	320	268	35	16	0.5
0.3 (weak scat.)	298	224	51	22	0.6
0.01 (very weak scat.)	239	170	53	16	0.4

While the OPL and l_{sup} decrease with the decrease in $\mu'_{s\text{-}CSF}$, l_{CSF} and l_{gray} increase with the decrease in $\mu'_{s\text{-}CSF}$. The results can be understood phenomenologically as the following. When $\mu'_{s\text{-}CSF}$ = 1.0 mm^{-1}, the CSF layer exhibits strong scattering similarly to that of soft tissue, and the light propagation pattern inside the head model can be schematically described as in Figure 2a where the thickness of the CSF layer is exaggerated to see its effect on light propagation. The propagation path from the source, S, to the detector, D, depicts a so-called banana shape as a semitransparent yellow region. In this case, the typical propagation path expressed by the red zig-zag lines crosses the interface between the superficial and CSF layers at positions "a" and "d" with a relatively short distance between them.

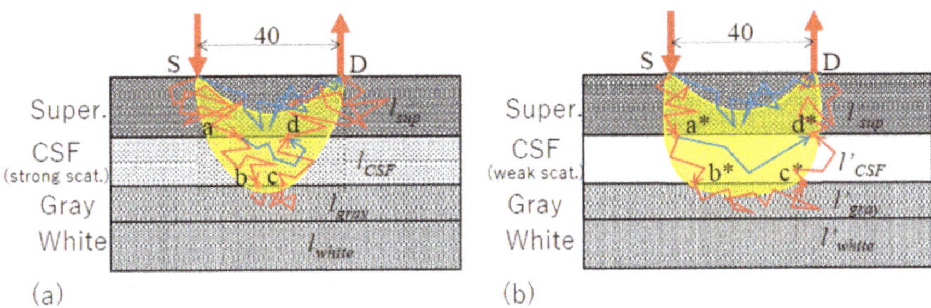

Figure 2. Patterns of light propagation in the head model with (**a**) $\mu'_{s\text{-}CSF}$ = 1.0 mm^{-1} (strong scattering as soft tissue) and (**b**) $\mu'_{s\text{-}CSF}$ = 0.01 mm^{-1} (very weak scattering).

When $\mu'_{s\text{-}CSF}$ = 0.01 mm^{-1}, the CSF layer exhibits very weak (or almost no) scattering, and the light propagation pattern can be described as in Figure 2b. The propagation pattern in Figure 2b is widened from the banana shape in Figure 2a due to the presence of the almost non-scattering CSF layer, and the typical propagation path crosses the interface between the superficial and CSF layers at positions "a*" and "d*" with a longer distance between them than the distance between "a" and "d". Because light in the CSF layer propagates a long distance without being scattered, a part of light propagating in the superficial layer can circumvent the strong scattering superficial layer by going through the CSF layer to reach the detector, D. Resultantly, l_{sup} decreases while l_{CSF} increases although the decrease in l_{sup} is much larger than the increase in l_{CSF}. This is the reason why the OPL decreases with the decrease in $\mu'_{s\text{-}CSF}$ in Table 6.

The results of Table 6 do not describe the relationship between the OPL and $\mu'_{s\text{-}CSF}$ but that between the OPL and the extent of VOI atrophy. However, the decrease in the $\mu'_{s\text{-}CSF}$ can be understood to be equivalent to the increase in the extent of Volume of interest (VOI) atrophy. Nevertheless, there is still a possibility that more precise simulation of light propagation in the human head with increasing or decreasing thickness of the CSF layer may provide the opposite results to those described above, i.e., the OPL increases with the increase in the thickness of the CSF layer. Further investigation about the relation between the OPL and the thickness of the CSF layer is necessary.

5. Conclusions

The present study demonstrated that TRS may be applicable to assessment of cognitive dysfunction, since Hb concentrations measured by TRS at rest in the PFC correlated with cognitive functions evaluated by the MMSE. It should be emphasized that TRS measurements at resting condition may be useful in aged people, particularly subjects with cognitive dysfunction who cannot perform cognitive tasks. In contrast to activation methods, the present method does not allow us to investigate the type of cognitive function being impaired by changing the type of task. However, TRS may be applicable to screening test of cognitive impairment; our preliminary study demonstrated that deep learning allows to predict the MMSE scores based on the TRS parameters [25,29].

Author Contributions: K.S. designed the study, and wrote the initial draft of the manuscript. L.H. and N.O. contributed to analysis and interpretation of data, assisted in the preparation of the manuscript. Y.Y. contributed to interpretation of TRS data, and critically reviewed the manuscript. All authors approved the final version of the manuscript, and agree to be accountable for all aspects of the work in insuring that questions related to the accuracy of integrity of any part of the work are appropriately investigated and resolved.

Funding: This work was supported in part by a Grant-in-Aid from the Ministry of Education, Culture, Sports, Sciences and Technology of Japan (Strategic Research Foundation Grant-aided Project for Private Universities S1411017), JSPS Grant-in-Aid for Young Scientists (B, 16K16077), from MEXT of Japan, and Ministry of Economy, Trade and Industry (Groundwork Project for Creating Industrial Models Using IoT)

Conflicts of Interest: The authors declare no conflict of interest.

References

1. Livingston, G.; Sommerlad, A.; Orgeta, V.; Costafreda, S.G.; Huntley, J.; Ames, D.; Ballard, C.; Banerjee, S.; Burns, A.; Cohen-Mansfield, J.; et al. Dementia prevention, intervention, and care. *Lancet* **2017**, *390*, 2673–2734. [CrossRef]
2. Folstein, M.F.; Folstein, S.E.; McHugh, P.R. "Mini-mental state". A practical method for grading the cognitive state of patients for the clinician. *J. Psychiatr. Res.* **1975**, *12*, 189–198. [CrossRef]
3. Arevalo-Rodriguez, I.; Smailagic, N.; Roqué, I.; Figuls, M.; Ciapponi, A.; Sanchez-Perez, E.; Giannakou, A.; Pedraza, O.L.; Bonfill Cosp, X.; Cullum, S. Mini-Mental State Examination (MMSE) for the detection of Alzheimer's disease and other dementias in people with mild cognitive impairment (MCI). *Cochrane Database Syst. Rev.* **2015**, *3*, CD010783.
4. Scheltens, P.; Blennow, K.; Breteler, M.M.; de Strooper, B.; Frisoni, G.B.; Salloway, S.; Van der Flier, W.M. Alzheimer's disease. *Lancet* **2016**, *388*, 505–517. [CrossRef]
5. Jöbsis, F.F. Noninvasive, infrared monitoring of cerebral and myocardial oxygen sufficiency and circulatory parameters. *Science* **1977**, *198*, 1264–1267. [CrossRef]
6. Reynolds, E.O.; Wyatt, J.S.; Azzopardi, D.; Delpy, D.T.; Cady, E.B.; Cope, M.; Wray, S. New non-invasive methods for assessing brain oxygenation and hemodynamics. *Br. Med. Bull.* **1988**, *44*, 1052–1075. [CrossRef]
7. Chance, B.; Leigh, J.S.; Miyake, H.; Smith, D.S.; Nioka, S.; Greenfeld, R.; Finander, M.; Kaufmann, K.; Levy, W.; Young, M. Comparison of time-resolved and -unresolved measurements of deoxyhemoglobin in brain. *Proc. Natl. Acad. Sci. USA* **1988**, *85*, 4971–4975. [CrossRef] [PubMed]
8. Patterson, M.S.; Chance, B.; Wilson, C. Time resolved reflectance and transmittance for the non-invasive measurement of tissue optical properties. *Appl. Opt.* **1989**, *28*, 2331–2336. [CrossRef] [PubMed]
9. Katagiri, A.; Dan, I.; Tuzuki, D.; Okamoto, M.; Yokose, N.; Igarashi, K.; Hoshino, T.; Fujiwara, T.; Katayama, Y.; Yamaguchi, Y.; et al. Mapping of optical pathlength of human adult head at multi-wavelengths in near infrared spectroscopy. *Adv. Exp. Med. Biol.* **2010**, *662*, 205–212.
10. Yokose, N.; Sakatani, K.; Murata, Y.; Awano, T.; Igarashi, T.; Nakamura, S.; Hoshino, T.; Katayama, Y. Bedside monitoring of cerebral blood oxygenation and hemodynamics after aneurysmal subarachnoid hemorrhage by quantitative time-resolved near-infrared spectroscopy. *World Neurosurg.* **2010**, *73*, 508–513. [CrossRef]
11. Oda, M.; Nakano, T.; Suzuki, A.; Shimomura, F.; Suzuki, T. Near infrared time-resolved spectroscopy system for tissue oxygenation monitor. *SPIE* **2000**, *4160*, 204–210.
12. Okada, E.; Delpy, D.T. Near-infrared light propagation in an adult head model. I. Modeling of low-level scattering in the cerebrospinal fluid layer. *Appl. Opt.* **2003**, *42*, 2906–2914. [CrossRef]
13. Brayne, C.; Calloway, P. The association of education and socioeconomic status with the Mini Mental State Examination and the clinical diagnosis of dementia in elderly people. *Age Ageing* **1990**, *19*, 91–96. [CrossRef]
14. Crum, R.M.; Anthony, J.C.; Bassett, S.S.; Folstein, M.F. Population-based norms for the Mini-Mental State Examination by age and educational level. *JAMA* **1993**, *269*, 2386–2391. [CrossRef] [PubMed]
15. Tombaugh, T.N.; McIntyre, N.J. The mini-mental state examination: A comprehensive review. *J. Am. Geriatr. Soc.* **1992**, *40*, 922–935. [CrossRef] [PubMed]
16. Matsuda, H.; Mizumura, S.; Nemoto, K.; Yamashita, F.; Imabayashi, E.; Sato, N. Automatic voxel-based morphometry of structural MRI by SPM8 plus diffeomorphic anatomic registration through exponentiated lie algebra improves the diagnosis of probable Alzheimer Disease. *Am. J. Neuroradiol.* **2012**, *33*, 1109–1114. [CrossRef]

17. Matsuda, H. Voxel-based Morphometry of Brain MRI in Normal Aging and Alzheimer's Disease. *Aging Dis.* **2013**, *4*, 29–37. [PubMed]
18. Ohmae, E.; Ouchi, Y.; Oda, M. Cerebral hemodynamics evaluation by near-infrared time-resolved spectroscopy: Correlation with simultaneous positron emission tomography measurements. *NeuroImage* **2006**, *29*, 697–705. [CrossRef]
19. Sakatani, K.; Yamashita, D.; Yamanaka, T. Changes of cerebral blood oxygenation and optical pathlength during activation and deactivation in the prefrontal cortex measured by time-resolved near infrared spectroscopy. *Life Sci.* **2006**, *78*, 2734–2741. [CrossRef]
20. Machida, A.; Shirato, M.; Tanida, M.; Kanemaru, C.; Nagai, S.; Sakatani, K. Effects of Cosmetic Therapy on Cognitive Function in Elderly Women Evaluated by Time-Resolved Spectroscopy Study. *Adv. Exp. Med. Biol.* **2016**, *876*, 289–295. [PubMed]
21. Martin, A.J.; Friston, K.J.; Colebatch, J.G.; Frackowiak, R.S. Decreases in regional cerebral blood flow with normal aging. *J. Cereb. Blood Flow Metab.* **1991**, *11*, 684–689. [CrossRef]
22. Sze, V.; Chen, Y.H.; Yang, T.J.; Joel, S. Efficient Processing of Deep Neural Networks: A tutorial and Survey. *Proc. IEEE* **2017**, *105*, 2295–2329. [CrossRef]
23. Hosny, A.; Parmar, C.; Quackenbush, J.; Schwartz, L.H.; Aerts, H.J.W.L. Artificial intelligence in radiology. *Nat. Rev. Cancer* **2018**, *18*, 500–510. [CrossRef]
24. Liu, S.; Liu, S.; Cai, W.; Pujol, S.; Kikinis, R.; Feng, D. Early diagnosis of Alzheimer's disease with deep learning. In Proceedings of the 2014 IEEE 11th International Symposium on Biomedical Imaging (ISBI), Beijing, China, 29 April–2 May 2014; pp. 1015–1018.
25. Oyama, K.; Hu, L.; Sakatani, K. Prediction of MMSE score using time-resolved near-infrared spectroscopy. In Proceedings of the 45th Annual Meeting of the International Society on Oxygen Transport to Tissue (ISOTT), Halle, Germany, 19–23 August 2017.
26. Sato, Y.; Komuro, Y.; Lin, L.; Tang, Z.; Hu, L.; Kadowaki, S.; Ugawa, Y.; Yamada, Y.; Sakatani, K. Differences in Tissue Oxygenation, Perfusion and Optical Properties in Brain Areas Affected by Stroke: A Time-Resolved NIRS Study. *Adv. Exp. Med. Biol.* **2018**, *1072*, 63–67.
27. Minkova, L.; Habich, A.; Peter, J.; Kaller, C.P.; Eickhoff, S.B.; Klöppel, S. Gray matter asymmetries in aging and neurodegeneration: A review and meta-analysis. *Hum. Brain Mapp.* **2017**, *38*, 5890–5904. [CrossRef]
28. Koyama, T.; Iwasaki, A.; Ogoshi, Y.; Okada, E. Practical and adequate approach to modeling light propagation in an adult head with low-scattering regions by use of diffusion theory. *Appl. Opt.* **2005**, *44*, 2094–2103. [CrossRef]
29. Oyama, K.; Hu, L.; Sakatani, K. Prediction of MMSE score using time-resolved near-infrared spectroscopy. *Adv. Exp. Med. Biol.* **2018**, *1072*, 145–150.

 *applied sciences*

Article

Relationship of Total Hemoglobin in Subcutaneous Adipose Tissue with Whole-Body and Visceral Adiposity in Humans

Miyuki Kuroiwa [1,†], **Sayuri Fuse** [1,†], **Shiho Amagasa** [2], **Ryotaro Kime** [1], **Tasuki Endo** [1], **Yuko Kurosawa** [1] **and Takafumi Hamaoka** [1,*]

[1] Department of Sports Medicine for Health Promotion Tokyo Medical University, 6-1-1 Shinjuku, Shinjuku ku, Tokyo 160-8402, Japan; mkuroiwa@tokyo-med.ac.jp (M.K.); fuse@tokyo-med.ac.jp (S.F.); kime@tokyo-med.ac.jp (R.K.); endo-tasuki-fv@ynu.jp (T.E.); kurosawa@tokyo-med.ac.jp (Y.K.)

[2] Department of Preventive Medicine and Public Health Tokyo Medical University, 6-1-1 Shinjuku, Shinjuku ku, Tokyo 160-8402, Japan; shiho.ama@gmail.com

* Correspondence: kyp02504@nifty.com; Tel.: +81-3-3351-6141

† The authors contributed equally to this work.

Received: 6 April 2019; Accepted: 11 June 2019; Published: 14 June 2019

Abstract: High whole-body and visceral adiposity are risk factors that can cause metabolic diseases. We hypothesized that the total hemoglobin concentration (total-Hb) in abdominal subcutaneous adipose tissue (SAT_{ab}), an indicator of white adipose tissue (WAT) vascularity, correlates negatively with risk factors for developing metabolic diseases, such as whole-body and visceral adiposity. We tested the optical characteristics of abdominal tissue in 140 participants (45 men and 95 women) who were apparently healthy individuals with a median age of 39 years. They also had a median body fat percentage of 25.4%, a visceral fat area of 50.4 cm^2, and a SAT_{ab} thickness of 1.05 cm. These tests were conducted using near-infrared time-resolved spectroscopy (NIR_{TRS}) with a 2-cm optode separation. To distinguish the segments of SAT_{ab} (Seg_{SAT}) and the mixture of muscle and SAT_{ab} ($Seg_{SAT+Mus}$), the threshold was analyzed using the slopes of (total-Hb) against the thickness of SAT_{ab} using the least-squares mean method. According to the results from the logistic regression analysis, the percentage of body fat and visceral fat area remained significant predictors of the (total-Hb) ($p = 0.005$ and $p = 0.043$, respectively) in the data for Seg_{SAT} (no influence from the SAT_{ab} thickness). We conclude that simple, rapid, and noninvasive NIR_{TRS}-determined (total-Hb) in WAT could be a useful parameter for evaluating risk factors for metabolic diseases.

Keywords: near-infrared time-resolved spectroscopy; noninvasive; subcutaneous white adipose tissue; tissue total hemoglobin

1. Introduction

White adipose tissue (WAT), which is constantly remodeled by metabolic challenges, is one of the most plastic tissues in multicellular beings. The capillary density of WAT varies for individual organs depending on their metabolic rate, e.g., the capillary density of the prenatal depot is much higher than in the subcutaneous one [1]. The vascular network in the subcutaneous adipose tissue in the abdomen (SAT_{ab}) of a non-obese group is greater than that for an obese group. Along with the increase in vascularity owing to the increase in energy demand due to exercise, mitochondrial gene expression in WAT can also shift to a metabolically active brown and/or beige type [2,3]. Thus, long-term adaptation or remodeling of the vascular network in adipocytes is needed for maintaining energy homeostasis in WAT [4].

To evaluate vascularity, an invasive sampling of WAT is needed. Invasive sampling prohibits widespread research on humans. Compared with visible light wavelengths, near-infrared (NIR)

wavelengths in the range 700–3000 nm show less scattering and as a result they show better penetration into biological tissue. However, light absorption by water limits tissue penetration above the 900 nm wavelength, thus, the 650–850 nm range is suitable for measurements [5]. Several types of near-infrared spectroscopy (NIRS) allow for noninvasive monitoring of tissue oxygenation and hemodynamics in vivo [6–10]. Among them, NIR time-resolved spectroscopy (NIR_{TRS}) is a method employing picosecond light pulse emissions from the skin surface for measuring the time distribution of the photons scattered and/or absorbed in tissue several centimeters away from the point of light emission. It noninvasively quantifies a range of tissue optical properties, including the absorption coefficient (μ_a), reduced scattering coefficient (μ_s'), and light path length, and allows for the calculation of tissue oxygenated hemoglobin concentration (oxy-Hb), deoxygenated hemoglobin concentration (deoxy-Hb), total hemoglobin concentration (total-Hb), and oxygen saturation (StO_2) [5,10,11]. It was reported that (total-Hb) is an indicator of tissue vascularity and μ_s' is a mitochondria parameter in the brown adipose and muscle tissues [12,13]. Although there is a difference in μ_s' between muscle and WAT, the μ_s' of WAT is unrelated to its mitochondrial content [13]. Thus, among the NIR_{TRS} parameters, we have only chosen and used total-Hb as an indicator of tissue vascularity.

We hypothesize that total-Hb in the SAT_{ab} is an indicator of WAT vascularity. These indicators correlate negatively with risk factors for developing metabolic diseases, such as whole-body and visceral adiposity. Thus, the purpose of this study was to confirm the relationship between the vascularity of a localized WAT and the risk factors for metabolic diseases.

2. Materials and Methods

2.1. Subjects and Study Design

For this study, 140 participants over 20 years of age were recruited (Table 1). Volunteers were recruited via poster advertisements in the Kanto region in Japan. The participants arrived at the laboratory and the following parameters were measured: (Total-Hb), (oxy-Hb), (deoxy-Hb), μ_a, μ_s', SAT_{ab}, percentage of whole-body fat (%BF), and visceral fat area (VFA). In this study, participants with different SAT_{ab} thicknesses were chosen for obtaining the physiological and optical properties of SAT_{ab}. The room temperature in the laboratory was regulated from 23 °C to 27 °C using an air-conditioner. The study design and protocols were approved by the institutional review boards of Tokyo Medical University (IRB 2017-199), in accordance with the ethical principles contained in the Declaration of Helsinki. Written informed consent was obtained from all the participants. This study was conducted in the summer season, July to August, in 2017 and 2018.

2.2. Measurements of Anthropometric Parameters

The SAT_{ab} thickness was monitored using B-mode ultrasonography (Vscan Dual Probe; GE Vingmed Ultrasound AS, Horten, Norway). The measurement points were fixed 1.0 cm dorsally and ventrally from the center of the NIR_{TRS} probe (the anterior axillary line across the umbilical height), which generally contains the thickest fat layer. The SAT_{ab} thickness was measured by the investigator using the attached distance measuring system and calculated as the mean value of two measurements.

The %BF was estimated using the multi-frequency bioelectric impedance method (InBody 720, InBody Japan, Tokyo, Japan). The InBody 720 measured impedance at various frequencies (1, 5, 250, 500, and 1000 kHz) across the legs, arms, and trunk. All four extremities were in contact with the electrodes, and the participant stood barefoot on the device until the completion of the test. Measurements of total body water, %BF, fat mass, fat-free mass, and lean body mass were obtained. The electrical resistance of fat is greater than that of other tissues, such as muscle; therefore, the fat value is estimated by the alternating current resistance value [14]. The InBody 720 is an excellent body component analyzer that can measure 30 impedance values and 15 reactance values. The %BF measured by the InBody 720 and dual energy X-ray absorptiometry (DXA) showed a significant correlation ($r^2 = 0.858$) [15].

The %BF measured by the InBody 720 and underwater weighing (a four-component model) showed a significant correlation (r = 0.85) [16].

The VFA, at the abdominal level of L4–L5, was estimated using a bioelectrical impedance analysis (EW-FA90; Panasonic, Osaka, Japan). The VFA can be calculated by applying the abdominal bioimpedance method and directly measuring the abdominal resistance due to VFA using impedance technology [17,18]. The VFA, as measured by bioimpedance and a CT examination, showed a significant correlation (r = 0.87) [17].

2.3. Measurements of (total-Hb), (oxy-Hb), (deoxy-Hb), μ_a, and μ_s'

As the center of the NIR_{TRS} probe was placed at the anterior axillary line across the umbilical height. The location of the light input and output was deviated 1.0 cm dorsally and ventrally from the center.

NIR_{TRS} can evaluate optical properties, such as the absorption coefficient (μ_a) and reduced scattering coefficient (μ_s'), and therefore it can be used to noninvasively quantify tissue oxygenated hemoglobin (oxy-Hb), deoxygenated Hb (deoxy-Hb), and total Hb (total-Hb) concentrations. The μ_a, μ_s', (total-Hb), (oxy-Hb), and (deoxy-Hb) in the abdominal region were measured for one minute using NIR_{TRS} (TRS-20; Hamamatsu Photonics K.K., Hamamatsu, Japan). After a five minute rest, the probes were placed on the skin of the abdomen and participants were required to remain in a sitting position during the measurements. The optode separation for NIR_{TRS} was 2 cm in this study.

The methods for calculating the μ_a, μ_s', (total-Hb), (oxy-Hb), and (deoxy-Hb) were as follows: The target tissue in the abdominal region was repeatedly irradiated using semiconductor pulse laser lights of three different wavelengths (760, 800, and 830 nm). This was done under the conditions of a full width at half maximum of 100 ps, a pulse rate of 5 MHz, and an average output power at the optical irradiation fiber's end of approximately 100 μW at each wavelength (the total average power level was 250–300 μW). The pulsed light that was scattered and absorbed inside the tissue was detected by a photomultiplier tube capable of single-photon detection. Time-resolved measurements were performed using the time-correlated single photon counting method. The values of μ_a and μ_s' for the obtained tissue were estimated by fitting the temporal profile of the flux (reflectance) derived from the analytical solution of the photon diffusion equation R (ρ,t) (Equation (1)) that convolved the instrument response function to the time-response properties of the samples.

Where (t) is the response time, (ρ) is the distance between the light source and detector, μ_a and μ_s' are the absorption coefficient and equivalent scattering coefficient, respectively, $\underline{D = 1/3\ \mu_s'}$ is the photon diffusion coefficient, c is the velocity of light inside the light scattering medium $\underline{(20\ cm\ ns^{-1})}$, Z_0 $(= 1/\mu_s')$ is the transport mean free path, and the average path length (L) $(= \int[R(\rho, t)\ t\ dt]\ c/\int [R\ (\rho, t)\ dt])$. (deoxy-Hb) and (oxy-Hb) were obtained using simultaneous equations with μ_a obtained from Equation (1) using the least-squares fitting method [19]. Then, the absolute (total-Hb) was calculated as the sum of (oxy-Hb) and (deoxy-Hb) [5,12] from the following calculation formula.

$$R\ (\rho, t) = (4\pi Dc)^{-3/2} \cdot Z0t^{-5/2}\ \exp(-\mu_a\ ct)\ \exp[-(\rho^2 + Z0^2)/4Dct] \qquad (1)$$

$$[\text{total-Hb}] = [\text{deoxy-Hb}] + [\text{oxy-Hb}] \qquad (2)$$

The data were collected every 10 seconds by the NIR_{TRS}. The coefficient of variation (SD/mean) for repeated measurements of (total-Hb) was 4.9% [12].

Maximum permissible exposure (MPE) is the highest power or energy density (in W/cm^2 or J/cm^2) of the light source that is considered safe. Based on this MPE value, five classes (1, 2, 3A, and 4) for indicating the hazard that a laser product represents were defined as per the accessible emission limit (AEL); Class 1 is the most secure. NIR_{TRS} was classified as Class 1 by both the Japanese Industrial Standards Committee (JIS) C 6802 and the International Electrotechnical Commission (IEC) 60825-1. Thus, the safety of using NIR_{TRS} has been established.

2.4. Data Analysis for the Threshold by Analyzing the Slopes of (total-Hb) against the Thickness of the Subcutaneous Adipose Tissue in the Abdomen (SAT$_{ab}$)

To identify the thickness threshold of SAT$_{ab}$ for evaluating the SAT$_{ab}$ without the influence of the underlying muscle layer, the threshold was analyzed using the slopes of (total-Hb) against the thickness of SAT$_{ab}$ by the least-square mean method using the R software (v.3.5.2, R Foundation for Statistical Computing, Vienna, Austria, 2018). In this study, we attempted to identify the break-point location where the linear relation changes and in the relevant regression parameters (slopes of straight lines). The break-points were appropriately calculated using the 'segmented' Package. The two different slopes of (total-Hb) against the SAT$_{ab}$ thickness in the two segments primarily resulted from different (total-Hb) values between muscle and SAT$_{ab}$. One has a steeper slope for a segment with lower values of SAT$_{ab}$ thickness; the other has a lower slope for a segment with higher SAT$_{ab}$ thicknesses (Figure 1). The former segment comprises data derived from the mixture of muscle and SAT$_{ab}$ (Seg$_{SAT+Mus}$) and the latter, the specific SAT$_{ab}$ (Seg$_{SAT}$). Thus, the threshold of the SAT$_{ab}$ thickness, or the *x*-value at the intersection of the two slopes, indicates the need to evaluate SAT$_{ab}$ without the influence of the underlying muscle layer.

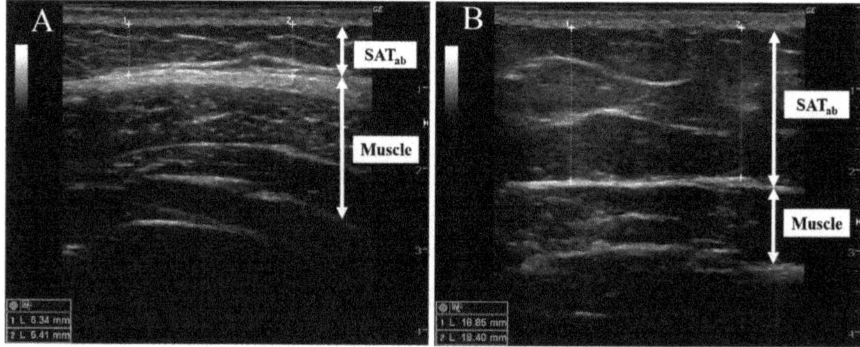

Figure 1. Typical ultrasonic images of the subcutaneous adipose tissue in the abdominal region (SAT$_{ab}$).

2.5. Statistical Analyses

Data are expressed as a median (first and third quartile) ± standard deviation (SD). The Pearson's product moment correlation coefficient was used to analyze the relationship between each parameter. Logistic regression analysis was conducted to assess the factors influencing (total-Hb). The predictor variables were median SAT$_{ab}$ thickness, %BF, and visceral fat area. To compare the participants' profiles between Seg$_{SAT+Mus}$ and Seg$_{SAT}$, we used the independent *t*-test or the Mann–Whitney test, as required. The analyses were performed using SPSS (IBM SPSS Statistics 25, IBM Japan, Tokyo, Japan, 2017), and $P < 0.05$ was considered statistically significant.

3. Results

3.1. Participant Profiles

As shown in Table 1, there were 140 participants. The median age was 39 years old (the age range being 22–67), the median %BF was 25.4 ± 7.12%, and the visceral fat area was 50.4 ± 37.6 cm^2. In the abdominal region, the median (total-Hb) was 19.9 ± 15.8 μM, (oxy-Hb) was 13.4 ± 11.7 μM, (deoxy-Hb) was 6.88 ± 6.31 μM, μ_a was 0.054 ± 0.032 cm^{-1}, and μ_s' was 9.09 ± 1.45 cm^{-1}. The minimum value of SAT$_{ab}$ thickness was 0.100 cm and the maximum value was 5.41 cm. The median SAT$_{ab}$ thickness was 1.50 ± 0.795 cm.

The SAT$_{ab}$ in the abdomen was measured using B-mode ultrasonography (Vscan Dual Probe; GE Vingmed Ultrasound AS, Hort e, Norway), whereas the thickness was measured using the attached distance measuring system by an investigator and calculated as the mean value of two measurements. Part A of Figure 1 is the ultrasonic image of the abdominal region in a 31-year-old woman. The mean layer thickness of the SAT_{ab} was 0.638 cm. B is the ultrasonic image of the abdominal region in a 40-year-old woman. The mean layer thickness of the SAT_{ab} was 1.91 cm.

Table 1. Participant profiles.

$n = 140$ (45 Men/95 Women)			
	Mean		**SD**
Age (Year)	**39.3**	±	**7.59**
SAT_{ab} thickness (cm)	1.50	±	0.80
%BF (%)	25.4	±	7.12
visceral fat area (cm^2) in the abdomen	50.4	±	37.6
(total-Hb) (µM)	19.9	±	15.8
(oxy-Hb) (µM)	13.4	±	11.7
(deoxy-Hb) (µM)	6.88	±	6.31
μ_a (cm^{-1})	0.054	±	0.032
$\mu_s{}'$ (cm^{-1})	9.09	±	1.45

Values are expressed as median ± standard deviation (SD). %BF, percentage of whole-body fat; (total-Hb), total hemoglobin; (oxy-Hb), oxy hemoglobin; (deoxy-Hb), deoxy hemoglobin; μ_a, absorption coefficient; and μ_s', reduced scattering coefficient of the subcutaneous adipose tissue in the abdomen (SAT_{ab}) were measured. The distance between transmission and detection with near infrared time resolved spectroscopy (NIR_{TRS}) in the abdomen was 2 cm.

3.2. The Thickness Threshold of the Subcutaneous Adipose Tissue in the Abdomen (SAT_{ab})

From the threshold analysis of the slope of (total-Hb) against the SAT_{ab} thickness, the respective thresholds of the SAT_{ab} thickness was 1.45 cm (Figure 2).

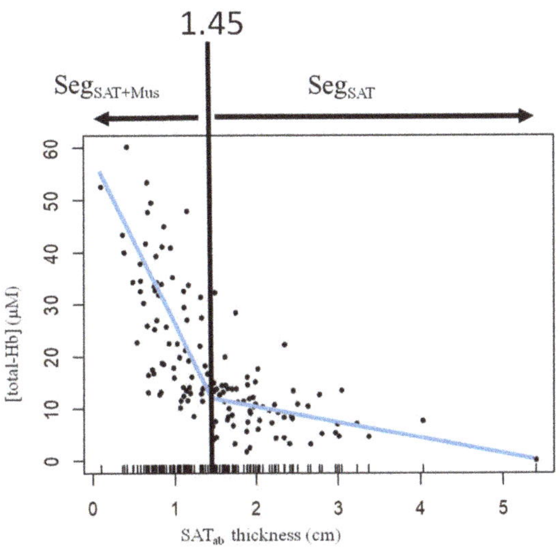

Figure 2. The relationship between total hemoglobin concentration (total-Hb) and the thickness of the subcutaneous adipose tissue (SAT_{ab}).

To identify the threshold of the thickness of SAT_{ab}, a segment regression analysis was conducted using the R software. The regression equations were $y = -31.5x + 57.8$ and $y = -3.0x + 16.6$. The inflection point between the (total-Hb) and the SAT_{ab} thickness. The left segment from the threshold line was defined as $Seg_{SAT+Mus}$, while the right, Seg_{SAT}.

3.3. Participant Profiles: $Seg_{SAT+Mus}$ and Seg_{SAT}

Using the value of 1.45 cm from the threshold (Section 3.2), the participants were divided into two groups: $Seg_{SAT+Mus}$, with SAT_{ab} less than 1.45 cm and the data derived from the SAT_{ab} and muscle, and $Seg_{SAT.}$, with SAT_{ab} greater than 1.45 cm and the data derived from the SAT_{ab}. Participant profiles for $Seg_{SAT+Mus}$ and Seg_{SAT} are presented in Table 2. A significant difference was found in all items, when $Seg_{SAT+Mus}$ and Seg_{SAT} were compared.

Table 2. Participant profiles.

A $Seg_{SAT+Mus}$			B Seg_{SAT}			*p*-Value
n = 72 (26 men/46 women)			*n* = 68 (19 men/49 women)			
	Mean	SD		Mean	SD	
age (year)	37.5 ±	7.14	age (year)	41.2 ±	7.63	= 0.004 [b]
SAT_{ab} thickness (cm)	0.911 ±	0.312	SAT_{ab} thickness (cm)	2.12 ±	0.67	<0.001 [a]
%BF (%)	21.4 ±	6.04	%BF (%)	29.6 ±	5.63	<0.001 [a]
visceral fat area (cm^2) in the abdomen	39.1 ±	28.8	visceral fat area (cm^2) in the abdomen	62.3 ±	41.8	<0.001 [b]
(total-Hb) (μM)	29.1 ±	16.8	(total-Hb) (μM)	10.2 ±	5.63	<0.001 [b]
(oxy-Hb) (μM)	20.2 ±	12.4	(oxy-Hb) (μM)	6.14 ±	3.99	<0.001 [b]
(deoxy-Hb) (μM)	9.47 ±	7.46	(deoxy-Hb) (μM)	4.15 ±	2.96	<0.001 [b]
μ_a (cm^{-1})	0.073 ±	0.0337	μ_a (cm^{-1})	0.035 ±	0.0119	<0.001 [b]
$\mu_s{'}$ (cm^{-1})	9.75 ±	1.30	$\mu_s{'}$ (cm^{-1})	8.40 ±	1.27	<0.001 [a]

Values are expressed as median ± standard deviation (SD). $Seg_{SAT+Mus}$, participants whose SAT_{ab} was less than 1.45 cm with the data derived from the SAT_{ab} and muscle; Seg_{SAT}, participants whose SAT_{ab} was greater than 1.45 cm with the data derived from the SAT_{ab}; %BF, percentage of whole-body fat; (total-Hb), total hemoglobin; (oxy-Hb), oxy hemoglobin; (deoxy-Hb), deoxy hemoglobin; μ_a, absorption coefficient; $\mu_s{'}$, reduced scattering coefficient; and SAT_{ab}, subcutaneous adipose tissue in the abdomen are shown. The distance between transmission and detection with NIR_{TRS} in the abdomen was 2 cm. $Seg_{SAT+Mus}$ is data derived from the mixture of SAT_{ab} and muscle; Seg_{SAT} is data derived from the SAT_{ab}. To compare the participants' profiles between $Seg_{SAT+Mus}$ and Seg_{SAT}, we used independent *t*-test or the Mann–Whitney test, as appropriate. A: Independent *t*-test and b: Mann–Whitney test.

3.4. Predictor Analysis for (total-Hb) Using Body Indicators (SAT_{ab} thickness, %BF, and Visceral Fat Area)

According to the results of the logistic regression analysis, the SAT_{ab} thickness ($p = 0.001$) and %BF ($p < 0.001$) remained as significant predictors of the (total-Hb) in the data for $Seg_{SAT+Mus}$ (Table 3A). The %BF and visceral fat area remained as significant predictors of the (total-Hb) ($p = 0.005$ and $p = 0.043$, respectively) in the data for Seg_{SAT} (no influence by the SAT_{ab} thickness) (Table 3B).

Table 3. Logistic regression analysis.

A	$Seg_{SAT+Mus}$ *n* = 75				
				95% C.I. for EXP (B)	
		p	Exp(B)	Lower	Upper
	(total-Hb)				
	SAT_{ab} thickness	0.009 **	0.022	0.001	0.384
	%BF	0.001 **	0.739	0.622	0.878
	visceral fat area	0.030 *	0.956	0.917	0.996
B	Seg_{SAT} *n* = 65				
				95% C.I. for EXP (B)	
		p	Exp(B)	Lower	Upper
	(total-Hb)				
	SAT_{ab} thickness	-	-	-	-
	%BF	0.004 **	0.830	0.731	0.941
	visceral fat area	0.044 *	0.982	0.966	1.000

Relationship between (total-Hb) in the abdomen and body indicators, such as subcutaneous adipose tissue in the abdomen (SAT_{ab}) thickness, %BF, and visceral fat area. Data derived from SAT_{ab}. $Seg_{SAT+Mus}$, SAT_{ab} thickness is less than 1.45 cm; Seg_{SAT}, SAT_{ab} thickness is over 1.45 cm.

4. Discussion

We found from the SAT_{ab} (Seg_{SAT}) measurements that a significant correlation exists between microvascular density evaluated by (total-Hb) and whole body and visceral adiposity without influenced by the thickness of SAT_{ab}.

We successfully discriminated data obtained from Seg_{SAT} (thickness of $SAT_{ab} \geqq$ approximately 1.5 cm) from those of $Seg_{SAT+Mus}$ (thickness of $SAT_{ab} <$ approximately 1.5 cm) by analyzing the different slopes of (total-Hb) against the thickness of SAT_{ab}. However, for both inhomogeneous $Seg_{SAT+Mus}$ and homogeneous Seg_{SAT}, (total-Hb) decreases as SAT_{ab} thickness increases. The correlation between the (total-Hb) and risk factors for metabolic diseases is specific to data obtained from Seg_{SAT} without being influenced by the thickness of SAT_{ab}. The question arises as to why microvascular density in the Seg_{SAT} tends to vary in each individual [4]. It may be attributed to the findings that, due to changes in metabolic rates in adipocytes, adaptation or remodeling of the vascular network and mitochondrial phenotype or density are needed for maintaining energy homeostasis in WAT [4]. Previous studies also demonstrated that remodeling of the vascular bed occurs through close interaction of adipocytes with metabolic changes via angiogenic factors released by the adipocytes themselves (paracrinologic action) or by other organs (endocrinologic action) [20].

In a previous study, optical and physiological properties were measured using diffuse optical spectroscopic imaging for three months in overweight or obese individuals under calorie-restricted diets. This study found alterations in tissue structure, determined by optical scattering signals, that possibly correlate with reductions in adipose cell volume, improved SAT_{ab} perfusion, and oxygen extraction determined by water and hemoglobin dynamics [21]. Our result is in accordance with the data, in that microvascular density is high in individuals with lower visceral and whole-body adiposity.

The most common, commercially available continuous wave NIRS (NIR_{CWS}) provides only relative values of tissue oxygenation. The main reason for this method's inability to provide quantitative data is the unknown path of NIR light through biological tissues [8–10]. It is suggested that the depth of light penetration is approximately 15 mm with a 30 mm optode separation for NIR_{CWS} [9]. In contrast, NIR_{TRS} measures the time distribution of the photons scattered and/or absorbed in tissue several centimeters from the point of light emission. It can also provide absolute values for tissue hemodynamics. Furthermore, according to a recent study [22], the mean depth of light penetration would be greater (approximately two-thirds of optode separation) and more homogeneous when NIR_{TRS} is used. The difference between this model and the previous model is the following: μ_a and μ_s' were 0.023 cm^{-1} and 10 cm^{-1}, respectively, with the 3.0 cm separation optode at the 807 nm wavelength on the intralipid phantom in the previous study [22] and were 0.035 cm^{-1} and 8.404 cm^{-1}, respectively, with the 2 cm optode separation at 800 nm in this study. The greater μ_a observed in the current study might allow for shallower penetration of photons.

The limitations of this study are the following: As this is a cross-sectional study, we could not draw a conclusive remark on whether the findings are true for a single individual encountering changes in weight. A longitudinal study is needed. In addition, only a part of the body, SAT_{ab}, was measured, and the other parts of the body have not been considered. It is difficult to make explicit connections between the observed data and microscopic tissue properties (microvascular and mitochondrial densities) without accompanying histopathology evidence. We realize that obtaining this in a human subject study is challenging. Despite this, data can be obtained from an animal model to test these hypotheses. Furthermore, as the participants of this study were healthy people, it is necessary to examine whether the current results hold true for people with various risk factors. Moreover, as this study was conducted in the summer, seasonal differences might have an influence on the results. As the data determined by NIR_{TRS} are derived based on a single-layered model, we should be careful to interpret current data with the multiple-layered (skin/fat/muscle) models. As the skin thickness is less than 0.2 cm and the inter-individual variation is small, we do not usually consider the effect of skin thickness. However, we should consider the effect of the overlying skin layer on the SAT_{ab} measurement in the future. The distance between the transmitter and receiver was 2 cm with NIR_{TRS} in this study; however, for measuring people with less adiposity, it would be more appropriate to use a 1 cm optode separation to specifically observe the optical and physiological characteristics of WAT.

5. Conclusions

We conclude that simple, rapid, and noninvasive NIR_{TRS}-determined (total-Hb) in SAT_{ab} could be a useful parameter for evaluating risk factors for metabolic diseases.

Author Contributions: Conceptualization, T.H.; formal analysis, M.K. and S.F.; investigation, M.K., S.F., S.A., T.H., R.K., T.E., and Y.K.; data curation, M.K. and S.F.; writing—original draft preparation, M.K.; writing—review and editing, Y.K., and T.H.; funding acquisition, T.H.

Funding: This work was supported by a Grant-in-Aid for Scientific Research from the Ministry of Education, Culture, Sports, Science, and Technology of Japan (15H03100).

Acknowledgments: This work was supported by a Grant-in-Aid for Scientific Research from the Ministry of Education, Culture, Sports, Science, and Technology of Japan (15H03100).

Conflicts of Interest: The authors declare that there are no conflicts of interest.

References

1. Crandall, D.L.; Hausman, G.J.; Kral, J.G. A review of the microcirculation of adipose tissue: Anatomic, metabolic, and angiogenic perspectives. *Microcirculation* **1997**, *4*, 211–232. [CrossRef] [PubMed]
2. Otero-Diaz, B.; Rodriguez-Flores, M.; Sanchez-Munoz, V.; Monraz-Preciado, F.; Ordonez-Ortega, S.; Becerril-Elias, V.; Baay-Guzman, G.; Obando-Monge, R.; Garcia-Garcia, E.; Palacios-Gonzalez, B.; et al. Exercise Induces White Adipose Tissue Browning Across the Weight Spectrum in Humans. *Front. Physiol.* **2018**, *9*, 1781. [CrossRef] [PubMed]
3. Wang, H.; Lee, J.H.; Tian, Y. Critical Genes in White Adipose Tissue Based on Gene Expression Profile Following Exercise. *Int. J. Sports Med.* **2019**, *40*, 57–61. [CrossRef] [PubMed]
4. Kadomatsu, T.; Oike, Y. Angiogenesis in adipose tissue and obesity. *J. Clin. Exp. Med. (IGAKU NO AYUMI)* **2014**, *250*, 761–765.
5. Hamaoka, T.; McCully, K.K.; Quaresima, V.; Yamamoto, K.; Chance, B. Near-infrared spectroscopy/imaging for monitoring muscle oxygenation and oxidative metabolism in healthy and diseased humans. *J. Biomed. Opt.* **2007**, *12*, 062105. [CrossRef] [PubMed]
6. Jobsis, F.F. Noninvasive, infrared monitoring of cerebral and myocardial oxygen sufficiency and circulatory parameters. *Science* **1977**, *198*, 1264–1267. [CrossRef]
7. Delpy, D.T.; Cope, M. Quantification in tissue near-infrared spectroscopy. *Philos. Trans. R. Soc. B Biol. Sci.* **1997**, *352*, 649–659. [CrossRef]
8. Ferrari, M.; Mottola, L.; Quaresima, V. Principles, techniques, and limitations of near infrared spectroscopy. *Can. J. Appl. Physiol.* **2004**, *29*, 463–487. [CrossRef]
9. Chance, B.; Dait, M.T.; Zhang, C.; Hamaoka, T.; Hagerman, F. Recovery from exercise-induced desaturation in the quadriceps muscles of elite competitive rowers. *Am. J. Physiol.* **1992**, *262*, C766–C775. [CrossRef]
10. Chance, B.; Nioka, S.; Kent, J.; McCully, K.; Fountain, M.; Greenfeld, R.; Holtom, G. Time-resolved spectroscopy of hemoglobin and myoglobin in resting and ischemic muscle. *Anal. Biochem.* **1988**, *174*, 698–707. [CrossRef]
11. Hamaoka, T.; Katsumura, T.; Murase, N.; Nishio, S.; Osada, T.; Sako, T.; Higuchi, H.; Kurosawa, Y.; Shimomitsu, T.; Miwa, M.; et al. Quantification of ischemic muscle deoxygenation by near infrared time-resolved spectroscopy. *J. Biomed. Opt.* **2000**, *5*, 102–105. [CrossRef] [PubMed]
12. Nirengi, S.; Yoneshiro, T.; Sugie, H.; Saito, M.; Hamaoka, T. Human brown adipose tissue assessed by simple, noninvasive near-infrared time-resolved spectroscopy. *Obesity (Silver Spring)* **2015**, *23*, 973–980. [CrossRef] [PubMed]
13. Beauvoit, B.; Chance, B. Time-resolved spectroscopy of mitochondria, cells and tissues under normal and pathological conditions. *Mol. Cell Biochem.* **1998**, *184*, 445–455. [CrossRef] [PubMed]
14. Lukaski, H.C.; Johnson, P.E.; Bolonchuk, W.W.; Lykken, G.I. Assessment of fat-free mass using bioelectrical impedance measurements of the human body. *Am. J. Clin. Nutr.* **1985**, *41*, 810–817. [CrossRef] [PubMed]

15. Lim, J.S.; Hwang, J.S.; Lee, J.A.; Kim, D.H.; Park, K.D.; Jeong, J.S.; Cheon, G.J. Cross-calibration of multi-frequency bioelectrical impedance analysis with eight-point tactile electrodes and dual-energy X-ray absorptiometry for assessment of body composition in healthy children aged 6-18 years. *Pediatr. Int.* **2009**, *51*, 263–268. [CrossRef] [PubMed]

16. Gibson, A.L.; Holmes, J.C.; Desautels, R.L.; Edmonds, L.B.; Nuudi, L. Ability of new octapolar bioimpedance spectroscopy analyzers to predict 4-component-model percentage body fat in Hispanic, black, and white adults. *Am. J. Clin. Nutr* **2008**, *87*, 332–338. [CrossRef] [PubMed]

17. Ryo, M.; Nakamura, T.; Nishida, M.; Takahashi, M.; Hotta, K.; Matsuzawa, Y. Development of Visceral-fat Measuring Apparatus Using Abdominal Bioelectrical Impedance. *J. Japan Soc. Stud. Obes.* **2003**, 136–142.

18. Ryo, M.; Maeda, K.; Onda, T.; Katashima, M.; Okumiya, A.; Nishida, M.; Yamaguchi, T.; Funahashi, T.; Matsuzawa, Y.; Nakamura, T.; et al. A new simple method for the measurement of visceral fat accumulation by bioelectrical impedance. *Diabetes Care* **2005**, *28*, 451–453. [CrossRef] [PubMed]

19. Ohmae, E.; Oda, M.; Suzuki, T.; Yamashita, Y.; Kakihana, Y.; Matsunaga, A.; Kanmura, Y.; Tamura, M. Clinical evaluation of time-resolved spectroscopy by measuring cerebral hemodynamics during cardiopulmonary bypass surgery. *J. Biomed. Opt.* **2007**, *12*, 062112. [CrossRef]

20. Lemoine, A.Y.; Ledoux, S.; Queguiner, I.; Calderari, S.; Mechler, C.; Msika, S.; Corvol, P.; Larger, E. Link between adipose tissue angiogenesis and fat accumulation in severely obese subjects. *J. Clin. Endocrinol. Metab.* **2012**, *97*, E775–E780. [CrossRef]

21. Ganesan, G.; Warren, R.V.; Leproux, A.; Compton, M.; Cutler, K.; Wittkopp, S.; Tran, G.; O'Sullivan, T.; Malik, S.; Galassetti, P.R.; et al. Diffuse optical spectroscopic imaging of subcutaneous adipose tissue metabolic changes during weight loss. *Int. J. Obes. (Lond)* **2016**, *40*, 1292–1300. [CrossRef] [PubMed]

22. Gunadi, S.; Leung, T.S.; Elwell, C.E.; Tachtsidis, I. Spatial sensitivity and penetration depth of three cerebral oxygenation monitors. *Biomed. Opt. Express* **2014**, *5*, 2896–2912. [CrossRef] [PubMed]

 *applied sciences*

Article

Comparison of Lipid and Water Contents by Time-domain Diffuse Optical Spectroscopy and Dual-energy Computed Tomography in Breast Cancer Patients

Etsuko Ohmae [1,*,†], Nobuko Yoshizawa [2,†], Kenji Yoshimoto [1], Maho Hayashi [2], Hiroko Wada [1], Tetsuya Mimura [1], Yuko Asano [3], Hiroyuki Ogura [3], Yutaka Yamashita [1], Harumi Sakahara [2] and Yukio Ueda [1]

[1] Central Research Laboratory, Hamamatsu Photonics K.K., Hamamatsu 434-8601, Japan; k-yoshimoto@crl.hpk.co.jp (K.Y.); hiroko.wada@crl.hpk.co.jp (H.W.); tetsuya.mimura@crl.hpk.co.jp (T.M.); yutaka@crl.hpk.co.jp (Y.Y.); ueda@crl.hpk.co.jp (Y.U.)

[2] Department of Diagnostic Radiology & Nuclear Medicine, Hamamatsu University School of Medicine, Hamamatsu 431-3192, Japan; nonchan@hama-med.ac.jp (N.Y.); mh1318241@gmail.com (M.H.); sakahara@hama-med.ac.jp (H.S.)

[3] Department of Breast Surgery, Hamamatsu University School of Medicine, Hamamatsu 431-3192, Japan; D17003@hama-med.ac.jp (Y.A.); h.ogura@hama-med.ac.jp (H.O.)

* Correspondence: etuko-o@crl.hpk.co.jp; Tel.: +81-53-586-7111

† Both of these authors contributed equally to this work.

Received: 31 January 2019; Accepted: 4 April 2019; Published: 9 April 2019

Abstract: We previously compared time-domain diffuse optical spectroscopy (TD-DOS) with magnetic resonance imaging (MRI) using various water/lipid phantoms. However, it is difficult to conduct similar comparisons in the breast, because of measurement differences due to modality-dependent differences in posture. Dual-energy computed tomography (DECT) examination is performed in the same supine position as a TD-DOS measurement. Therefore, we first verified the accuracy of the measured fat fraction of fibroglandular tissue in the normal breast on DECT by comparing it with MRI in breast cancer patients (n = 28). Then, we compared lipid and water signals obtained in TD-DOS and DECT from normal and tumor-tissue regions (n = 16). The TD-DOS breast measurements were carried out using reflectance geometry with a source–detector separation of 3 cm. A semicircular region of interest (ROI), with a transverse diameter of 3 cm and a depth of 2 cm that included the breast surface, was set on the DECT image. Although the measurement area differed between the modalities, the correlation coefficients of lipid and water signals between TD-DOS and DECT were rs = 0.58 ($p < 0.01$) and rs = 0.90 ($p < 0.01$), respectively. These results indicate that TD-DOS captures the characteristics of the lipid and water contents of the breast.

Keywords: diffuse optical spectroscopy; time-resolved spectroscopy; breast cancer

---

1. Introduction

Near-infrared spectroscopy (NIRS), in the range of 650–1000 nm, is widely used to noninvasively measure the concentration of light absorbing substances such as hemoglobin, water, and lipids in living tissues. In breast cancer studies, NIRS has been used to monitor tumor response to neoadjuvant chemotherapy (NAC), as hemoglobin, water, and lipid proportions reflect microvasculature, cellular metabolism, angiogenesis, edema, hypoxia, and necrosis [1–7]. These parameters are significantly altered in tumor growth and regression.

The validity of NIRS systems has generally been verified using tissue simulating phantoms such as intralipid-based aqueous phantoms [8,9] and resin-based hard phantoms [10], while phantoms with various lipid and water contents have been proposed in previous studies [11–14]. Merrit et al. and Ohmae et al. used such phantoms to compare lipid and water measurements between a diffuse optical spectroscopy (DOS) system and magnetic resonance imaging (MRI), which was defined as the "gold standard" technique [11,14]. In those studies, the authors confirmed that MRI and DOS measurements of water/lipid ratios in phantoms were almost identical. Studies using breast phantoms [13] and simulations [15] have also been conducted, but the structure of the living body is more complicated. Several studies have reported on the relationship between DOS parameters and breast density on MRI and their link with breast cancer risk [3,16,17]. Although previous studies have compared DOS with other modalities such as MRI, there have been no reports comparing DOS with other modalities in the direct measurement of lipid and water signals in the human breast. Although we previously compared lipid and water measurements of the contents of a phantom between MRI and time-domain diffuse optical spectroscopy (TD-DOS) [14], comparison of these parameters in living humans is difficult, because the measurement postures required in the two modalities differ: MRI measurement is conducted with the subject in a prone (downward facing) position, while the subject is generally in a supine (upward facing) or standing position for the DOS measurement.

Computed tomography (CT) is a cross-sectional, high-resolution, three-dimensional diagnostic imaging modality that generally uses single-energy polychromatic X-rays. Its recently increased clinical utility is primarily attributed to significantly increased scan speed as a synergic effect of increased gantry rotation speed and increased longitudinal detector coverage, as well as the development of various radiation-lowering techniques to improve patient risk/benefit ratios [18]. In many cases, chest CT scans are performed as the first choice of investigation to stage breast cancers in our hospital. Although CT has an inherent limitation in terms of soft tissue differentiation because the pixel value or CT number depends entirely on the linear attenuation coefficient, which has considerable overlap between different body materials, dual-energy CT (DECT) can improve material differentiation by using two different X-ray energy spectra.

Applications of DECT can largely be divided into exploration of material-nonspecific and material-specific energy-dependent information. Both types of evaluations can be qualitative or quantitative. The former includes virtual monoenergetic imaging, effective atomic mapping, and electron density mapping. The latter includes material decomposition, material labeling, and material highlighting [18].

In this study, we evaluated the quantitative correlations between in vivo TD-DOS and DECT signals to examine the validity of TD-DOS. The CT examination was performed in the same supine position as the TD-DOS measurement was to facilitate comparisons between the two modalities. Lipid and water contents were measured on TD-DOS by assessing tissue absorption due to the vibrational overtones of the lipid C–H bonds (930 nm) and the water O–H bond (978 nm) [11]. With a three-material decomposition algorithm, DECT can quantify the fat fraction according to the attenuation differences of tissues at varying energy levels [19]. We applied the Liver Virtual Non-contrast (VNC) application program (Siemens Healthcare, Forchheim, Germany) to measure the fat fraction in the breast. The MRI-based proton density fat fraction (PDFF) technique has been shown to provide accurate quantification of the hepatic fat fraction [20], and can also be applied to breast density measurement [21]. In the first part of this study, we confirm the accuracy of DECT by comparing the MRI-measured fat fraction of fibroglandular tissue in the normal breast with that measured using DECT, with the subjects in different postures for MRI and DECT. In the second part of this study, the lipid and water contents of normal and tumor tissues of patients with breast cancer were measured by TD-DOS, and compared with the fat and fat-free fractions measured by DECT in the supine position. The fat-free fraction of the breast is predominantly water [22].

2. Materials and Methods

2.1. Patients

DECT was used to measure the fat fraction of breast cancer and the contralateral normal breasts of 71 patients between August 2017 and August 2018.

In the first part of this study, the DECT-measured fat fraction of the normal breasts was compared with our MRI measurements. We excluded 35 patients who had not undergone an MRI examination in our hospital, five patients in whom a part of the breasts was out of the field of view on the CT, one patient whose fibroglandular tissue had a massive calcification, one patient whose MRI images were of poor quality, and one other patient who had bilateral cancer. Finally, 28 patients were included in this study. Their median age was 66 years (age range: 42–78 years).

In the second part of this study, we compared TD-DOS with DECT for a measurement of the fat and water contents of breast cancer and the contralateral normal breast. Twenty-six patients underwent both DECT and TD-DOS. We excluded four patients in whom a part of the breasts was out of the field of view on CT, one patient with small breasts that prevented the region of interest (ROI) from being placed solely on the breast, four patients who had undergone chemotherapy or hormonal therapy, and one patient for whom the optical probe could not be placed in a proper position. Finally, 16 patients with a median age of 65.5 years (with an age range of: 41–78 years), were included in the second part of the study. Twelve patients had invasive ductal carcinoma, two had ductal carcinoma in situ, one had invasive lobular carcinoma, and another had mucinous carcinoma. The mean tumor thickness measured using an ultrasonography system (US; EUB-7500, Hitachi Medical Corporation, Tokyo, Japan) with a linear probe (EUP-L65, Hitachi Medical Corporation) attached to the spectroscopic probe was 9.8 mm (range: 2–19 mm). Ten patients were included in both the first and second parts of the study.

The study protocol was approved by the Ethical Review Committee of Hamamatsu University School of Medicine. All patients signed a written informed consent form.

2.2. Magnetic Resonance Imaging (MRI) Measurement

All breast MRI examinations were performed on a 3-T magnet (Discovery 750w, GE Healthcare, Waukesha, WI, USA) using an 8-channel dedicated phased-array breast coil. The patients were examined in the prone position with both breasts symmetrically positioned in the coil. The PDFF [23,24] pulse sequence used for this study was provided by the manufacturer of the MR unit (IDEAL IQ, GE Healthcare). The pulse sequence used the multipoint Dixon technique, which uses a low flip angle to limit T1 bias, acquires six echoes to correct for T2* effects, and uses a multipeak fat model. The parameters of this sequence included: Repetition time, 8.3 ms; shortest echo time, 3.2 ms; field of view, 35 × 35 cm; matrix, 160 × 160; bandwidth, 111.1 kHz; flip angle, 4°; section thickness, 7 mm; and a single dimensional image with 24 sections. The images were processed using the software provided by the manufacturer (IDEAL IQ, GE Healthcare) to instantaneously create water, fat, R2*, and fat fraction maps.

To measure the fat fraction of fibroglandular tissue in a normal breast, a ROI was placed over the entire fibroglandular tissue shown as an intermediate intensity region at the level of the nipple on a transverse image of the fat fraction maps (Figure 1a). The fat fraction was calculated using a medical imaging system (SYNAPSE version 4.1, Fujifilm Medical, Tokyo, Japan).

2.3. Dual-Energy Computed Tomography (DECT) Measurement

All CT examinations were performed on a dual-source multidetector-row scanner (SOMATOM Definition FLASH system, Siemens Healthcare). All patients lay in the supine position with hands raised on the scanner table. After the acquisition of lateral and anteroposterior topograms, each patient was scanned in the craniocaudal direction with their breath held after deep inspiration. Dual-energy images were acquired with an automatic exposure control system (Care dose 4D; Siemens Healthcare; Table 1).

Image data with two different tube potentials (100 kVp and 140 kVp) were processed with dual-energy software (Syngovia version VB10B; Siemens Healthcare) running on a workstation. The Liver VNC application program (Siemens Healthcare) consists of a three-material decomposition algorithm analyzing soft tissue, fat, and iodine, and their relative changes of attenuation at different peak voltages [19]. This application normally uses the CT number of the liver as the CT number of soft tissue. In this study, the CT number of breast cancer was used instead of that of the liver, with this being determined in a previous study [25]. Assuming that every voxel in the field is composed of fat, soft tissue, and iodine, this spectral algorithm generates a map that encodes not only the iodine distribution, but also the fat distribution.

In the first part of the study, a ROI was placed over the entire fibroglandular tissue, shown as an intermediate density region at the level of the nipple on a transverse image, to measure the fat fraction of fibroglandular tissue in normal breast by DECT (Figure 1b).

To measure the fat content of breast cancer by DECT in the second part of the study, an image showing the largest diameter of the tumor was selected. In general, the measurement depth of the NIRS is stated as half of the source–detector separation distance. However, Yoshizawa et al. reported that the influence of the chest wall was small, and that the tHb concentration was constant, when the skin-to-chest wall distance was 2 cm or more for the TD-DOS source–detector separation of 3 cm [26]. According to the relationship between the skin-to-chest wall distance and tHb, a semicircular ROI with a transverse diameter of 3 cm and a depth of 2 cm was placed over the tumor (Figure 1c). In the second part of the study, the fat fraction of normal breast was measured by placing the same ROI in a symmetrical area in the contralateral breast (Figure 1d). The remaining components of the fat-free fraction were calculated by subtracting the fat fraction from 100%. Ding et al. measured twenty pairs of postmortem breast samples using chemical analysis as a gold standard [22]. They reported that the breast consisted mainly of water, lipid, and protein, whose measured values were about 10–70%, 25–90%, and 0–10%, respectively. In this study, the fat-free fraction of the breast is considered to be a parameter related to water content, as the breast is low in protein.

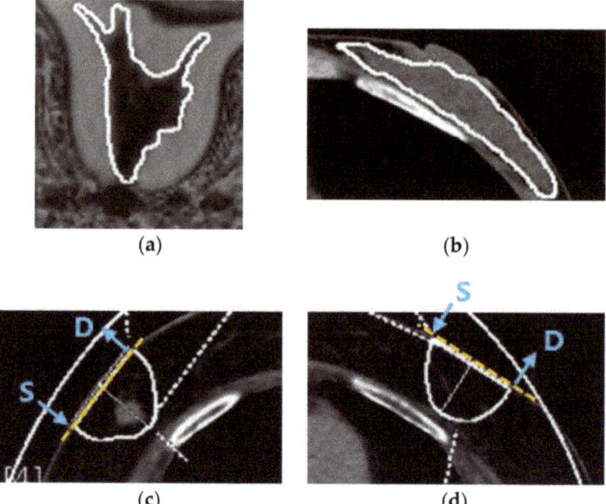

Figure 1. ROIs on MRI and DECT. (**a**) ROI on an MRI image of fibroglandular tissue in a normal breast in the first part of the study. (**b**) ROI on a DECT image of fibroglandular tissue in a normal breast in the first part of the study. (**c**) ROI on a DECT image of breast cancer in the second part of the study. (**d**) ROI on a DECT image in a symmetrical area of the contralateral breast in the second part of the study. Yellow dashed lines show the measurement axes for TD-DOS. Blue arrows indicate the position of the source (S) and detector (D) fiber bundles.

Table 1. DECT examination protocol.

	Plain Dual-Energy-CT
Tube voltage (kV)	100/Sn140
Tube current	CT-auto exposure control
Quality ref. (mAs)	250/193
Rotation time (s)	0.5
Detector configuration (mm)	32 × 0.6 = 19.2
Pitch factor	0.8
Recon kernel	D30f medium smooth
Slice (mm)	1
Increment (mm)	1
Field of view (mm)	332

2.4. Time-Domain Diffuse Optical Spectroscopy (TD-DOS) Measurement

Our TD-DOS system (TRS-21-6W) uses the time-correlated single photon counting (TCSPC) method to measure the temporal profile of the detected light. The TRS-21-6W consists of two light source units, two photodetector units, a single photon counting (SPC) circuit, and optical fiber bundles. The light source units (custom-designed, Hamamatsu Photonics K.K., Hamamatsu, Japan) contain three laser diodes per unit (762 nm, 802 nm, and 838 nm for unit 1; 908 nm, 936 nm, and 976 nm for unit 2). Each laser diode emits a pulse 100 ps to 200 ps wide (full width at half maximum) at a repetition rate of 5 MHz. The specimen is irradiated with the pulsed laser through a 1-mm diameter source optical fiber bundle with a numerical aperture (NA) of 0.29 (Sumita Optical Glass, Inc., Saitama, Japan). The light propagating through the specimen is collected by a 3-mm diameter detection fiber bundle with an NA of 0.29 (Sumita Optical Glass, Inc.), and is guided to two photodetector units containing different photomultiplier tubes (GaAs and InGaAs PMT, Hamamatsu Photonics K.K.). The detected light is converted into an electrical signal by the photodetector units, and processed with a custom-designed SPC circuit consisting of a constant fraction discriminator, time-to-amplitude converter, an analog-to-digital converter, and histogram memory. The processed signal is acquired as a temporal profile of detected light.

The TRS-21-6W can quantify the optical properties of a specimen measured in the reflectance and transmittance geometry by analyzing the temporal profile according to photon diffusion theory. An estimated profile is calculated by convolving the instrument response function (IRF) with an analytical solution of the photon diffusion equation for a homogeneous medium, and then the μ'_s and μ_a of the specimen are determined by fitting the estimated profile to the measured profile with a non-linear least squares method based on the Levenberg-Marquardt algorithm. The details of analyzing the temporal profiles have been described previously [14].

The major chromophores in breast tissue at near-infrared wavelengths are oxy-hemoglobin (O_2Hb), deoxy-hemoglobin (HHb), water, and lipid [27], and the absorption coefficient at the wavelength λ is expressed as follows:

$$\mu_a(\lambda) = \varepsilon_{O_2Hb}(\lambda)C_{O_2Hb} + \varepsilon_{HHb}(\lambda)C_{HHb} + \mu_{a,water}^{100\%}(\lambda)V_{water} + \mu_{a,lipid}^{100\%}(\lambda)V_{lipid}, \tag{1}$$

where $\varepsilon_m(\lambda)$ and C_m are the molar extinction coefficient at the wavelength λ, and the concentration of the substance m, respectively, and $\mu_{a,n}^{100\%}(\lambda)$ and V_n are the absorption coefficient of the pure solution at the wavelength λ and the volume fraction of the substance n, respectively. Since $\varepsilon_m(\lambda)$ and $\mu_{a,n}^{100\%}(\lambda)$ can be measured with a spectrophotometer, the concentrations and volume fractions are determined from $\mu_a(\lambda)$ by solving the system of linear equations in Equation (1). Total hemoglobin (tHb) is derived as the sum of O_2Hb and HHb. Tissue oxygen saturation (StO_2) is defined as the ratio of O_2Hb to tHb.

The molar extinction coefficients of O_2Hb and HHb were taken from the literature by Matcher et al. [28]. The absorption spectrum of lipid that we used was reported by van Veen et al. [29]. They measured the absorption spectrum of clear purified oil obtained from lard above 36 °C. Differences in

153

absorption spectrum have been reported according to the type of oil (e.g., lard, soybean oil, mineral oil, or cod liver oil), and this affects the estimation of its lipid content. We used the absorption spectrum of oil from mammals close to humans. The absorption spectrum of free water was measured with a spectrophotometer (U-3500, Hitachi High-Technologies Corporation, Tokyo, Japan). The spectrum of water in vivo depends on its bonding state (i.e., free water vs. bound water) [30]. Chung et al. used this slight spectral difference to show the degree of in vivo water binding as the bound water index (BWI), and reported a positive correlation between this BWI and tumor histopathological grade. Therefore, the volume fractions of water and lipid measured by TRS-21-6W are relative values compared with those of pure solutions, and not exact values [31]. Figure 2 shows the normalized absorption spectra of O_2Hb, HHb, pure water and purified lard in the wavelength range of 700–1000 nm.

The breast measurements were carried out using reflectance geometry with a source–detector separation of 3 cm. To obtain the optical properties and ultrasound image of the breast simultaneously, we attached the optical fiber bundles to the ultrasound probe, as the resulting optical path is orthogonal to the ultrasound image. The optical properties along the transverse axis, and the ultrasound image in the sagittal plane, were obtained using the combined probe (Figure 1c,d). The measurements were performed at a position just above the tumor and in the symmetrical position on the healthy breast.

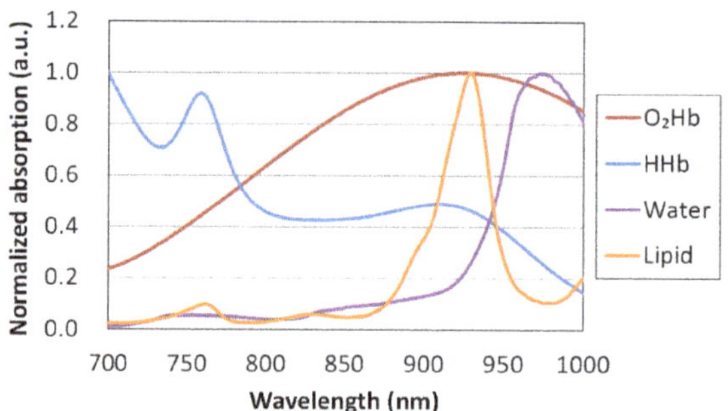

Figure 2. Absorption spectra of oxy-hemoglobin (O_2Hb), deoxy-hemoglobin (HHb), pure water, and purified lard.

2.5. Statistical Analysis

All statistical tests were performed using the Statistical Package for the Social Sciences (SPSS) for Microsoft Windows version 19 (IBM Corp., Armonk, NY, USA). Data normality was tested with the Shapiro-Wilk test. The relationship between the lipid content values of fibroglandular tissue in the normal breast obtained by MRI and DECT was evaluated using Pearson's correlation coefficients. Correlations of lipid and water contents between the DECT and TD-DOS parameters were evaluated using Spearman's correlation coefficients. Correlation coefficients between the lipid and water contents were compared using the Fisher z-transformation. The Wilcoxon signed rank test was used to compare quantitative data between normal and tumor tissue. A value of $p < 0.05$ was considered statistically significant.

3. Results

3.1. Comparison of the Fat Fraction of Fibroglandular Tissue in the Normal Breast between MRI and DECT in Different Positions

A positive correlation was found between the fat fraction measured by DECT and the fat fraction measured by MRI in the fibroglandular tissue of the normal breast. The slope and intercept of the regression line were 0.94 and 0.4, respectively (n = 28, $R^2 = 0.89$, $p < 0.01$; Figure 3).

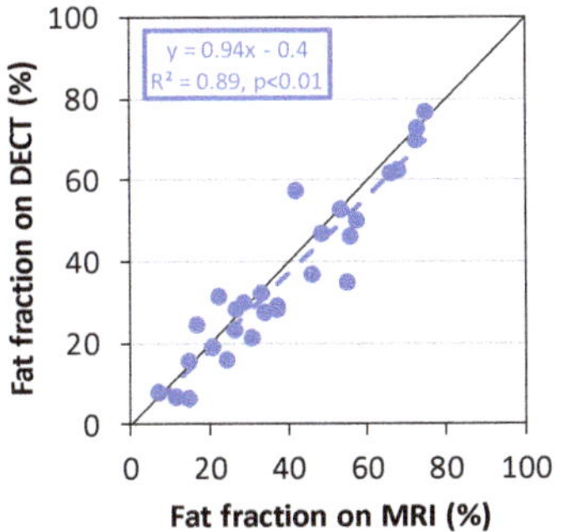

Figure 3. The fat fraction measured by DECT plotted against the fat fraction measured by MRI in the fibroglandular tissue of the normal-side breast.

3.2. Measurement by TD-DOS and DECT in the Supine Position

3.2.1. Comparison of Lipid and Water Contents between TD-DOS and DECT

Figure 4 shows an example of IRFs, measured temporal profiles, and the estimated profiles at 6 wavelengths obtained from TD-DOS measurement of the normal breast in a 41-year-old patient. The estimated profiles fit well with the measured temporal profiles. In this measurement, the concentration of O_2Hb was 14.6 µM, that of HHb was 3.9 µM, the water content was 15.4%, and the lipid content was 63.9%.

The relationship between the lipid content estimated by TD-DOS and the fat fraction estimated by DECT measured at ROIs in normal and tumor tissue is shown in Figure 5a. A positive correlation between the lipid contents obtained by the two methods was observed, with a slope of 0.58 (95% Confidence interval (CI) = 0.40–0.76) and an intercept of 23.27 (95% CI = 9.9–36.6) (n = 16, rs = 0.58, $p < 0.01$). The correlation between the water and fat-free content (Figure 5b) has a slope of 0.52 (95% CI = 0.43–0.61), an intercept of 5.36 (95% CI = 2.0–8.7) (n = 16, rs = 0.90, $p < 0.01$), and a higher correlation coefficient than that for lipid content ($p < 0.05$).

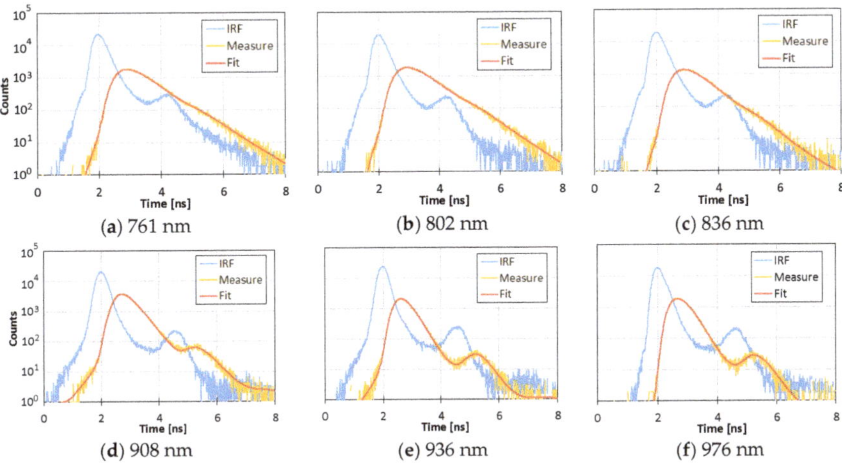

Figure 4. Examples of IRFs (blue lines), measured temporal profiles (yellow lines), and the fitting results (red lines) at 6 wavelengths (761 nm (**a**), 802 nm (**b**), 836nm (**c**), 908 nm (**d**), 936 nm (**e**), and 976 nm (**f**)) obtained from TD-DOS measurement of normal breast tissue with O_2Hb = 14.6 µM, HHb = 3.9 µM, water content = 15.4%, and lipid content = 63.9%.

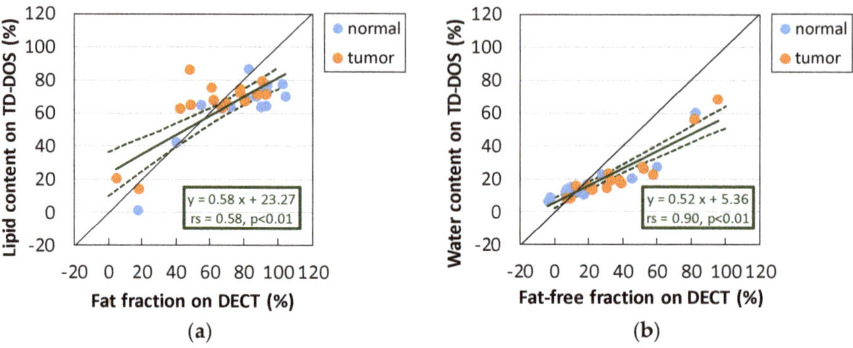

Figure 5. Measurement of normal breast tissue and breast tumor tissue in the supine position. (**a**) Relationship between lipid content estimated by TD-DOS and fat fraction estimated by DECT. (**b**) Relationship between the water content obtained by TD-DOS and the fat-free fraction obtained by DECT. The solid line on the inset is a regression line; the dashed lines represent the 95% confidence limits. (n = 16)

3.2.2. Normal Breast Tissue vs. Tumor Tissue

Figure 6 shows box plots of the various parameters obtained by TD-DOS and DECT for normal and tumor tissue. DECT fat measurements were significantly different between normal and tumor tissue ($p < 0.01$). However, TD-DOS showed a significant difference in water content and StO_2, but no difference in lipid content and tHb between normal and tumor tissue.

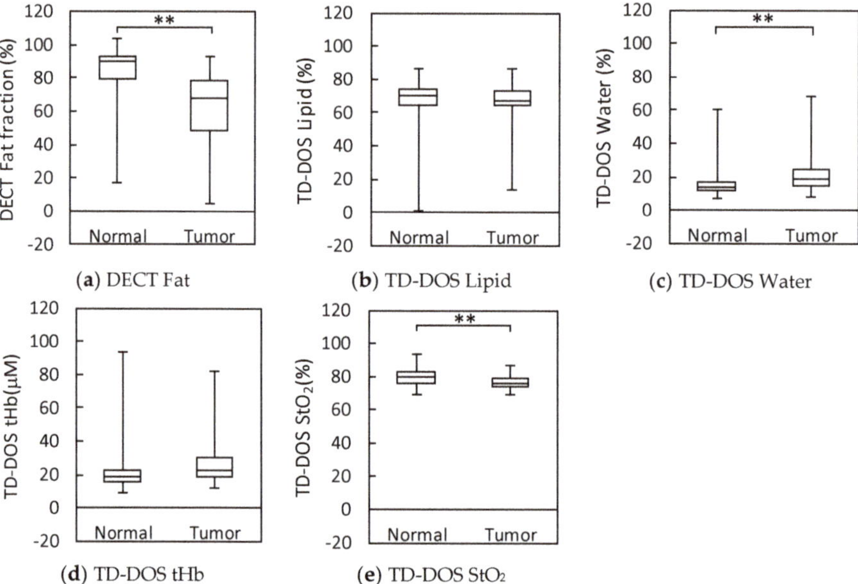

Figure 6. Box plots of parameter measurements by DECT and TD-DOS in normal and tumor tissues. (n = 16) (**a**) Fat fraction by DECT. (**b**) Lipid content by TD-DOS. (**c**) Water content by TD-DOS. (**d**) tHb by TD-DOS. (**e**) StO_2 by TD-DOS, ** $p < 0.01$ by the Wilcoxon signed rank test. Boxplots show the median (horizontal lines within the boxes), 75th and 25th percentiles (upper and lower edges of the boxes, respectively), and maximum and minimum values (upper and lower bars, respectively).

4. Discussion

The fat fractions of fibroglandular tissue in the normal breast measured by DECT and MRI in the first part of the study were comparable (Figure 3). The differences between the two measurements could be caused by differences in the body position: DECT was acquired with the subject in a supine position, whereas MRI was acquired with the subject in the prone position. Unlike MRI, DECT can be measured in the same supine position as that used in TD-DOS. Therefore, DECT was adopted as the method for comparison with TD-DOS in the second part of the study. The tumors of patients with breast cancer and the contralateral normal breasts were measured by both methods, to assess the validity of the in vivo TD-DOS lipid and water content measurements. There were two patients whose fat fractions measured by DECT were slightly over 100%. The fat fraction can show values under 0% or over 100% due to the limitation of the calculation algorithm when the CT number measurement by DECT is close to the that in pure soft tissue or in pure fat.

Our TD-DOS system measures hemoglobin, lipid, and water contents, while the DECT measures fat fraction and the remaining components as fat-free content. The fat-free fraction of living tissue is composed of water, protein, carbohydrates, and minerals, but in this breast study, the fat-free fraction was regarded as a parameter related to water content [22]. The correlation between TD-DOS and DECT was moderate for lipid content and very strong for water content (Figure 5). These results show that TD-DOS captures the characteristics of the lipid and water contents of the breast. Although the correlation for lipid was lower than that for water content, a moderate correlation was still found for the lipid. However, the 95% CI of the slope and the intercept for the lipid were wider than those for water due to the small number of patients with very low lipid levels (n = 3). The absorption peaks of lipid and water are around 936 and 976 nm, respectively. The accuracy for lipid may be lower than that for water because the absorption coefficient at 936 nm is affected by water absorption, which is also high at this wavelength. A similar trend was also shown in the absorption coefficient results from phantoms with various proportions of water and lipid in our previous study [14]. The absorption coefficient at 976 nm reflected the relative amount of water in the phantom, whereas the absorption coefficient at 936 nm did not necessarily reflect the relative amount of lipid in the phantom. Taroni et al. used seven wavelengths from 635 to 1060 nm to measure collagen as an important biomarker in a pathological classification [32]. They stated that 635 and 1060 nm were very effective wavelengths for collagen identification. They also reported that the presence or absence of collagen affects the calculation of other parameters [33]. The level of collagen varies with age, body mass index (BMI), and breast density [17,34], and is reported to be involved in tumor development and progression [32]. It is difficult to measure collagen with the wavelengths of 760–1000 nm used in this study, as the absorption of collagen is lower than that of water and lipid at those wavelengths, and the spectral peak positions of lipid and collagen are similar within this wavelength range [33]. In future study, we need to investigate whether or not individual differences in collagen affect the lipid results calculated from similar spectra using water/lipid phantoms with added collagen.

There was much less dynamic range in TD-DOS than DECT for both lipid and water. The regression slope between the TD-DOS and DECT measurements of the breast tissues was 0.5–0.6, lower than that of a water/lipid phantom [14]. There are several possible reasons for this. For DECT measurement, a semicircular ROI with depth 2 cm, including the breast surface, was set on a 1-mm-thick DECT image. In contrast, the optical path of TD-DOS is spatial and banana-shaped [35], and the measurement region differs from the ROI used on the DECT images. Furthermore, the TD-DOS method is less affected by the superficial layer, but more sensitive to deeper regions than other DOS methods [36–38]. In this study, the depth of ROI was determined as 2 cm based on the relationship between the skin-to-chest wall distance and the tHb [26]. In the case of breast measurement, the absorption in the 900 nm region, which reflects the characteristics of water and lipid, is fairly high compared with the absorption at 750–850 nm reflecting the characteristics of hemoglobin. It causes difficulty in detection of the photons which have passed through the deeper region in the 900 nm range. Therefore, it is expected that the water and lipid contents obtained by TD-DOS reflect the information within the region shallower than 2 cm. In addition, the fat and fat-free fractions obtained by DECT change according to the size of ROI due to the heterogeneity of the breast tissue. For that reason, the regression slope in Figure 5 will be improved by employing a ROI that has more appropriate depth.

Several reports have compared DOS-measured physiological parameters between normal and tumor breast tissue [4,27,39–41]. In tumor tissue, tHb and water are higher than those in normal tissue, while lipid is lower. A significant reduction in StO_2 in tumor caused by tissue hypoxia due to metabolically active tumor cells has been reported [39,41], but results showing no significant differences have also been reported [27]. High tHb reflects angiogenesis as a tumor grows, and Ueda et al. described the association between tHb level estimated by TD-DOS and mitotic count score evaluated as a proliferation marker, or SUV_{max} on FDG PET/CT [4]. High water content and low lipid content of tumors compared with normal tissue has also been reported using magnetic resonance spectroscopy (MRS) [42]. The high water content indicates edema and increased cellularity [43]. As breast tumors grow, adipose tissue is extruded, so a decrease in lipid content is predicted. Our results showed a significant difference in water and StO_2 between tumor and normal tissue. Although tHb and lipid did not show a significant difference, tHb tended to be higher in the tumor ($p = 0.063$). In this study, differences in these parameters may not have been observed because the tumors were smaller in size than those in the previous report or because the number of cases was small. When all 32 patients, including ones only measured by TD-DOS (untreated patients, excluding one case with the probe resting on the nipple, and 2 cases with a hematoma near the tumor) were analyzed using the Wilcoxon signed rank test, and significant differences in the parameters became clear as the number of cases increased. We confirmed a significant decrease in lipid ($p = 0.0043$) and an increase in tHb ($p = 0.0032$) in tumor, as described in previous reports [27,41].

The present study has some limitations. Since the number of patients was small, especially the number of patients with very low lipid content, the regression line between TD-DOS and DECT for lipid might be affected by this scarcity of lipid data. Furthermore, due to the small tumor size, some parameters (tHb and lipid) showed no significant difference between normal and tumor, unlike other cases reported in the literature.

It has been reported that skin-to-chest wall distance, depth of the tumor, and thickness of the tumor, all affect the measurement values of breast cancer [26,44]. Yoshizawa et al. proposed tHb_{net} as an index of tHb for tumors in order to improve the evaluation of tumors by reducing the effect of the skin-to-chest wall distance, which differs among individuals. This measurement used a standard curve of tHb and skin-to-chest wall distance for a normal breast [44]. In addition, a tissue optical index (TOI) combining water, lipid, and HHb has also been proposed as a contrast function for cancer and normal tissue [27]. It has been reported that a TOI used during the intermediate period of treatment is an effective predictor of clinical response to NAC [5]. In this study, we used DECT to confirm that TD-DOS reflects the characteristics of lipid and water contents in breasts in vivo. We plan to conduct a further study monitoring the effects of chemotherapy on breast cancer patients. This evaluation was conducted using single-point measurements, but it is expected that the contrast between tumor and normal tissue will be clearer using two- or three-dimensional images and multipoint measurements [4,45].

Author Contributions: Conceptualization, N.Y. and E.O.; software, K.Y., H.W., and T.M.; formal analysis, E.O., N.Y., K.Y., and M.H.; investigation, N.Y., H.W., and T.M.; resources, Y.A. and H.O.; writing—original draft preparation, E.O. and N.Y.; writing—review and editing, H.W. and K.Y.; supervision, Y.Y., H.S., and Y.U.; All authors discussed the results and contributed to the final manuscript.

Funding: This research was partially funded by Japan Society for the Promotion of Science (JSPS) KAKENHI Grant Numbers (JP15K19781 and JP17H03591).

Acknowledgments: We are grateful to Toshiaki Saeki and Shigeto Ueda, Department of Breast Oncology, International medical Center, Saitama Medical University, for useful discussions. We would like to thank Yuki Hirai and Naoko Hyodo, Department of Diagnostic Radiology & Nuclear Medicine, Hamamatsu University School of Medicine, for acquiring the consent forms from the patients for DECT and MRI.

Conflicts of Interest: The authors declare no conflict of interest.

References

1. Roblyer, D.; Ueda, S.; Cerussi, A.; Tanamai, W.; Durkin, A.; Mehta, R.; Hsiang, D.; Butler, J.A.; McLaren, C.; Chen, W.-P.; et al. Optical imaging of breast cancer oxyhemoglobin flare correlates with neoadjuvant chemotherapy response one day after starting treatment. *Proc. Natl. Acad. Sci. USA* **2011**, *108*, 14626–14631. [CrossRef] [PubMed]
2. Ueda, S.; Roblyer, D.; Cerussi, A.; Durkin, A.; Leproux, A.; Santoro, Y.; Xu, S.; O'Sullivan, T.D.; Hsiang, D.; Mehta, R.; et al. Baseline Tumor Oxygen Saturation Correlates with a Pathologic Complete Response in Breast Cancer Patients Undergoing Neoadjuvant Chemotherapy. *Cancer Res.* **2012**, *72*, 4318–4328. [CrossRef]
3. O'Sullivan, T.D.; Leproux, A.; Chen, J.-H.; Bahri, S.; Matlock, A.; Roblyer, D.; McLaren, C.E.; Chen, W.-P.; Cerussi, A.E.; Su, M.-Y.; et al. Optical imaging correlates with magnetic resonance imaging breast density and revealscomposition changes during neoadjuvant chemotherapy. *Breast Cancer Res.* **2013**, *15*, R14. [CrossRef]
4. Ueda, S.; Nakamiya, N.; Matsuura, K.; Shigekawa, T.; Sano, H.; Hirokawa, E.; Shimada, H.; Suzuki, H.; Oda, M.; Yamashita, Y.; et al. Optical imaging of tumor vascularity associated with proliferation and glucose metabolism in early breast cancer: Clinical application of total hemoglobin measurements in the breast. *BMC Cancer* **2013**, *13*, 514. [CrossRef]
5. Tromberg, B.J.; Zhang, Z.; Leproux, A.; O'Sullivan, T.D.; Cerussi, A.E.; Carpenter, P.M.; Mehta, R.S.; Roblyer, D.; Yang, W.; Paulsen, K.D.; et al. Predicting Responses to Neoadjuvant Chemotherapy in Breast Cancer: ACRIN 6691 Trial of Diffuse Optical Spectroscopic Imaging. *Cancer Res.* **2016**, *76*, 5933–5944. [CrossRef]
6. Ueda, S.; Saeki, T.; Takeuchi, H.; Shigekawa, T.; Yamane, T.; Kuji, I.; Osaki, A. In vivo imaging of eribulin-induced reoxygenation in advanced breast cancer patients: A comparison to bevacizumab. *Br. J. Cancer* **2016**, *114*, 1212–1218. [CrossRef] [PubMed]
7. Xu, C.; Vavadi, H.; Merkulov, A.; Li, H.; Erfanzadeh, M.; Mostafa, A.; Gong, Y.; Salehi, H.; Tannenbaum, S.; Zhu, Q. Ultrasound-Guided Diffuse Optical Tomography for Predicting and Monitoring Neoadjuvant Chemotherapy of Breast Cancers: Recent Progress. *Ultrason. Imaging* **2016**, *38*, 5–18. [CrossRef]
8. Spinelli, L.; Martelli, F.; Farina, A.; Pifferi, A.; Torricelli, A.; Cubeddu, R.; Zaccanti, G. Calibration of scattering and absorption properties of a liquid diffusive medium at NIR wavelengths. Time-resolved method. *Opt. Express* **2007**, *15*, 6589. [CrossRef] [PubMed]
9. Spinelli, L.; Botwicz, M.; Zolek, N.; Kacprzak, M.; Milej, D.; Sawosz, P.; Liebert, A.; Weigel, U.; Durduran, T.; Foschum, F.; et al. Determination of reference values for optical properties of liquid phantoms based on Intralipid and India ink. *Biomed. Opt. Express* **2014**, *5*, 2037. [CrossRef]
10. Pogue, B.W.; Patterson, M.S. Review of tissue simulating phantoms for optical spectroscopy, imaging and dosimetry. *J. Biomed. Opt.* **2006**, *11*, 041102. [CrossRef] [PubMed]
11. Merritt, S.; Gulsen, G.; Chiou, G.; Chu, Y.; Deng, C.; Cerussi, A.E.; Durkin, A.J.; Tromberg, B.J.; Nalcioglu, O. Comparison of Water and Lipid Content Measurements Using Diffuse Optical Spectroscopy and MRI in Emulsion Phantoms. *Technol. Cancer Res. Treat.* **2003**, *2*, 563–569. [CrossRef] [PubMed]
12. Nachabé, R.; Hendriks, B.H.W.; Desjardins, A.E.; van der Voort, M.; van der Mark, M.B.; Sterenborg, H.J.C.M. Estimation of lipid and water concentrations in scattering media with diffuse optical spectroscopy from 900 to 1600 nm. *J. Biomed. Opt.* **2010**, *15*, 037015. [CrossRef] [PubMed]
13. Michaelsen, K.E.; Krishnaswamy, V.; Shenoy, A.; Jordan, E.; Pogue, B.W.; Paulsen, K.D. Anthropomorphic breast phantoms with physiological water, lipid, and hemoglobin content for near-infrared spectral tomography. *J. Biomed. Opt.* **2014**, *19*, 026012. [CrossRef]
14. Ohmae, E.; Yoshizawa, N.; Yoshimoto, K.; Hayashi, M.; Wada, H.; Mimura, T.; Suzuki, H.; Homma, S.; Suzuki, N.; Ogura, H.; et al. Stable tissue-simulating phantoms with various water and lipid contents for diffuse optical spectroscopy. *Biomed. Opt. Express* **2018**, *9*, 5792. [CrossRef] [PubMed]
15. Prince, S.; Malarvizhi, S. Monte Carlo simulation of NIR diffuse reflectance in the normal and diseased human breast tissues. *BioFactors* **2007**, *30*, 255–263. [CrossRef] [PubMed]
16. Brooksby, B.; Pogue, B.W.; Jiang, S.; Dehghani, H.; Srinivasan, S.; Kogel, C.; Tosteson, T.D.; Weaver, J.; Poplack, S.P.; Paulsen, K.D. Imaging breast adipose and fibroglandular tissue molecular signatures by using hybrid MRI-guided near-infrared spectral tomography. *Proc. Natl. Acad. Sci. USA* **2006**, *103*, 8828–8833. [CrossRef]

17. Taroni, P.; Pifferi, A.; Quarto, G.; Spinelli, L.; Torricelli, A.; Abbate, F.; Villa, A.; Balestreri, N.; Menna, S.; Cassano, E.; et al. Noninvasive assessment of breast cancer risk using time-resolved diffuse optical spectroscopy. *J. Biomed. Opt.* **2010**, *15*, 060501. [CrossRef] [PubMed]

18. Goo, H.W.; Goo, J.M. Dual-Energy CT: New Horizon in Medical Imaging. *Korean J. Radiol.* **2017**, *18*, 555–569. [CrossRef]

19. Ascenti, G.; Mazziotti, S.; Lamberto, S.; Bottari, A.; Caloggero, S.; Racchiusa, S.; Mileto, A.; Scribano, E. Dual-Energy CT for Detection of Endoleaks After Endovascular Abdominal Aneurysm Repair: Usefulness of Colored Iodine Overlay. *Am. J. Roentgenol.* **2011**, *196*, 1408–1414. [CrossRef]

20. Idilman, I.S.; Aniktar, H.; Idilman, R.; Kabacam, G.; Savas, B.; Elhan, A.; Celik, A.; Bahar, K.; Karcaaltincaba, M. Hepatic Steatosis: Quantification by Proton Density Fat Fraction with MR Imaging versus Liver Biopsy. *Radiology* **2013**, *267*, 767–775. [CrossRef]

21. Ding, J.; Stopeck, A.T.; Gao, Y.; Marron, M.T.; Wertheim, B.C.; Altbach, M.I.; Galons, J.-P.; Roe, D.J.; Wang, F.; Maskarinec, G.; et al. Reproducible automated breast density measure with no ionizing radiation using fat-water decomposition MRI. *J. Magn. Reson. Imaging JMRI* **2018**, *48*, 971–981. [CrossRef]

22. Ding, H.; Ducote, J.L.; Molloi, S. Measurement of breast tissue composition with dual energy cone-beam computed tomography: A postmortem study: Breast composition measurement with dual energy CBCT. *Med. Phys.* **2013**, *40*, 061902. [CrossRef]

23. Serai, S.D.; Dillman, J.R.; Trout, A.T. Proton Density Fat Fraction Measurements at 1.5- and 3-T Hepatic MR Imaging: Same-Day Agreement among Readers and across Two Imager Manufacturers. *Radiology* **2017**, *284*, 244–254. [CrossRef]

24. Hernando, D.; Sharma, S.D.; Aliyari Ghasabeh, M.; Alvis, B.D.; Arora, S.S.; Hamilton, G.; Pan, L.; Shaffer, J.M.; Sofue, K.; Szeverenyi, N.M.; et al. Multisite, multivendor validation of the accuracy and reproducibility of proton-density fat-fraction quantification at 1.5T and 3T using a fat-water phantom: Proton-Density Fat-Fraction Quantification at 1.5T and 3T. *Magn. Reson. Med.* **2017**, *77*, 1516–1524. [CrossRef]

25. Okamura, Y.; Yoshizawa, N.; Yamaguchi, M.; Kashiwakura, I. Application of Dual-Energy Computed Tomography for Breast Cancer Diagnosis. *Int. J. Med. Phys. Clin. Eng. Radiat. Oncol.* **2016**, *05*, 288–297. [CrossRef]

26. Yoshizawa, N.; Ueda, Y.; Nasu, H.; Ogura, H.; Ohmae, E.; Yoshimoto, K.; Takehara, Y.; Yamashita, Y.; Sakahara, H. Effect of the chest wall on the measurement of hemoglobin concentrations by near-infrared time-resolved spectroscopy in normal breast and cancer. *Breast Cancer* **2016**, *23*, 844–850. [CrossRef]

27. Cerussi, A.; Shah, N.; Hsiang, D.; Durkin, A.; Butler, J.; Tromberg, B.J. In vivo absorption, scattering, and physiologic properties of 58 malignant breast tumors determined by broadband diffuse optical spectroscopy. *J. Biomed. Opt.* **2006**, *11*, 044005. [CrossRef]

28. Matcher, S.J.; Elwell, C.E.; Cooper, C.E.; Cope, M.; Delpy, D.T. Performance comparison of several published tissue near-infrared spectroscopy algorithms. *Anal. Biochem.* **1995**, *227*, 54–68. [CrossRef]

29. van Veen, R.L.P.; Sterenborg, H.J.C.M.; Pifferi, A.; Torricelli, A.; Chikoidze, E.; Cubeddu, R. Determination of visible near-IR absorption coefficients of mammalian fat using time- and spatially resolved diffuse reflectance and transmission spectroscopy. *J. Biomed. Opt.* **2005**, *10*, 054004. [CrossRef]

30. Chung, S.H.; Cerussi, A.E.; Klifa, C.; Baek, H.M.; Birgul, O.; Gulsen, G.; Merritt, S.I.; Hsiang, D.; Tromberg, B.J. *In vivo* water state measurements in breast cancer using broadband diffuse optical spectroscopy. *Phys. Med. Biol.* **2008**, *53*, 6713–6727. [CrossRef]

31. Cerussi, A.E.; Tanamai, V.W.; Hsiang, D.; Butler, J.; Mehta, R.S.; Tromberg, B.J. Diffuse optical spectroscopic imaging correlates with final pathological response in breast cancer neoadjuvant chemotherapy. *Philos. Trans. R. Soc. Math. Phys. Eng. Sci.* **2011**, *369*, 4512–4530. [CrossRef]

32. Taroni, P.; Paganoni, A.M.; Ieva, F.; Pifferi, A.; Quarto, G.; Abbate, F.; Cassano, E.; Cubeddu, R. Non-invasive optical estimate of tissue composition to differentiate malignant from benign breast lesions: A pilot study. *Sci. Rep.* **2017**, *7*, 40683. [CrossRef]

33. Taroni, P.; Comelli, D.; Pifferi, A.; Torricelli, A.; Cubeddu, R. Absorption of collagen: Effects on the estimate of breast composition and related diagnostic implications. *J. Biomed. Opt.* **2007**, *12*, 014021. [CrossRef]

34. Taroni, P.; Quarto, G.; Pifferi, A.; Abbate, F.; Balestreri, N.; Menna, S.; Cassano, E.; Cubeddu, R. Breast Tissue Composition and Its Dependence on Demographic Risk Factors for Breast Cancer: Non-Invasive Assessment by Time Domain Diffuse Optical Spectroscopy. *PLoS ONE* **2015**, *10*, e0128941. [CrossRef]

35. Gratton, G.; Maier, J.S.; Fabiani, M.; Mantulin, W.W.; Gratton, E. Feasibility of intracranial near-infrared optical scanning. *Psychophysiology* **1994**, *31*, 211–215. [CrossRef]

36. Sato, C.; Yamaguchi, T.; Seida, M.; Ota, Y.; Yu, I.; Iguchi, Y.; Nemoto, M.; Hoshi, Y. Intraoperative monitoring of depth-dependent hemoglobin concentration changes during carotid endarterectomy by time-resolved spectroscopy. *Appl. Opt.* **2007**, *46*, 2785–2792. [CrossRef]

37. Gunadi, S.; Leung, T.S.; Elwell, C.E.; Tachtsidis, I. Spatial sensitivity and penetration depth of three cerebral oxygenation monitors. *Biomed. Opt. Express* **2014**, *5*, 2896. [CrossRef]

38. Koga, S.; Poole, D.C.; Kondo, N.; Oue, A.; Ohmae, E.; Barstow, T.J. Effects of increased skin blood flow on muscle oxygenation/deoxygenation: Comparison of time-resolved and continuous-wave near-infrared spectroscopy signals. *Eur. J. Appl. Physiol.* **2015**, *115*, 335–343. [CrossRef]

39. Ntziachristos, V.; Yodh, A.G.; Schnall, M.D.; Chance, B. MRI-Guided Diffuse Optical Spectroscopy of Malignant and Benign Breast Lesions. *Neoplasia* **2002**, *4*, 347–354. [CrossRef]

40. Grosenick, D.; Wabnitz, H.; Moesta, K.T.; Mucke, J.; Schlag, P.M.; Rinneberg, H. Time-domain scanning optical mammography: II. Optical properties and tissue parameters of 87 carcinomas. *Phys. Med. Biol.* **2005**, *50*, 2451–2468. [CrossRef]

41. Wang, J.; Pogue, B.W.; Jiang, S.; Paulsen, K.D. Near-infrared tomography of breast cancer hemoglobin, water, lipid, and scattering using combined frequency domain and cw measurement. *Opt. Lett.* **2010**, *35*, 82. [CrossRef]

42. Thakur, S.B.; Brennan, S.B.; Ishill, N.M.; Morris, E.A.; Liberman, L.; Dershaw, D.D.; Bartella, L.; Koutcher, J.A.; Huang, W. Diagnostic usefulness of water-to-fat ratio and choline concentration in malignant and benign breast lesions and normal breast parenchyma: An in vivo 1H MRS study. *J. Magn. Reson. Imaging* **2011**, *33*, 855–863. [CrossRef]

43. Chung, S.H.; Yu, H.; Su, M.-Y.; Cerussi, A.E.; Tromberg, B.J. Molecular imaging of water binding state and diffusion in breast cancer using diffuse optical spectroscopy and diffusion weighted MRI. *J. Biomed. Opt.* **2012**, *17*, 071304. [CrossRef]

44. Yoshizawa, N.; Ueda, Y.; Mimura, T.; Ohmae, E.; Yoshimoto, K.; Wada, H.; Ogura, H.; Sakahara, H. Factors affecting measurement of optic parameters by time-resolved near-infrared spectroscopy in breast cancer. *J. Biomed. Opt.* **2018**, *23*, 1–6.

45. Yoshimoto, K.; Ohmae, E.; Yamashita, D.; Suzuki, H.; Homma, S.; Mimura, T.; Wada, H.; Suzuki, T.; Yoshizawa, N.; Nasu, H.; et al. Development of time-resolved reflectance diffuse optical tomography for breast cancer monitoring. In *Proceedings of the Optical Tomography and Spectroscopy of Tissue XII*; International Society for Optics and Photonics: San Diego, CA, USA, 2017; Volume 10059, p. 100590M.

Article

A Versatile Setup for Time-Resolved Functional Near Infrared Spectroscopy Based on Fast-Gated Single-Photon Avalanche Diode and on Four-Wave Mixing Laser

Laura Di Sieno [1], Alberto Dalla Mora [1], Alessandro Torricelli [1,2], Lorenzo Spinelli [2], Rebecca Re [1,2], Antonio Pifferi [1,2] and Davide Contini [1,*

[1] Politecnico di Milano, Dipartimento di Fisica, Piazza Leonardo da Vinci 32, I-20133 Milano, Italy;
laura.disieno@polimi.it (L.D.S.); alberto.dallamora@polimi.it (A.D.M.); alessandro.torricelli@polimi.it (A.T.);
rebecca.re@polimi.it (R.R.); antonio.pifferi@polimi.it (A.P.)

[2] Consiglio Nazionale delle Ricerche, Istituto di Fotonica e Nanotecnologie, Piazza Leonardo da Vinci 32,
20133 Milano-I, Italy; lorenzo.spinelli@polimi.it

*   Correspondence: davide.contini@polimi.it; Tel.: +39-02-23996148

Received: 15 April 2019; Accepted: 4 June 2019; Published: 10 June 2019

Abstract: In this paper, a time-domain fast gated near-infrared spectroscopy system is presented. The system is composed of a fiber-based laser providing two pulsed sources and two fast gated detectors. The system is characterized on phantoms and was tested in vivo, showing how the gating approach can improve the contrast and contrast-to-noise-ratio for detection of absorption perturbation inside a diffusive medium, regardless of source-detector separation.

Keywords: time-domain NIRS; null source-detector separation; brain; diffuse optics

1. Introduction

Among the fascinating applications of lasers, the non-invasive/non-destructive analysis of diffusive samples [1] (such as biological tissues, fruit, chemicals and wood [2–4]) has recently gained a lot of interest for the promising exploitations in biomedicine, in the food sector and in industrial fields. Photons injected in a diffusive medium undergo many scattering events due to refractive index mismatches at the micro and meso scale. Moreover, if light absorption is weak, like what happens in the so-called "therapeutic window" (600 to 1100 nm), where biological tissues are almost transparent to red and near infrared light, diffusing photons can easily be reemitted at the medium boundaries after having travelled deep in the tissue (e.g., >1 cm below skin in biological tissue in a reflectance configuration). The information carried by diffusing photons relates to the structural properties (linked to the scattering coefficient) and to the chemical properties (linked to the absorption coefficient) of the medium. For biomedical applications, the main chromophores of interest in the range between 600–850 nm are oxygenated and deoxygenated hemoglobin (O_2Hb and HHb, respectively) [5]. For this reason, diffuse optical spectroscopy is a promising tool to assess brain and muscle oxygenation. In particular, functional near infrared spectroscopy (fNIRS) exploits the difference in absorption spectrum between HHb and O_2Hb to recover their concentration over time, thus enabling the ability to see variation in haemodynamics, which can be a fingerprint of a brain activation thanks to the neurovascular coupling mechanism.

While the use of steady state light sources and detectors enables the so-called continuous wave (CW) approach to fNIRS, the time-resolved (TR) approach relies on the injection in the sample of short laser pulses (few tens of ps duration) and on the recording of the arrival time of re-emitted photons (distribution of time of flight, DTOF) at a given distance, the so-called source-detector distance

(SDD). It has been demonstrated that the TR-technique provides a higher information content than CW, in particular when a single SDD is used [6]. Indeed, it is possible to recover the optical properties (absorption coefficient -μ_a- and reduced scattering coefficient -μ_s'), after fitting of the DTOF with a proper theoretical model [7]. Moreover, when using the TR technique in the so-called "reflectance geometry" (i.e., injection and collection put on the same side of the sample), the depth reached by photons is intrinsically encoded in their arrival time—the later the photon is re-emitted, the longer is its pathlength, thus increasing significantly the probability of the photon to have probed a deeper region [8]. It has also been demonstrated that the mean penetration depth of photons is independent of the SDD. For this reason, the null-SDD approach was proposed some years ago [9]. The use of short (or even null) SDD improves the confinement of photons, thus resulting in a higher spatial resolution and number of photons recorded, as well as a larger contrast at all times. To apply this approach, however, a fast-gated detector (i.e., a detector that can be turned on/off in a few hundred ps) is needed to prevent the detector saturation (or its damage). To meet this requirement, fast-gated (FG) single-photon avalanche diodes (SPADs) have been proposed in the literature [10,11]. The use of a FG-detector permits also the ability to reconstruct the DTOF with a larger dynamic range [12] and this great potentiality can be exploited if the laser source is powerful enough to increase the signal at late times, thus allowing the ability to probe deeper regions [13]. Additionally, the possibility to collect photons only at a given portion of the DTOF (by using a classical time-correlated single photon counting board or with simpler devices [14]) allows for the ability to increase the number of late photons, thus improving the visibility of possible deep inhomogeneities in the medium which affect the DTOF shape.

Different works based on FG-SPADs have been published, demonstrating experimentally that an improved spatial resolution, contrast and even chromophore quantification can be achieved using a gated detector [15–17]. However, to the best of our knowledge, the FG-SPAD detectors have never been adopted to demonstrate the improvement given by the use of short-SDD in fNIRS measurements. Indeed, the higher spatial resolution allows for the ability to retrieve more precisely the area wherein the activation occurs and the capability to collect photons in a given portion of the DTOF, improving the robustness of the detectability of a brain activation.

In this paper, we propose an instrument [18] based on a prototypal dual-channel (710 and 820 nm) high power (around 100 mW output for each channel) laser which allows for the ability to exploit at its best capacity the high-dynamic range acquisition technique. The laser light was spread over a sufficiently large area in order to guarantee skin safety by means of a spacer, while eye safety was guaranteed by means of the use of proper protection. Two state-of-the-art fast-gated SPAD modules (featuring a diffusion tail of about 90 ps [19]) were adopted to monitor both wavelengths, thus allowing for the recovery of the concentration of both O_2Hb and HHB simultaneously. The proposed system was firstly characterized using some tests from two internationally-agreed protocols for performance assessment of TR-instrument/oximeter (such as the Instrument Response Function/IRF) from the "Basic Instrumental Protocol" (BIP) protocol [20], and contrast and contrast-to-noise ratio (CNR) from "nEUROPt" ones [21]). We then exploited the system for fNIRS measurements with the aim to demonstrate the improvement in terms of detectability of brain activation with respect to the classical oximetry instrument [22,23], simulating a free-running (i.e., not-gated) acquisition.

2. Materials and Methods

2.1. Experimental Setup

The setup used in the measurements is schematically depicted in Figure 1a. The laser source was a prototypal four-wave mixing laser (Fianium Ltd, Southampton, UK), providing two separate laser beams (710 and 820 nm) with a temporal duration of about 25 ps full width at half maximum (FWHM) at the repetition rate of 40 MHz. Each beam was properly attenuated by means of the variable optical attenuator (VOA, Edmund Optics Inc., Barrington, NJ, USA) and light was collimated into a 400 μm fiber by means of a collimator. Collimated beams were then injected into the sample and retro-diffused photons were collected through two 1 mm core fibers (with a NA of 0.37) at a given source-detector distance (5 or 30 mm).

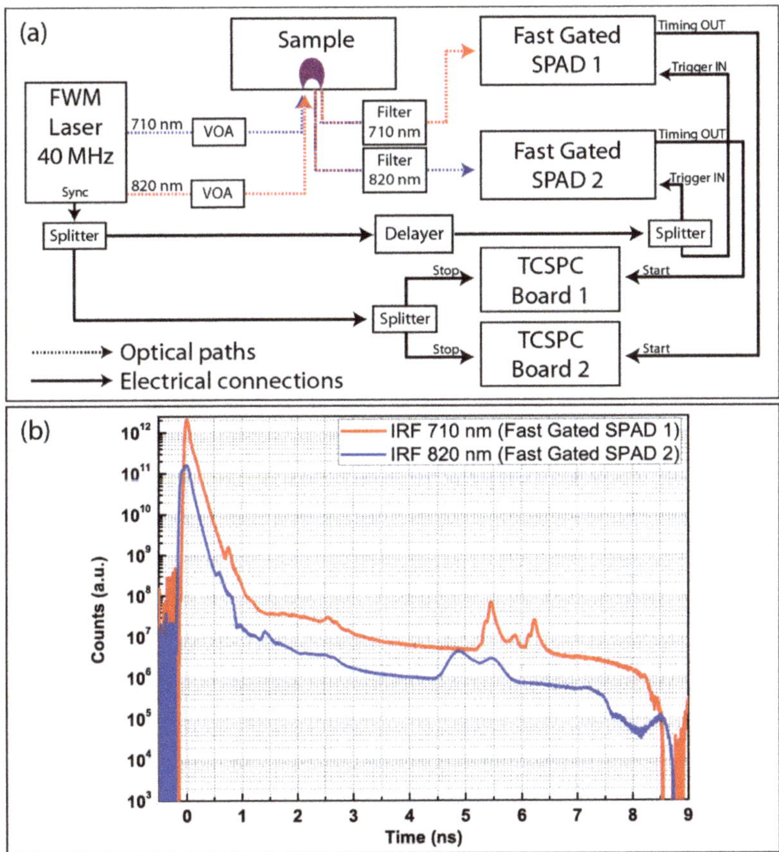

Figure 1. (a) Schematics of the setup used for the in vivo experiment. (b) Instrumental response function recorded for both fast-gated single-photon avalanche diode (SPAD) modules at their operating wavelengths (710 nm for fast-gated SPAD (FG-SPAD) 1—red line; and 820 nm for FG-SPAD 2—blue line). Time bin size: 3.05 ps. IRF = instrument response function, VOA = variable optical attenuator, TCSPC = time-correlated single photon counting. FWM: Four Wave Mixing.

In the in vivo measurements, it was fundamental to disentangle the contribution of the two wavelengths in order to recover the concentration of both oxygenated and deoxygenated hemoglobin. For this reason, an interferential filter (centered at 710 or 820 nm) was put at the other tip of each collection fiber. Such a solution allowed for the ability to separate the two beams and send each of them to a different fast-gated SPAD (FG-SPAD) module, thus having a parallel acquisition of the two wavelengths. The arrangement of the launching and detection fiber was conceived so as to probe the same area with both source-detector couples.

The FG-SPAD modules were state-of-the-art time-gated detectors, having an active area of 100 µm diameter and nearly 200 ps of raising edge for the gate opening. The full description of the FG-SPAD modules can be found in Boso et al. [10].

Concurrently with the optical pulse, the laser provided a synchronism signal, which was split into two paths. The former was fed to the FG-SPAD modules to trigger the enabling of the detector for a given period of time (5 ns). In order to acquire different "slices" of the DTOF, the detector was enabled at given delays with respect to the peak of the reflectance curve by means of a home-made delayer. It was a transmission-line based one, and it provided delays from 0 up to 6375 ps at a step of 25 ps.

On the other side, the second path of the split sync was sent to the time-correlated single photon counting (TCSPC) electronics, feeding the "stop" signal for both acquiring boards (SPC-130, Becker&Hickl GmbH, Berlin, Germany). The "start" signal for the TCSPC board was provided by each FG-SPAD module. Indeed, each module generated a pulse (timing OUT signal in Figure 1a) when a photon was detected within each SPAD, thus triggering an avalanche.

2.2. Performance Assessment

In order to impartially assess the performances of the proposed setup, we made use of some parameters of two internationally agreed protocols for diffuse optics instrumentation (the "Basic Instrumental Protocol" (BIP) [20] and "nEUROPt" [21] ones). Their aim is to characterize the TR setup for brain imaging.

For what concerns the BIP protocol, we acquired the IRF of the proposed setup at both wavelengths, coupling the launching and collection fiber with a diffusive medium (a thin film of Teflon^{TM}) so as to excite all propagation modes of the detection fibers. In order to have a high-dynamic range measurement, we adopted the technique explained in Dalla Mora et al. and Tosi et al. [12,24], acquiring the re-emitted photons at different delays (from 0 up to 3450 ps at step of 25 ps) with an integration time of each delay of 10 s. Before rescaling the acquisitions and reconstructing the IRF, we subtracted the constant background (the so-called "dark count", which is not correlated with laser light) to all curves. We finally computed the FWHM and the dynamic range of the reconstructed IRF, as suggested by the BIP protocol.

We then characterized the proposed setup at 820 nm using two features of the "nEUROPt" protocol (contrast and contrast-to-noise ratio (CNR)) since they are meant to define quantitatively the capability of detecting a small absorption change within the probed volume (e.g., absorption changes in the cortex due to brain activation).

For both tests, we made use of a liquid phantom composed of a mix of water, Intralipid 20%, and Indian ink in such a proportion to have an absorption coefficient (μ_a) of 0.1 cm^{-1} and a reduced scattering one of 10 cm^{-1} at 820 nm. In order to simulate the absorption changes related to a localized brain activation, we made use of solid black cylindrical inclusions (height: 4.5 mm; diameter: 5.32 mm). In Martelli et al. [25], the equivalence relation between a realistic absorption change in a finite volume and the perturbation produced by a small black object had been demonstrated. In our case, we performed all tests using the 100 mm^3 totally absorbing inclusion (which corresponded to a $\Delta\mu_a$ of 0.15 cm^{-1} over a 1 cm^3 volume). The inclusion was posed beneath the surface of the phantom and its position was moved from 2 up to 30 mm in depth (the distance between the centroid of the inclusion and the surface). For each depth, the time-resolved curve was acquired in difference slices, delaying the opening of the gate from 0 ns (situation where the first re-emitted photon was recorded) up to 3250 ps at step of 250 ps. The VOA was set in order to have nearly 1 Mcps for each delay (except for the latest where it was not possible to increase the injected power). For each delay, we acquired five repetitions (1 s each) of the distribution of time-of-flight of re-emitted photons.

We computed contrast and CNR according the definition given in Wabnitz et al. [21] and briefly summarized in the following section.

The contrast is a primary assessment of the effect of variation in absorption coefficient. In the case of time-resolved reflectance measurements, often the analysis is performed using time-windows, meaning that the distribution of time-of-flight is subdivided into portions and contrast (as well as other features) and can be computed within those subdivisions of the curve. In case of the so-called "time-window" analysis, the contrast in each temporal window ($C(t_W)$) was computed by the following equation:

$$C^{(t_w)} = \frac{N_0^{(t_w)} - N^{(t_w)}}{N_0^{(t_w)}}, \tag{1}$$

where $N_0^{(t_w)}$ and $N^{(t_w)}$ represent the overall number of counts computed in the time-window (t_w) in the homogeneous or perturbed (e.g., when the inclusion mimicking the perturbation is inserted into the probed volume) state, respectively. In our case, each recorded curve was subdivided in time-windows of 500 ps in width. For the homogeneous acquisition, the inclusion was posed 30 mm aside from the source-detector couple, where its effect on the time-resolved curve was considered negligible.

The contrast-to-noise ratio affects the detection of a small variation in absorption coefficient. Indeed, CNR was defined in the case of "time-window" analysis as

$$CNR^{(t_w)} = \frac{N^{(t_w)} - N_0^{(t_w)}}{\sigma\left(N_0^{(t_w)}\right)}, \tag{2}$$

where $\sigma\left(N_0^{(t_w)}\right)$ is the fluctuation of counts in the unperturbed state (i.e., the inclusion far from the source-detector couple) computed over different repetitions. Usually, this standard deviation is mostly due to photonic noise, but also instrumental fluctuations (such as laser instability) can affect this value.

Both contrast and CNR were computed with and without the exploitation of the FG technique. Indeed, when analyzing the "non-gated" data, we used the measurements where the gate was opened before the first re-emitted photons were detected, and closed when the last ones were collected, thus mimicking the acquisitions obtained with a "classical" single-photon detector operating in the so-called "free-running" mode. On the other hand, for "gated" results, we analyzed the curve acquired enabling the FG-SPAD module with a given delay with respect to the peak of the reflectance curve.

Contrast and CNR were computed using the same portion of the curve (the same time-window defined with respect to the peak of the IRF). In this way the photons analyzed had reached the same mean depth (which was encoded in time for time-resolved measurements), thus leading to a fair comparison between the "gated" and "non-gated" technique.

2.3. In Vivo Measurements

Preliminary in vivo experiments were performed in order to test the performance of the system in a real case. Different finger-tapping experiments were performed to check system sensitivity in functional studies of the brain. Two healthy volunteers were recruited from the lab and informed consent was obtained (Subject one: male, 30 years old. Subject two: male, 43 years old). The competent ethics committee authorized the study.

The optical probe was placed over the left motor area and it was centered (according to the international 10–20 system for the Electro Encephalography -EEG- (electrode placement) at the C3 point. The measurement protocol was composed of 20 s of baseline, 20 s of motor task (finger opposition at a frequency rate as high as possible) and 20 s of recovery. This sequence was repeated five times with the right hand (controlateral hand) and five times with the left hand (ipsilateral hand). The entire protocol was repeated four times for each subject—using a short source-detector separation ($\rho = 6$ mm for both subjects) with and without the gating approach and using a long source-detector separation ($\rho = 20$ mm for subject 1 and $\rho = 30$ mm for subject 2) with and without the gating approach. The short source-detector separation distance was limited to 6 mm because of the size of the fiber terminator at the probe side, while the long source-detector separation was chosen in order to guarantee a photon count as closer as possible to 1 Mcps. The same criterion was followed for the choice of the applied gating. Measurements were performed in a dimmed room to decrease the amount of background light.

In all the measurements, reemitted photons at both wavelengths were collected together with an integration time of 1 s. The photon time of flight distributions were divided into 25 time windows of 500 ps each. For each time window, we estimated the value of the quantity $\Delta\mu_a vt$ at both the wavelengths [24], where $\Delta\mu_a$ is the variation of the absorption coefficient in the activated brain cortex, while v is the speed of light in the medium and t, the time spent by photons in the activated brain cortex. From this quantity at both wavelengths, we derived the changes in the amount of oxygenated hemoglobin ($\Delta O_2 Hb$) and deoxygenated hemoglobin (ΔHHb), expressed in an arbitrary unit in the activated brain cortex. As mentioned in the phantom measurements section, we used the same time window for both gated and non-gated experiments in order to consider photons investigating the same depth. From the $\Delta O_2 Hb$ and ΔHHb curves, we can estimate contrast and CNR of the changes in the hemoglobin species with respect to the baseline.

3. Results and Discussion

3.1. Characterization of the Setup (Phantom Measurements)

Figure 1b represents the IRF of the proposed setup for the two channels. For both wavelengths the dynamic range was about 5.5 decades, while the computed FWHM was about 120 and 180 ps, respectively, for 710 and 820 nm. Some reflections were visible at 5.5 ns or 4.8 ns after the main peak (depending on the wavelength). Those reflections were most probably due to a second order of Fresnel reflections at the tip of the fibers. When conceiving the setup, the fibers' length was chosen in order to delay reflections at least some ns from the peak. Indeed, reflections did not represent a problem for the measurements, since they were quite late and several decades (about 4.5) under the main peak. Other small reflections presented at a shorter time were not significant since their amplitude was small enough to not distort the curve or saturate the dynamic range of the gated measurements.

Figure 2 reports the contrast and CNR computed at both the 5 and 30 mm source-detector distance (first and second row respectively) in the case of the fast-gating technique being applied ("gated", blue curve) or not ("non-gated", red lines). For all features, the time-window used for the analysis was opened 3 ns after the peak of the IRF (i.e., 3 ns after the injection of the light in the sample).

When a 5 mm source-detector distance was used, the fast-gating technique was fundamental to retrieve information about a buried perturbation. Indeed, the contrast recorded at 5 mm source-detector distance in the non-gated mode was low and very noisy, while, for the gated acquisitions, the contrast was much higher and its standard deviation (computed on the several repetitions) was nearly negligible. The improvement given by the fast-gating technique was clearly noticeable also in the CNR graph. Indeed, the curve relative to the gated acquisition was always higher than 10, even at the deepest position of the inclusion. On the other hand, the CNR for the non-gated acquisition was always less than 10 and, in most cases, also smaller than 1, meaning that the contrast was more sensitive to the noise of the measurements (e.g., photonic noise) than to the presence of an inclusion.

As predicted by the theory [9,26], the improvement given by the fast-gating technique was fundamental when using a small interfiber distance. Indeed, when the distance between the launching and detection fiber was shortened, a huge increase of the scarcely diffused photons (the so-called "early photons") was experienced. If a non-gated detector was used, this peak of "early photons" caused the saturation of its dynamic range, thus limiting the number of late photons detected and lowering the capability to discriminate a perturbation in depth.

The improvements given by the use of a fast-gated acquisition was less evident if a larger SDD was used, as it was clearly noticeable from curves reported in the last row of Figure 2. Indeed, the contrast obtained exploiting the fast-gating technique was slightly higher than what was achievable in the "non-gated" situation. The CNR graph clearly shows that in the gated mode there was a constant improvement of about a factor of 4.

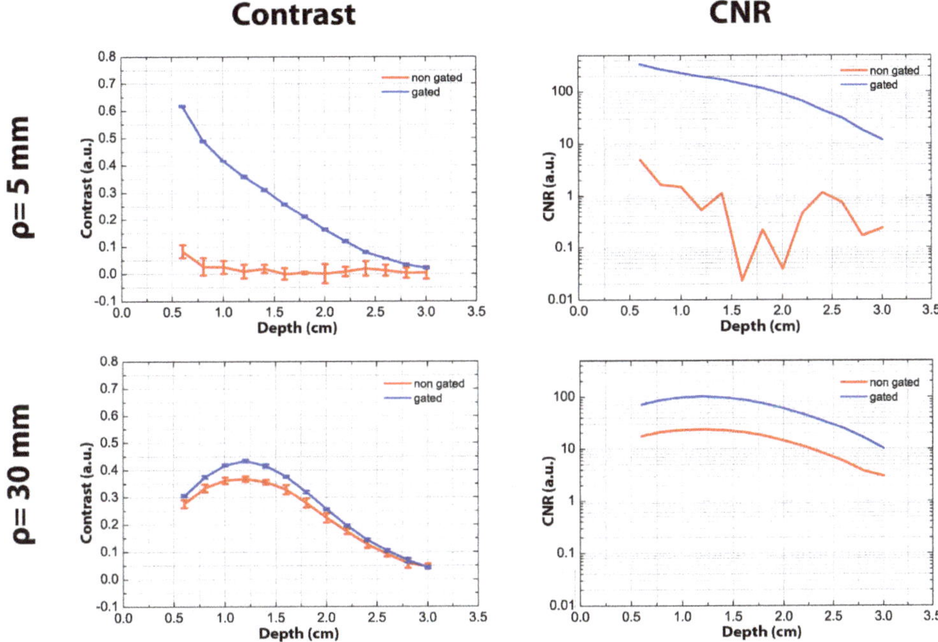

Figure 2. Contrast (left column) and contrast-to-noise ratio (CNR) (right column) computed at two different source-detector distances (5 mm, top row; 30 mm, bottom row) with and without the exploitation of the fast-gating technique (blue and red curves, respectively).

This was mostly because, when using the fast-gating technique, we recorded photons enabling the FG-SPAD module with a delay of 2.25 ns with respect to what was done in the non-gated mode, thus rejecting the first part of the time-resolved curve. The delay was chosen in order to exploit the dynamic range of the detector with the latest possible delay. Since we collected the same number of photons (nearly 10^6 per acquisition) for both gated and not gated acquisition, it was obvious that in the former case there was a boost in the number of late photons detected, thus increasing the number of counts recorded in the region of the time-window used for analysis. This resulted in a decrease of the noise (which was also visible from the errorbar in the contrast graph) and in an increase of the CNR in the case of gated measurements.

Considering as a conservative threshold the 0.05 contrast and a CNR higher than 10, using the gated technique we could detect the inclusion until 2.6 cm and 3 cm in depth, with 5 and 30 mm source-detector distance, respectively.

This conclusion is in contrast with theoretical findings predicting a contrast that is always higher when using a small source-detector distance [26]. Such a disagreement can be due to both the effect of the "diffusion tail" of the detector and to the so-called "memory effect".

The former effect [19] is due to the non-perfectly efficient time-gating mechanism that causes a "leakage" of early photons in a temporal region of the curve where only the contribution of late photons are expected. This effect is as high as the source-detector distance decreases, since the peak of early photons increases. On the other side, the memory effect [27] causes a boost in the background level at longer delays that can hide the late photons, thus decreasing the maximum contrast achievable. Since the memory effect is proportional to the number of photons impinging on the detector when it is in the OFF state, its consequences are more evident when using a small source-detector distance.

3.2. In Vivo Measurements

Table 1 represents the summary of the in vivo experiments. Only data regarding the finger tapping with the contralateral hand were proposed because of the absence of any activation during finger tapping with the ipsilateral hand due to the lateralization of the motor area. For measurements at short SDD, contrast and CNR are higher for gated measurements, confirming that the time gated approach is necessary in order to perform time resolved functional measurements at short SDD. For long SDD measurements, this finding was not confirmed. In fact, only for subject 1 and for HHb in subject 2 did we have a slight improvement in contrast and CNR.

Figures 3 and 4 show the changes in hemoglobin species (Hb) concentrations averaged over the five repetitions of finger tapping with the contralateral hand for subject 1 and 2, respectively. Error bars represent the standard deviation over the five repetitions. Normally, brain activation is identified by a task related increase in O_2Hb concentration and a corresponding decrease in HHb. We can clearly see this pattern in the gated column of both Figures 3 and 4, demonstrating the capability of a time gated system in performing functional measurements independently of the SDD, while in the non-gated approach this trend can be recognized only for long SDD.

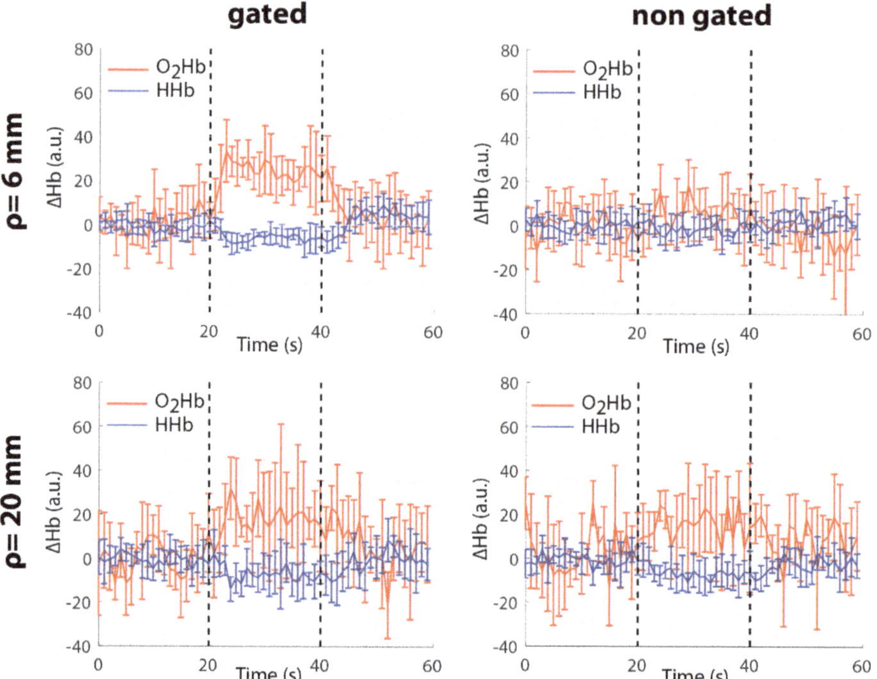

Figure 3. In vivo measurements of subject 1. Changes in O_2Hb (red) and HHb (blue) averaged over the five repetitions of finger tapping with the contralateral hand for the gated and non-gated approach (left and right column, respectively) at the short and long distance (upper and bottom lines, respectively). Error bars represent the standard deviation over the five repetitions.

Table 1. Summary of the in vivo measurements—finger tapping with the controlateral hand. Error for contrast and CNR is the standard deviation over the 5 finger tapping task repetitions. SDD = source-detector distance, O_2Hb = oxygenated hemoglobin, HHb = deoxygenated hemoglobin.

Subject	SDD	Gated	Time Window	Contrast O_2Hb	CNR O_2Hb	Contrast HHb	CNR HHb
Subject 1	6 mm	Yes	2.5–3 ns	21.9 ± 4.4	2.1 ± 0.6	4.5 ± 2.8	1.1 ± 0.7
Subject 1	6 mm	No	2.5–3 ns	5.1 ± 4.7	0.4 ± 0.4	1.7 ± 0.5	0.3 ± 0.1
Subject 2	6 mm	Yes	3–3.5 ns	23 ± 6.3	3.7 ± 2.5	11.2 ± 6.2	3.2 ± 3.2
Subject 2	6 mm	No	3–3.5 ns	6.9 ± 6.5	0.6 ± 0.5	1.7 ± 3	0.3 ± 0.5
Subject 1	20 mm	Yes	2.5–3 ns	18.1 ± 7	1.3 ± 0.7	6.3 ± 3.9	1 ± 0.7
Subject 1	20 mm	No	2.5–3 ns	12.4 ± 3	0.7 ± 0.2	5.4 ± 1	0.7 ± 0.2
Subject 2	30 mm	Yes	4.5–5 ns	44.8 ± 14.7	2.6 ± 1.2	25.4 ± 4.9	2.5 ± 1.1
Subject 2	30 mm	No	4.5–5 ns	50.2 ± 13.6	3.6 ± 1.1	13.5 ± 16.8	2.3 ± 3.2

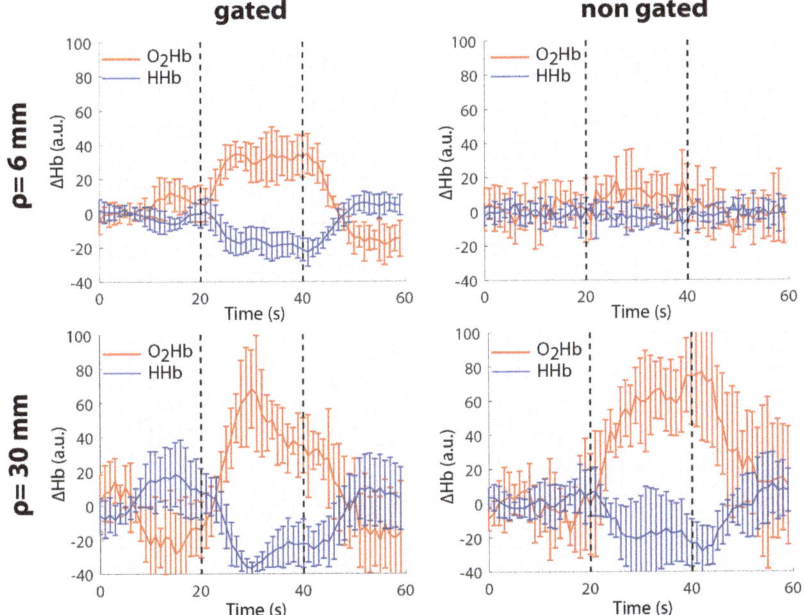

Figure 4. In vivo measurements of subject 2. Changes in O_2Hb (red) and HHb (blue) averaged over the five repetitions of finger tapping with the contralateral hand for the gated and non-gated approach (left and right column, respectively) at the short and long distance (upper and bottom lines, respectively). Error bars represent the standard deviation over the five repetitions.

4. Conclusions

The double-wavelength time-gated system proposed can perform measurements at SDD of few millimeters, showing the improvements theoretically predicted—increased contrast and CNR as highlighted by phantom and in vivo tests. Furthermore, the simultaneous acquisition of photon counts in a temporal window at two wavelengths allowed for the estimation of the changes in O_2Hb and HHb concentration during brain activation. The use of time gating—necessary for short SDD—however, can be useful also for long SDD measurements. In fact, as demonstrated with phantom measurements,

CNR for an absorption perturbation was always higher in time gated measurements. This was true because, in the presence of enough laser power, time gating allowed the exploitation of the entire dynamic range of the acquisition chain of the system to collect more late photons—which brought about information about the deeper structure in the medium—than without time gating. This last aspect was demonstrated with phantom measurements, but not completely with in vivo measurements. It is noteworthy that the in vivo measurements at different gates and at different distances were performed at different times; it is therefore plausible that their comparison is not completely reliable because of the intrinsic variability of the brain activation.

Author Contributions: Conceptualization, A.D.M., A.T., L.S., A.P. and D.C.; data curation, L.D.S., A.D.M. and D.C.; formal analysis, L.D.S., A.T., L.S., R.R. and D.C.; validation, L.D.S., A.D.M., A.T., L.S., A.P. and D.C.; writing—original draft, L.D.S., A.D.M., A.T., L.S., A.P. and D.C.; writing—review and editing, L.D.S., A.D.M., A.T., L.S., R.R., A.P. and D.C.

Funding: This research was partially supported from the EU project nEUROPt (grant agreement 201076).

Conflicts of Interest: The authors declare no conflict of interest.

References

1. Yodh, A.; Chance, B. Spectroscopy and imaging with diffusing light. *Phys. Today* **1995**, *48*, 34–41. [CrossRef]
2. Wolf, M.; Ferrari, M.; Quaresima, V. Progress of near-infrared spectroscopy and topography for brain and muscle clinical applications. *J. Biomed. Opt.* **2007**, *12*, 062104. [CrossRef] [PubMed]
3. Durduran, T.; Choe, R.; Baker, W.B.; Yodh, A.G. Diffuse optics for tissue monitoring and tomography. *Rep. Prog. Phys.* **2010**, *73*, 076701. [CrossRef] [PubMed]
4. Bellincontro, A.; Taticchi, A.; Servili, M.; Esposto, S.; Farinelli, D.; Mencarelli, F. Feasible Application of a Portable NIR-AOTF Tool for On-Field Prediction of Phenolic Compounds during the Ripening of Olives for Oil Production. *J. Agric. Food Chem.* **2012**, *60*, 2665–2673. [CrossRef] [PubMed]
5. Scholkmann, F.; Kleiser, S.; Metz, A.J.; Zimmermann, R.; Pavia, J.M.; Wolf, U.; Wolf, M. A review on continuous wave functional near-infrared spectroscopy and imaging instrumentation and methodology. *Neuroimage* **2014**, *85*, 6–27. [CrossRef] [PubMed]
6. Torricelli, A.; Contini, D.; Pifferi, A.; Caffini, M.; Re, R.; Zucchelli, L.; Spinelli, L. Time domain functional NIRS imaging for human brain mapping. *Neuroimage* **2014**, *85*, 28–50. [CrossRef] [PubMed]
7. Martelli, F.; del Bianco, S.; Ismaelli, A.; Zaccanti, G. *Light Propagation through Biological Tissue and Other Diffusive Media: Theory, Solutions, and Software*; SPIE: Bellingham, WA, USA, 2009.
8. Martelli, F.; Binzoni, T.; Pifferi, A.; Spinelli, L.; Farina, A.; Torricelli, A. There's plenty of light at the bottom: Statistics of photon penetration depth in random media. *Sci. Rep.* **2016**, *6*, 27057. [CrossRef]
9. Pifferi, A.; Torricelli, A.; Spinelli, L.; Contini, D.; Cubeddu, R.; Martelli, F.; Zaccanti, G.; Tosi, A.; Dalla Mora, A.; Zappa, F.; et al. Time-Resolved Diffuse Reflectance Using Small Source-Detector Separation and Fast Single-Photon Gating. *Phys. Rev. Lett.* **2008**, *100*, 138101. [CrossRef]
10. Boso, G.; Dalla Mora, A.; della Frera, A.; Tosi, A. Fast-gating of single-photon avalanche diodes with 200 ps transitions and 30 ps timing jitter. *Sens. Actuators A Phys.* **2013**, *191*, 61–67. [CrossRef]
11. Saha, S.; Lu, Y.; Weyers, S.; Sawan, M.; Lesage, F. Compact Fast Optode-Based Probe for Single-Photon Counting Applications. *IEEE Photonics Technol. Lett.* **2018**, *30*, 1515–1518. [CrossRef]
12. Dalla Mora, A.; Tosi, A.; Zappa, F.; Cova, S.; Contini, D.; Pifferi, A.; Spinelli, L.; Torricelli, A.; Cubeddu, R. Fast-gated single-photon avalanche diode for wide dynamic range near infrared spectroscopy. *Sel. Top. Quantum Electron.* **2010**, *16*, 1023–1030. [CrossRef]
13. Behera, A.; di Sieno, L.; Pifferi, A.; Martelli, F.; Dalla Mora, A. Instrumental, optical and geometrical parameters affecting time-gated diffuse optical measurements: A systematic study. *Biomed. Opt. Express* **2018**, *9*, 5524–5542. [CrossRef] [PubMed]
14. Di Sieno, L.; Dalla Mora, A.; Boso, G.; Tosi, A.; Pifferi, A.; Cubeddu, R.; Contini, D. Diffuse optics using a dual window fast-gated counter. *Appl. Opt.* **2014**, *53*, 7394–7401. [CrossRef] [PubMed]
15. Puszka, A.; di Sieno, L.; Dalla Mora, A.; Pifferi, A.; Contini, D.; Planat-Chrétien, A.; Koenig, A.; Boso, G.; Tosi, A.; Hervé, L.; et al. Spatial resolution in depth for time-resolved diffuse optical tomography using short source-detector separations. *Biomed. Opt. Express* **2015**, *6*, 1–10. [CrossRef] [PubMed]

16. Zouaoui, J.; di Sieno, L.; Hervé, L.; Pifferi, A.; Farina, A.; Dalla Mora, A.; Derouard, J.; Dinten, J.-M. Quantification in time-domain diffuse optical tomography using Mellin-Laplace transforms. *Biomed. Opt. Express* **2016**, *7*, 4346–4363. [CrossRef] [PubMed]

17. Di Sieno, L.; Wabnitz, H.; Pifferi, A.; Mazurenka, M.; Hoshi, Y.; Dalla Mora, A.; Contini, D.; Boso, G.; Becker, W.; Martelli, F.; et al. Characterization of a time-resolved non-contact scanning diffuse optical imaging system exploiting fast-gated single-photon avalanche diode detection. *Rev. Sci. Instrum.* **2016**, *87*, 035118. [CrossRef] [PubMed]

18. Di Sieno, L.; Contini, D.; Dalla Mora, A.; Torricelli, A.; Spinelli, L.; Cubeddu, R.; Tosi, A.; Boso, G.; Pifferi, A. Functional near-infrared spectroscopy at small source-detector distance by means of high dynamic-range fast-gated SPAD acquisitions: First in-vivo measurements. *Proc. SPIE* **2013**, 880402–880406.

19. Contini, D.; Dalla Mora, A.; Spinelli, L.; Farina, A.; Torricelli, A.; Cubeddu, R.; Martelli, F.; Zaccanti, G.; Tosi, A.; Boso, G.; et al. Effects of time-gated detection in diffuse optical imaging at short source-detector separation. *J. Phys. D Appl. Phys.* **2015**, *48*, 45401. [CrossRef]

20. Wabnitz, H.; Taubert, D.R.; Mazurenka, M.; Steinkellner, O.; Jelzow, A.; Macdonald, R.; Milej, D.; Sawosz, P.; Kacprzak, M.; Liebert, A.; et al. Performance assessment of time-domain optical brain imagers, part 1: Basic instrumental performance protocol. *J. Biomed. Opt.* **2014**, *19*, 86010. [CrossRef] [PubMed]

21. Wabnitz, H.; Jelzow, A.; Mazurenka, M.; Steinkellner, O.; Macdonald, R.; Milej, D.; Zolek, N.; Kacprzak, M.; Sawosz, P.; Maniewski, R.; et al. Performance assessment of time-domain optical brain imagers, part 2: nEUROPt protocol. *J. Biomed. Opt.* **2014**, *19*, 86012. [CrossRef] [PubMed]

22. Kirilina, E.; Jelzow, A.; Heine, A.; Niessing, M.; Wabnitz, H.; Brühl, R.; Ittermann, B.; Jacobs, A.M.; Tachtsidis, I. The physiological origin of task-evoked systemic artefacts in functional near infrared spectroscopy. *Neuroimage* **2012**, *61*, 70–81. [CrossRef] [PubMed]

23. Gerega, A.; Milej, D.; Weigl, W.; Kacprzak, M.; Liebert, A. Multiwavelength time-resolved near-infrared spectroscopy of the adult head: Assessment of intracerebral and extracerebral absorption changes. *Biomed. Opt. Express* **2018**, *9*, 2974–2993. [CrossRef] [PubMed]

24. Tosi, A.; Dalla Mora, A.; Zappa, F.; Gulinatti, A.; Contini, D.; Pifferi, A.; Spinelli, L.; Torricelli, A.; Cubeddu, R. Fast-gated single-photon counting technique widens dynamic range and speeds up acquisition time in time-resolved measurements. *Opt. Express* **2011**, *19*, 10735–10746. [CrossRef] [PubMed]

25. Martelli, F.; Pifferi, A.; Contini, D.; Spinelli, L.; Torricelli, A.; Wabnitz, H.; Macdonald, R.; Sassaroli, A.; Zaccanti, G. Phantoms for diffuse optical imaging based on totally absorbing objects, part 1: Basic concepts. *J. Biomed. Opt.* **2013**, *18*, 066014. [CrossRef] [PubMed]

26. Torricelli, A.; Pifferi, A.; Spinelli, L.; Cubeddu, R.; Martelli, F.; del Bianco, S.; Zaccanti, G. Time-Resolved Reflectance at Null Source-Detector Separation: Improving Contrast and Resolution in Diffuse Optical Imaging. *Phys. Rev. Lett.* **2005**, *95*, 078101. [CrossRef] [PubMed]

27. Dalla Mora, A.; Tosi, A.; Contini, D.; di Sieno, L.; Boso, G.; Villa, F.; Pifferi, A. Memory effect in silicon time-gated single-photon avalanche diodes. *J. Appl. Phys.* **2015**, *117*, 114501. [CrossRef]

 *applied sciences*

Article

A Hybrid Inversion Scheme Combining Markov Chain Monte Carlo and Iterative Methods for Determining Optical Properties of Random Media

Yu Jiang [1], Yoko Hoshi [2], Manabu Machida [2,*] and Gen Nakamura [3]

[1] School of Mathematics, Shanghai University of Finance and Economics, Shanghai 200433, China
[2] Institute for Medical Photonics Research, Hamamatsu University School of Medicine, Hamamatsu 431-3192, Japan
[3] Department of Mathematics, Hokkaido University, Sapporo 060-0810, Japan
* Correspondence: machida@hama-med.ac.jp

Received: 31 July 2019; Accepted: 21 August 2019; Published: 24 August 2019

Abstract: Near-infrared spectroscopy (NIRS) including diffuse optical tomography is an imaging modality which makes use of diffuse light propagation in random media. When optical properties of a random medium are investigated from boundary measurements of reflected or transmitted light, iterative inversion schemes such as the Levenberg–Marquardt algorithm are known to fail when initial guesses are not close enough to the true value of the coefficient to be reconstructed. In this paper, we investigate how this weakness of iterative schemes is overcome using Markov chain Monte Carlo. Using time-resolved measurements performed against a polyurethane-based phantom, we present a case that the Levenberg–Marquardt algorithm fails to work but the proposed hybrid method works well. Then, with a toy model of diffuse optical tomography we illustrate that the Levenberg–Marquardt method fails when it is trapped by a local minimum but the hybrid method can escape from local minima by using the Metropolis–Hastings Markov chain Monte Carlo algorithm until it reaches the valley of the global minimum. The proposed hybrid scheme can be applied to different inverse problems in NIRS which are solved iteratively. We find that for both numerical and phantom experiments, optical properties such as the absorption and reduced scattering coefficients can be retrieved without being trapped by a local minimum when Monte Carlo simulation is run only about 100 steps before switching to an iterative method. The hybrid method is compared with simulated annealing. Although the Metropolis–Hastings MCMC arrives at the steady state at about 10,000 Monte Carlo steps, in the hybrid method the Monte Carlo simulation can be stopped way before the burn-in time.

Keywords: near-infrared spectroscopy; diffuse light; inverse problems; optical tomography

---

1. Introduction

In near-infrared spectroscopy (NIRS), we estimate optical properties of biological tissue by solving inverse diffuse problems [1,2]. Such inverse problems are commonly solved by means of iterative methods. In the case of a homogeneous medium, absorption and reduced scattering coefficients of the medium can be obtained with iterative methods such as the Levenberg–Marquardt algorithm [3,4] from time-resolved measurements (for example, see the review article [5]). Neuroimaging for the human brain by NIRS through the neurovascular coupling has been developed and is called functional NIRS (fNIRS) [6–9]. Since the region of interest on the head can be small, it is possible to assume a simple geometry of the half space. Quantitative measurements of inter-regional differences in neuronal activity requires accurate estimates of optical properties.

In Ref. [10], the Nelder–Mead simplex method was used to retrieve optical parameters in layered tissue. Heterogeneity of optical properties can be obtained by diffuse optical tomography [1,11,12].

Iterative numerical schemes are used to minimize the cost function when solving these inverse problems. A gradient-based approach [13] was used to detect breast cancer [14]. The brain activity of a newborn infant was investigated [15] with diffuse optical tomography by TOAST (temporal optical absorption and scattering tomography) [16,17], in which iterative algorithms such as the nonlinear conjugate gradients, damped Gauss–Newton method, and Levenberg–Marquardt method are implemented. Diffuse optical tomography was performed on human lower legs and a forearm with the algebraic reconstruction algorithm in the framework of the modified generalized pulse spectrum technique [18]. See Refs. [19,20] for numerical techniques for diffuse optical tomography. For these iterative numerical schemes to work, it is important to choose a good initial guess for the initial value of the iteration.

Solving inverse problems by the Bayesian approach has been sought as an alternative way. In Ref. [21], a novel use of the Bayesian approach was considered to take modeling error into account. The Bayesian inversion with the Metropolis–Hastings Markov chain Monte Carlo was used in Refs. [22,23]. The Bayesian approach was used to determine optical parameters of the human head [24]. Although the Markov chain Monte Carlo (MCMC) approach is in principle able to escape from local minima, it is computationally time consuming. Hence, despite the above-mentioned efforts, the use of Monte Carlo for inverse problems in NIRS has been extremely limited.

In this paper, we shed light on Markov chain Monte Carlo once again by combining it with an iterative scheme, and investigate the use of it in NIRS. In particular, we test a hybrid numerical scheme of Markov chain Monte Carlo and an iterative method. In the proposed hybrid scheme, the Markov chain Monte Carlo algorithm is first used to provide an initial guess using jumps in the landscape of the cost function, and then an iterative method is used after the initial large fluctuation. Thus, the proposed method realizes a fast reconstruction while, in the beginning, obtained values at Monte Carlo steps jump around in the landscape of the cost function. The MCMC simulation is necessary only for the first 100 steps. Then the hybrid method starts to use an iterative scheme and reconstructs optical properties by searching the global minimum. The computation time of the iterative scheme is negligible compared with that of the Monte Carlo simulation. Since the Monte Carlo simulation can be stopped way before the burn-in time for the hybrid method, the proposed scheme is at least ten times faster than simulated annealing, which is implemented by the naive use of the Metropolis–Hastings Markov chain Monte Carlo.

The rest of the paper is organized as follows. We develop diffusion theory in Section 2. A polyurethane-based phantom and numerical phantom are also described in Section 2. Section 3 is devoted to experimental and numerical results. Finally, discussion is given in Section 4.

2. Materials and Methods

2.1. Diffusion Theory

2.1.1. Diffuse Light in Three Dimensions

Let us suppose that a random medium occupies the three-dimensional half space. We assume that the random medium is characterized by the diffusion coefficient D and absorption coefficient μ_a. We have $D = 1/(3\mu_s')$, where μ_s' is the reduced scattering coefficient. Position in the half space $(-\infty < x < \infty, -\infty < y < \infty, 0 < z < \infty)$ is denoted by $\mathbf{r} = (\boldsymbol{\rho}, z), \boldsymbol{\rho} = (x, y)$. Let t denote time. Let c be the speed of light in the medium. We assume an incident beam on the boundary at the origin in the x-y plane. The energy density u of light in the medium obeys the following diffusion equation.

$$\left(\frac{1}{c} \frac{\partial}{\partial t} - D\Delta + \mu_a \right) u(\mathbf{r}, t) = 0, \tag{1}$$

with the boundary condition

$$-\ell\frac{\partial}{\partial z}u(\mathbf{r},t) + u(\mathbf{r},t) = \delta(x)\delta(y)q(t), \tag{2}$$

and the initial condition $u(\mathbf{r},0) = 0$. The right-hand side of the boundary condition is the incident laser beam which illuminates the medium at the origin $(0,0,0)$ with the temporal profile $q(t)$. In the phantom experiment described below, the incident light enters the phantom in the positive z direction. The extrapolation distance ℓ is a nonnegative constant. If we consider the diffuse surface reflection, we have [25]

$$\ell = 2D\frac{1+r_d}{1-r_d}, \tag{3}$$

where the internal reflection r_d is given by [26] $r_d = -1.4399n^{-2} + 0.7099n^{-1} + 0.6681 + 0.0636n$. Let us consider the corresponding surface Green's function $G_s(\mathbf{r},t;\rho',s)$, which satisfies Equation (1) and the boundary condition

$$\left(-\ell\frac{\partial}{\partial z} + 1\right) G_s(\mathbf{r},t;\rho',s) = \delta(x-x')\delta(y-y')\delta(t-s), \tag{4}$$

together with the initial condition $G_s(\mathbf{r},0;\rho',s) = 0$. We obtain [27–30]

$$\begin{aligned}
G_s(\mathbf{r},t;\rho',s) = \frac{cD}{\ell}\Bigg[& \frac{2e^{-\mu_a c(t-s)}}{(4\pi Dc(t-s))^{3/2}}e^{-\frac{(x-x')^2+(y-y')^2}{4Dc(t-s)}}e^{-\frac{z^2}{4Dc(t-s)}} \\
& -\frac{e^{-\mu_a c(t-s)}}{4\pi\ell Dc(t-s)}e^{-\frac{(x-x')^2+(y-y')^2}{4Dc(t-s)}}e^{\frac{z}{\ell}}e^{\frac{Dc(t-s)}{\ell^2}}\operatorname{erfc}\left(\frac{z+2Dc(t-s)/\ell}{\sqrt{4Dc(t-s)}}\right)\Bigg], \quad t > s,
\end{aligned} \tag{5}$$

with $G_s(\mathbf{r},t;\rho',s) = 0$ for $t \leq s$. The complementary error function is defined by $\operatorname{erfc}(x) = (2/\sqrt{\pi})\int_x^\infty \exp(-t^2)\,dt$. Let $\mathbf{r}_d$ be a point in the x-y plane with $|\mathbf{r}_d| > 0$. We obtain

$$\begin{aligned}
u(\mathbf{r}_d,t) &= \int_0^t G_s(\mathbf{r}_d,t;0,s)q(s)\,ds \\
&= \int_0^t e^{-\mu_a c(t-s)}e^{-\frac{|\mathbf{r}_d|^2}{4Dc(t-s)}}\frac{2q(s)}{(4\pi Dc(t-s))^{3/2}}\left[1 - \frac{\sqrt{4\pi Dc(t-s)}}{2\ell}e^{\frac{Dc(t-s)}{\ell^2}}\operatorname{erfc}\left(\frac{\sqrt{Dc(t-s)}}{\ell}\right)\right]ds.
\end{aligned} \tag{6}$$

This $u(\mathbf{r}_d,t)$ in Equation (6) is used for the calculation in Section 3.1.

2.1.2. Diffuse Light in Two Dimensions

For the later purpose of a numerical experiment, we here consider light propagation in two dimensions. Let us suppose that the two-dimensional half space of positive y is occupied with a random medium in the x-y plane. The energy density u of light at position $\rho = (x,y)$ in the medium due to the incident beam on the boundary (the x-axis) at ρ_s^i ($i = 1,\dots,M_s$) obeys the following diffusion equation.

$$\left(\frac{1}{c}\frac{\partial}{\partial t} - D\Delta + \mu_a(\rho)\right) u(\rho,t;\rho_s^i) = 0, \tag{7}$$

with the boundary condition

$$-\ell\frac{\partial}{\partial y}u(\rho,t;\rho_s^i) + u(\rho,t;\rho_s^i) = \delta(x-x_s^i)\delta(t), \tag{8}$$

and the initial condition $u(\rho, 0; \rho_s^i) = 0$. The incident laser beam on the right-hand side of the boundary condition was assumed to be a pulse at the position $\rho_s^i = (x_s^i, 0)$. We suppose that the diffusion coefficient D is a positive constant but μ_a varies in space.

Below we develop the Rytov approximation. Let us write $\mu_a(\rho) \geq 0$ as

$$\mu_a(\rho) = \mu_{a0} + \delta\mu_a(\rho), \tag{9}$$

where μ_{a0} is a constant and the perturbation $\delta\mu_a(\rho)$ spatially varies. We note the relation,

$$u(\rho, t; \rho_s^i) = u_0(\rho, t; \rho_s^i) - \int_0^t \int_0^\infty \int_{-\infty}^\infty G(\rho, t; \rho', s)\delta\mu_a(\rho')u(\rho', s; \rho_s^i)\, dx'dy'ds, \tag{10}$$

where $u_0(\rho, t; \rho_s^i)$ is the solution to the diffusion Equation (7) with the zeroth-order coefficient replaced by b_0. We introduce the Green's function $G(\rho, t; \rho', s)$, which satisfies

$$\left(\frac{1}{c}\frac{\partial}{\partial t} - D\Delta + \mu_{a0}\right) G = \delta(\rho - \rho')\delta(t - s), \tag{11}$$

with the boundary condition $-\ell\frac{\partial G}{\partial y} + G = 0$ at $y = 0$, and the initial condition $G = 0$ at $t = 0$. The above relation (10) can be directly verified. By recursive substitution, we obtain u as

$$u(\rho, t; \rho_s^i) = u_0(\rho, t; \rho_s^i) - \int_0^t \int_0^\infty \int_{-\infty}^\infty G(\rho, t; \rho', s)\delta\mu_a(\rho')u_0(\rho', s; \rho_s^i)\, dx'dy'ds + O((\delta\mu_a)^2). \tag{12}$$

By neglecting higher-order terms assuming that $\delta\mu_a$ is small, we arrive at the (first) Born approximation [31], in which u is given by

$$u(\rho, t; \rho_s^i) = u_0(\rho, t; \rho_s^i) - c \int_0^t \int_0^\infty \int_{-\infty}^\infty G(\rho, t; \rho', s)\delta\mu_a(\rho')u_0(\rho', s; \rho_s^i)\, dx'dy'ds. \tag{13}$$

We note that the Green's function is obtained as [27–30]

$$G(\rho, t; \rho', s) = \frac{e^{-\mu_{a0}c(t-s)}}{4\pi D(t-s)} e^{-\frac{(x-x')^2}{4Dc(t-s)}} \left[e^{-\frac{(y-y')^2}{4Dc(t-s)}} + e^{-\frac{(y+y')^2}{4Dc(t-s)}} \right.$$
$$\left. - \frac{\sqrt{4\pi Dc(t-s)}}{\ell} e^{-\frac{(y+y')^2}{4Dc(t-s)}} e^{\left(\frac{y+y'+2Dc(t-s)/\ell}{\sqrt{4Dc(t-s)}}\right)^2} \mathrm{erfc}\left(\frac{y+y'+2Dc(t-s)/\ell}{\sqrt{4Dc(t-s)}}\right) \right], \tag{14}$$

for $t > s$, and otherwise $G(\rho, t; \rho', s) = 0$. Moreover, we obtain

$$u_0(\rho, t; \rho_s^i) = \frac{e^{-\mu_{a0}ct}}{2\pi Dt} e^{-\frac{(x-x_s^i)^2+y^2}{4Dct}} \left[1 - \frac{\sqrt{\pi Dct}}{\ell} e^{\left(\frac{y+2Dct/\ell}{\sqrt{4Dct}}\right)^2} \mathrm{erfc}\left(\frac{y+2Dct/\ell}{\sqrt{4Dct}}\right) \right]. \tag{15}$$

The above expression of u_0 is similar to the formula in Equation (6) but does not have a time integral because in this case we assumed the delta function $\delta(t)$ for the incident beam.

To obtain the expression of the Rytov approximation, we introduce ψ_0, ψ_1 as [31]

$$u_0 = e^{\psi_0}, \qquad u = e^{\psi_0 + \psi_1}. \tag{16}$$

By plugging the above expressions of u, u_0 into the Born approximation and neglecting terms of $O((\delta\mu_a)^2)$, we obtain

$$
\begin{aligned}
e^{\psi_1} &= 1 - e^{-\psi_0} \int_0^t \int_0^\infty \int_{-\infty}^\infty G(\rho,t;\rho',s)\delta\mu_a(\rho')u_0(\rho',s;\rho_s^i)\,dx'dy'ds \\
&= \exp\left[-e^{-\psi_0} \int_0^t \int_0^\infty \int_{-\infty}^\infty G(\rho,t;\rho',s)\delta\mu_a(\rho')u_0(\rho',s;\rho_s^i)\,dx'dy'ds\right].
\end{aligned}
\tag{17}
$$

The (first) Rytov approximation is thus derived as

$$
u(\rho,t;\rho_s^i) = u_0(\rho,t;\rho_s^i)\exp\left[-\frac{1}{u_0(\rho,t;\rho_s^i)} \int_0^t \int_0^\infty \int_{-\infty}^\infty G(\rho,t;\rho',s)\delta\mu_a(\rho')u_0(\rho',s;\rho_s^i)\,dx'dy'ds\right].
\tag{18}
$$

Let us define

$$
\begin{aligned}
g(y,t;y',s) &= \frac{1}{4\pi D(t-s)}e^{-\frac{(y+y')^2}{4Dc(t-s)}}\left[1 + e^{\frac{(y+y')^2-(y-y')^2}{4Dc(t-s)}} - \frac{\sqrt{4\pi Dc(t-s)}}{\ell}\right. \\
&\quad \left. \times e^{\left(\frac{y+y'}{2\sqrt{Dc(t-s)}}+\frac{\sqrt{Dc(t-s)}}{\ell}\right)^2} \operatorname{erfc}\left(\frac{y+y'}{2\sqrt{Dc(t-s)}}+\frac{\sqrt{Dc(t-s)}}{\ell}\right)\right].
\end{aligned}
\tag{19}
$$

Then we have

$$
\int_0^t G(\rho,t;\rho',s)u_0(\rho',s)\,ds = e^{-\mu_{a0}ct} \int_0^t e^{-\frac{(x-x')^2}{4Dc(t-s)}} e^{-\frac{(x'-x_s^i)^2}{4Dcs}} g(y,t;y',s)g(y',s;0,0)\,ds.
\tag{20}
$$

Therefore, Equation (18) can be rewritten as

$$
\begin{aligned}
u(\rho,t;\rho_s^i) &= u_0(\rho,t;\rho_s^i)\exp\left[-\frac{e^{-\mu_{a0}ct}}{u_0(\rho,t;\rho_s^i)} \int_0^\infty \int_{-\infty}^\infty \delta\mu_a(\rho')\right. \\
&\quad \left. \times \left(\int_0^t e^{-\frac{(x-x')^2}{4Dc(t-s)}} e^{-\frac{(x'-x_s^i)^2}{4Dcs}} g(y,t;y',s)g(y',s;0,0)\,ds\right) dx'dy'\right].
\end{aligned}
\tag{21}
$$

This expression (21) is used to compute the forward data in Section 3.2.

2.2. Inverse Problems by an Iterative Scheme

We suppose that on the surface of a two- or three-dimensional random medium, for each source at ρ_s^i or $\mathbf{r}_s^i$ $(i = 1, \ldots, M_s)$ exiting light is detected at ρ_d^j or $\mathbf{r}_d^j$ $(j = 1, \ldots, M_d)$ and is measured at times t^k $(k = 1, \ldots, M_t)$. Let y_{ijk} be measured data whereas u is a solution to the diffusion equation. Let us suppose that u depends on a vector $\mathbf{a}$ which contains unknown parameters. We wish to reconstruct $\mathbf{a}$. In Section 3.1, $\mathbf{a} = (\mu_a, D)$, and a is a scalar (a one-dimensional vector) in Section 3.2. For example, in the former case the solution u depends on $\mathbf{a}$ since the calculated value of u depends on μ_a, D.

Let us introduce vectors $\mathbf{U}$ and $\mathbf{F}(\mathbf{a})$ of dimension $M_s M_d M_t$ as

$$
\{\mathbf{U}\}_{ijk} = y_{ijk}, \qquad \{\mathbf{F}(\mathbf{a})\}_{ijk} = u(\rho_d^j, t^k; \rho_s^i; \mathbf{a}) \text{ or } u(\mathbf{r}_d^j, t^k; \mathbf{r}_s^i; \mathbf{a}),
\tag{22}
$$

where we wrote $u(\rho_d^j, t^k; \rho_s^i) = u(\rho_d^j, t^k; \rho_s^i; \mathbf{a})$ and $u(\mathbf{r}_d^j, t^k; \mathbf{r}_s^i) = u(\mathbf{r}_d^j, t^k; \mathbf{r}_s^i; \mathbf{a})$ emphasizing that u depends on $\mathbf{a}$. We find optimal $\mathbf{a}$ by minimizing $\|\mathbf{U} - \mathbf{F}(\mathbf{a})\|_2^2$. Here we particularly consider the Levenberg–Marquardt method [3,4], i.e., the reconstructed value $\mathbf{a}^* = \lim_{k\to\infty} \mathbf{a}_k$ is computed by the iteration given by

$$
\mathbf{a}_{k+1} = \mathbf{a}_k + \left[F'(\mathbf{a}_k)^T F'(\mathbf{a}_k) + \lambda I\right]^{-1} F'(\mathbf{a}_k)^T (\mathbf{U} - \mathbf{F}(\mathbf{a}_k)),
\tag{23}
$$

where $F'(\mathbf{a})$ is the Jacobian matrix, which contains derivatives of $\mathbf{F}(\mathbf{a})$ with respect to $\mathbf{a}$, and the parameter λ is nonnegative. By modifying the original algorithm according to Ref. [32], our algorithm of the Levenberg–Marquardt method, which we call Algorithm 1, is described below.

Algorithm 1: Levenberg–Marquardt (LM)

---

1. Set $k = 0$ and $\lambda = 1$.
2. Calculate $\mathbf{F}(\mathbf{a}_k)$ and $F'(\mathbf{a}_k)$.
3. Calculate $S(\mathbf{a}_k) = \mathbf{d}^T\mathbf{d}$, where $\mathbf{d} = \mathbf{U} - \mathbf{F}(\mathbf{a}_k)$.
4. Prepare $A = F'(\mathbf{a}_k)^T F'(\mathbf{a}_k)$ and $\mathbf{v} = F'(\mathbf{a}_k)^T\mathbf{d}$.
5. Find δ from $(A + \lambda I)\delta = -\mathbf{v}$.
6. Obtain $S(\mathbf{a}_k + \delta)$ and $R = \frac{S(\mathbf{a}_k) - S(\mathbf{a}_k + \delta)}{-\delta^T(2\mathbf{v} + A\delta)}$.
7. If $R < 0.25$, then set $v = 10$ $(\alpha_c < 0.1)$, $v = 1/\alpha_c$ $(0.1 \le \alpha_c \le 0.5)$, or $v = 2$ $(\alpha_c > 0.5)$, where

   $\alpha_c = \left[2 - (S(\mathbf{a}_k + \delta) - S(\mathbf{a}_k))/\delta^T\mathbf{v}\right]^{-1}$. If $R < 0.25$ and $\lambda = 0$, set $\lambda = 1/\|A^{-1}\|$ and $v = v/2$.

   In the case of $R < 0.25$, we set $\lambda = v\lambda$. If $R > 0.75$, then we set $\lambda = \lambda/2$. If $R > 0.75$ and

   $\lambda < 1/\|A^{-1}\|$, set $\lambda = 0$. Otherwise when $0.25 \le R \le 0.75$, no update for λ.
8. If $S(\mathbf{a}_k + \delta) \ge S(\mathbf{a}_k)$, then return to Step 5.
9. If $S(\mathbf{a}_k + \delta) < S(\mathbf{a}_k)$, set $\mathbf{a}_{k+1} = \mathbf{a}_k + \delta$. Then put $k + 1 \to k$ and go back to Step 2. Repeat the

   above procedure until one of the stopping criteria $\|\delta\| < 10^{-4}$ and $S < 10^{-14}$ is fulfilled.

2.3. Inverse Problems by Markov Chain Monte Carlo

For simplicity in this section, we describe the Metropolis–Hastings Markov chain Monte Carlo algorithm (MH-MCMC) assuming a is the scalar appearing in Section 3.2. The extension of the derivation to the general case of vector $\mathbf{a}$ is straightforward.

Suppose that the coefficient μ_a is unknown and μ_a depends on a parameter a. We will reconstruct a within the framework of the Bayesian inversion algorithm [33,34], i.e., we will find the probability distribution $\pi(a|\mathbf{U})$ of a for measured data $\mathbf{U}$. Let $f_{\text{prior}}(a)$ be the prior probability density and $\pi(\mathbf{U}|a)$ be the likelihood density or the conditional probability density of $\mathbf{U}$ for a. Then the joint probability density $\pi(a, \mathbf{U})$ of $a, \mathbf{U}$ is given by $\pi(a, \mathbf{U}) = \pi(\mathbf{U}|a)f_{\text{prior}}(a)$. According to the Bayes formula, the conditional probability density $\pi(a|\mathbf{U})$ is given by

$$\pi(a|\mathbf{U}) = \frac{L(\mathbf{U}|a)f_{\text{prior}}(a)}{\int_{-\infty}^{\infty} L(\mathbf{U}|a)f_{\text{prior}}(a)\, da}, \tag{24}$$

where $L(\mathbf{U}|a)$ is a function proportional to $\pi(\mathbf{U}|a)$. Assuming Gaussian noise, we put

$$L(\mathbf{U}|a) = e^{-\frac{1}{2\sigma_e^2}\|\mathbf{U} - \mathbf{F}(a)\|_2^2}. \tag{25}$$

In this paper, we simply set $f_{\text{prior}}(a) = 1$, i.e., we can say $f_{\text{prior}}(a) \propto \mathbf{1}_{[a_{\min}, a_{\max}]}(a)$ and the interval $[a_{\min}, a_{\max}]$ is large enough so that all a's appearing in the Markov chain fall into this interval. Here, $\mathbf{1}_A(a)$ is the indicator function defined as $\mathbf{1}_A(a) = 1$ for $a \in A$ and $= 0$ for $a \notin A$. General uniform distributions can be used for $f_{\text{prior}}(a)$ if we use the prior-reversible proposal that satisfies $f_{\text{prior}}(a)q(a'|a) = f_{\text{prior}}(a')q(a|a')$ (see below for the proposal distribution $q(a'|a)$) [35]. Another possible choice of $f_{\text{prior}}(a)$ is a Gaussian distribution, which turns out to be the Tikhonov regularization term in the cost function.

Using the Metropolis–Hastings algorithm, we can evaluate $\pi(a|\mathbf{U})$ even when the normalization factor is not known and only the relation $\pi(a|\mathbf{U}) \propto L(\mathbf{U}|a)f_{\text{prior}}(a)$ is available [33,34]. We can find $\pi(a|\mathbf{U})$ using a sequence $p_1, p_2, \ldots$ as

$$\pi(a|\mathbf{U}) = \lim_{k \to \infty} p_k(a), \tag{26}$$

where $p_{k+1}(a)$ is obtained from $p_k(a)$ (Markov chain) as

$$p_{k+1}(a') = \int_{-\infty}^{\infty} K(a', a)p_k(a)\, da. \tag{27}$$

For all a, a', the transition kernel satisfies

$$K(a', a) \geq 0, \qquad \int_{-\infty}^{\infty} K(a', a)\, da' = 1. \tag{28}$$

Let us write $K(a', a)$ as

$$K(a', a) = \alpha(a', a)q(a'|a) + r(a)\delta(a' - a), \tag{29}$$

where $q(a'|a)$ is the proposal distribution and

$$r(a) = 1 - \int_{-\infty}^{\infty} \alpha(a', a)q(a'|a)\, da'. \tag{30}$$

In the Metropolis–Hastings algorithm we set $\alpha(a', a) = \min\{1, \pi(a'|\mathbf{U})q(a|a')/[\pi(a|\mathbf{U})q(a'|a)]\}$. With this choice of $\alpha(a', a)$, the detailed balance is satisfied and $K(a, a')\pi(a'|\mathbf{U}) = K(a', a)\pi(a|\mathbf{U})$. A common choice of $q(a'|a)$ is the normal distribution, i.e., $q(\cdot|a) = \mathcal{N}(a, \varepsilon^2)$ with the mean a and the standard deviation $\varepsilon > 0$. We note that $q(a'|a) = q(a|a')$ in this case. We have

$$\int_{-\infty}^{\infty} h(a)\pi(a|\mathbf{U})\, da = \lim_{k_{\max} \to \infty} \frac{1}{k_{\max}} \sum_{k=1}^{k_{\max}} h(a_k), \tag{31}$$

where $a_k \sim p_k(\cdot)$ and h is an arbitrary continuous bounded function.

Simulated annealing [36] is a type of the Metropolis–Hastings MCMC algorithm in which the *temperature* σ_e decreases during the simulation. The algorithm is summarized below as Algorithm 2. In this paper we consider two temperatures.

Algorithm 2: Two-temperature simulated annealing (SA)

---

1. Set large σ_e as a high temperature.
2. Generate $a' \sim q(\cdot|a_k) = \mathcal{N}(a_k, \varepsilon^2)$ with $\varepsilon > 0$ for given a_k.
3. Calculate $\alpha(a', a_k) = \min\{1, \pi(a'|\mathbf{U})/\pi(a_k|\mathbf{U})\}$.
4. Update a_k as $a_{k+1} = a'$ with probability $\alpha(a', a_k)$ but otherwise set $a_{k+1} = a_k$.
5. Continue while $k \leq k_b$.
6. Then decrease σ_e to a smaller value as a low temperature, and continue to update a_k.

Now we propose an MCMC-iterative hybrid method by combining Algorithms 1 with the Metropolis–Hastings Markov chain Monte Carlo. We first run the Markov chain Monte Carlo. Although the Metropolis–Hastings MCMC eventually gives the correct solution after about 10,000 steps when the chain reaches the steady state, we stop the Monte Carlo simulation way before the burn-in time. At the k_bth Monte Carlo step after initial rapid changes ceases, we record the obtained reconstructed value a_{k_b} and switch to Algorithm 1 with the initial value the recorded a_{k_b}. It is important that k_b is chosen after the initial rapid changes although k_b can be a significant way from the burn-in time. Otherwise, the final reconstructed results are quite robust against the choice of k_b. We refer to the following hybrid algorithm as Algorithm 3.

Algorithm 3: Hybrid

---

1. Choose an initial guess a_0, which is not necessarily close to the global minimum.
2. Generate $a' \sim q(\cdot|a_k) = \mathcal{N}(a_k, \varepsilon^2)$ with $\varepsilon > 0$ for given a_k.
3. Calculate $\alpha(a', a_k) = \min\{1, \pi(a'|U)/\pi(a_k|U)\}$.
4. Update a_k as $a_{k+1} = a'$ with probability $\alpha(a', a_k)$ but otherwise set $a_{k+1} = a_k$.
5. Obtain a reconstructed a_{k_b}.
6. Switch to Algorithm 1 with the initial guess a_{k_b}.

We close this subsection by the Gelman-Rubin convergence diagnostic, which uses the intra-chain variance and inter-chain variance [37,38]. We run M_{MC} different chains with different initial values. Let a_k^m denote the kth value in the mth chain ($k = 1, \ldots, k_b, m = 1, \ldots, M_{\text{MC}}$). We discard the first $k_a - 1$ steps before the initial rapid changes cease. Then we compute the following intra-chain average and variance.

$$\bar{a}_m = \frac{1}{k_b - k_a + 1} \sum_{k=k_a}^{k_b} a_k^m, \quad \sigma_m^2 = \frac{1}{k_b - k_a} \sum_{k=k_a}^{k_b} (a_k^m - \bar{a}_m)^2. \tag{32}$$

Next we compute the inter-chain mean and variance.

$$\bar{a} = \frac{1}{M_{\text{MC}}} \sum_{m=1}^{M_{\text{MC}}} \bar{a}_m, \quad B = \frac{k_b - k_a + 1}{M_{\text{MC}} - 1} \sum_{m=1}^{M_{\text{MC}}} (\bar{a}_m - \bar{a})^2. \tag{33}$$

Now we introduce

$$\hat{V} = \frac{k_b - k_a}{k_b - k_a + 1} W + \frac{1}{k_b - k_a + 1} B, \tag{34}$$

where $W = \frac{1}{M_{\text{MC}}} \sum_{m=1}^{M_{\text{MC}}} \sigma_m^2$. This is an unbiased estimator of the true variance. But W is also an unbiased estimate of the true variance if the chains converge. Therefore, we have $\sqrt{\hat{V}/W} \approx 1$ if converged. Below, we will see that we can choose k_b for which $\sqrt{\hat{V}/W}$ is not close to 1. In this paper, we set $M_{\text{MC}} = 10$.

2.4. TRS Measurements of a Polyurethane-Based Phantom

In this section, we consider time-resolved measurements for a phantom. The solid phantom (biomimic optical phantom) is made of polyurethane to simulate biological tissues (INO, Quebec, QC, Canada). The absorption coefficient and reduced scattering coefficient are $\mu_a = 0.0209\,\text{mm}^{-1}$ and $\mu_s' = 0.853\,\text{mm}^{-1}$ at wavelength 800 nm. The refractive index of the phantom is $n = 1.51$. The measurements were performed by the TRS (time-resolved spectroscopy) instrument (TRS-80, Hamamatsu Photonics K.K., Hamamatsu, Japan). Two optical fibers from TRS-80 are attached on the top of the phantom with the separation 13 mm ($M_s = M_d = 1$). The phantom is illuminated by near-infrared light of the wavelength 760 nm through one optical fiber and the reflected light is detected by the other optical fiber. The time interval of the measured data is 10 ps and we used the data from $t_1 = 2.00$ ns to $t_{M_t} = 8.00$ ns ($M_t = 601$). Measured counts, i.e., the number of photons, are shown in Figure 1. The upper panel of Figure 1 shows the instrument response function (IRF), which is given as a property of the experimental device. The measured reflected light is shown in the lower panel of Figure 1. The function $q(t^k)$ is the instrument response function divided by the maximum value of the measured reflected light. The peak of q is at 2.56 ns. $\{U\}_{11k} = y_{11k}$ is the measured reflected light normalized by its maximum value. In the lower panel of Figure 1, t_1 and t_{M_t} are marked by vertical dashed lines. The size of the phantom is large enough that we can assume the three-dimensional half space. Then the energy density of the detected light is computed by Equation (6). The parameter ℓ is obtained from Equation (3).

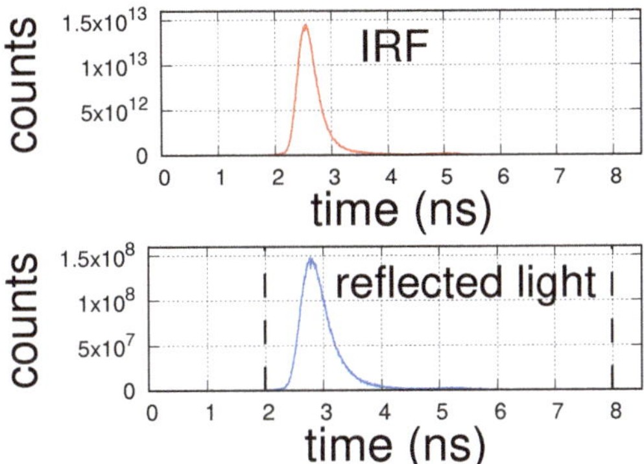

Figure 1. Measured data from TRS-80. The instrument response function and measured reflected light are shown in the upper and lower panels, respectively. In the lower panel, the dashed lines show $t_1 = 2.00\,\text{ns}$ and $t_{M_t} = 8.00\,\text{ns}$ ($M_t = 601$).

2.5. Numerical Phantom

To examine when the iterative scheme fails by being trapped by a local minimum and how the Markov chain Monte Carlo is capable of escaping from such local minima, a toy model is devised which is simple enough to explicitly understand the structure of the cost function but is complicated enough that the cost function has one local minimum and one global minimum.

We consider diffuse optical tomography in the two-dimensional space. In our toy model we suppose that the diffusion coefficient D is constant everywhere in the medium but there is absorption inhomogeneity and the absorption coefficient $\delta\mu_a$ is unknown. Moreover, we assume that $\delta\mu_a(\rho)$ is given by

$$\delta\mu_a(\rho) = \eta f_a(x)\delta(y - y_0),\tag{35}$$

where η, y_0 are given positive constants. Here, $f_a(x)$ is given by

$$f_a(x) = \left[a^3 + 3\left(1 + \frac{\tanh x^2}{10}\right)a^2\right]\left(1 - \tanh x^2\right),\tag{36}$$

where $a > 1.1$ is a constant. Thus, a is the parameter to be reconstructed in our toy inverse problem. As is shown in Figure 2, the function $f_a(x)$ has one peak at $x = 0$ and the maximum value is $f_a(0) = a^2(a + 3)$, f_a monotonically decays for $|x| > 0$, and $f_a \to 0$ as $|x| \to \infty$.

Now we describe how the forward data is computed. The unit of length and unit of time are taken to be mm and ps, respectively. On the x-axis, we place two sources ($M_s = 2$) at $\rho_s^1 = (-20,0)$, $\rho_s^2 = (20,0)$ and three detectors ($M_d = 3$) at $\rho_d^1 = (-40,0)$, $\rho_d^2 = (0,0)$, $\rho_d^3 = (40,0)$. We set $\mu_s' = 1$, $\mu_a = 0.02$, $n = 1.37$. Suppose that there is absorption inhomogeneity at depth 5. For $\delta\mu_a$, we put $\eta = 300/c$, $y_0 = 5$, and

$$a = 1.5.\tag{37}$$

To distinguish, hereafter the true value of a is denoted by $\bar{a}$. We assume 3% Gaussian noise and give the measured data as

$$y_{ijk} = u(\rho_d^j, t^k; \rho_s^i; \bar{a})(1 + e_{ijk}),\tag{38}$$

where $e_{ijk} \sim \mathcal{N}(0, 0.03^2)$. In this numerical experiment, $M_t = 500$ and $t^{k+1} - t^k = t^1 = 5$ ($k = 1, \ldots, M_t - 1$).

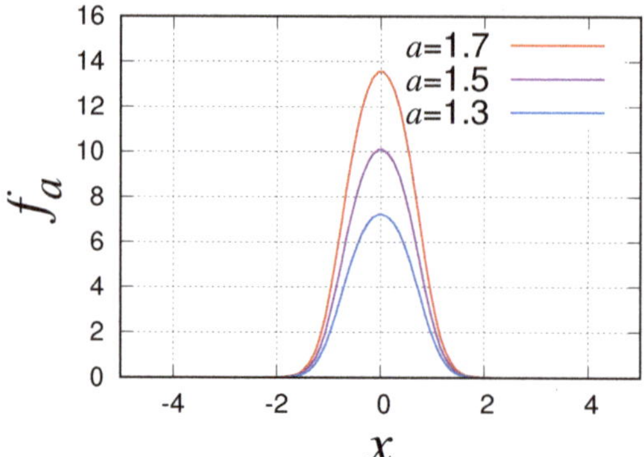

Figure 2. The function $f_a(x)$ in Equation (36) is plotted for $a = 1.3, 1.5,$ and 1.7.

By substituting the form (35) of δb in (21), we can express the energy density as

$$u(\rho_d^j, t^k; \rho_s^i; a) = u(\rho_d^j, t^k; \rho_s^i)$$

$$= u_0(\rho_d^j, t^k; \rho_s^i) \exp\left[-\frac{\eta e^{-\mu_{a0}ct^k}}{u_0(\rho_d^j, t^k; \rho_s^i)} \int_0^{t^k} g(0, t^k; y_0, s) \right.$$

$$\left. \times g(y_0, s; 0, 0) \left(\int_{-\infty}^{\infty} f_a(x') e^{-\frac{(x_d^j - x')^2}{4Dc(t^k-s)}} e^{-\frac{(x'-x_s^i)^2}{4Dcs}} dx' \right) ds \right],$$

(39)

where

$$u_0 = \frac{e^{-\mu_{a0}ct^k}}{2\pi Dt^k} e^{-\frac{(x_d^j - x_s^i)^2}{4Dct^k}} \left[1 - \frac{\sqrt{\pi Dct^k}}{\ell} e^{\left(\frac{\sqrt{Dct^k}}{\ell}\right)^2} \operatorname{erfc}\left(\frac{\sqrt{Dct^k}}{\ell}\right) \right].$$

(40)

3. Results

3.1. Determination of Optical Properties

Here we consider reconstructions from measured data in the phantom experiment. Let us first consider initial guesses below.

$$\text{Case 1:} \qquad \mu_a = 0.01\,\text{mm}^{-1}, \quad \mu_s' = 1.0\,\text{mm}^{-1}.$$

(41)

In this Case 1, the following μ_a, μ_s' are obtained both by Algorithm 1 (LM) and Algorithm 3 (hybrid).

$$\mu_a = 0.016\,\text{mm}^{-1}, \quad \mu_s' = 0.63\,\text{mm}^{-1}.$$

(42)

The results for μ_a and μ_s' are shown in Figures 3 and 4, respectively. In Figure 3, reconstructed values of μ_a are plotted against the iteration number k. In the top panel of Figure 3, we see that Algorithm 1 (LM) quickly converges. In Algorithms 2 and 3, we set $k_b = 99$. As is seen in the middle panel of Figure 3, the convergence of Algorithm 2 (SA) is slow. The *temperature* is decreased from $\sigma_e = 10^{-6}$ to $\sigma_e = 10^{-7}$ at the k_bth Monte Carlo step. Similarly, ε is changed from 0.1 to 0.001. Algorithm 2 returns the correct values after many Monte Carlo steps; we have $\sqrt{\hat{V}/W} = 1.02, 1.06$ for μ_a, μ_s', respectively, when $k_a = 9000$ and $k_b = 10000$. Finally, in the bottom panel of Figure 3, we see that the iteration of Algorithm 3 (hybrid) immediately converges after we switch from the Metropolis–Hastings MCMC ($\sigma_e = 10^{-6}, \varepsilon = 0.1$) to Algorithm 1 (LM). No convergence of the Monte Carlo chain is required, and we found $\sqrt{\hat{V}/W} = 8.6$ for μ_a and $= 11.0$ for μ_s' ($k_a = 49, k_b = 99$). A similar behavior is observed for the reconstruction of μ_s' in Figure 4. In the top panel of Figure 4, we find that Algorithm 1 (LM) works best and converges after a few iterations, whereas Algorithm 2 (SA) in the middle panel of Figure 4 has slow convergence and reconstructed values around at $k_b = 99$ are still away from μ'_s in (42). By switching from MH-MCMC to Algorithm 1 (LM) using Algorithm 3 (hybrid), convergence is easily obtained as shown in the bottom panel of Figure 4.

Now we start the simulation by setting the following initial values.

$$\text{Case 2:} \qquad \mu_a = 0.5\,\text{mm}^{-1}, \quad \mu'_s = 1.0\,\text{mm}^{-1}. \tag{43}$$

In this Case 2, Algorithm 1 (LM) fails and returns $\mu_a = 0.068\,\text{mm}^{-1}$ and $\mu'_s = 1.75\,\text{mm}^{-1}$ whereas Algorithm 3 (hybrid) still gives the correct values. The reconstructed values of μ_a and μ'_s at each iteration are shown in Figures 5 and 6, respectively. In Figure 5, the vertical axes of the left three panels are from 0 to 0.6 whereas the vertical axes of the right three panels are from 0 to 0.08. In Figure 6, the vertical axes of the left three panels are from 0 to 2 while the vertical axes of the right three panels are from 0 to 1.3. In the top panels of Figure 5, we see that the reconstruction of μ_a is unsuccessful by Algorithm 1 (LM). Algorithm 2 (SA) approaches μ_a in Equation (42) but still deviates from that value in the middle panels of Figure 5. As shown in the bottom panels of Figure 5, Algorithm 3 (hybrid) converges in a few iterations after switching to Algorithm 1 (LM). In the top panels of Figure 6, Algorithm 1 (LM) fails to reconstruct μ'_s. In the middle panels of Figure 6, reconstructed values by Algorithm 2 (SA) come close to μ'_s in Equation (42) but suffer from slow convergence. In the bottom panels of Figure 6, we see that Algorithm 3 (hybrid) arrives at μ'_s in Equation (42) without any problem.

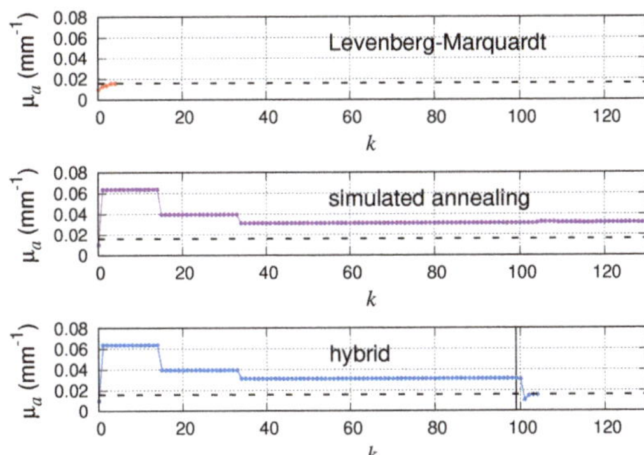

Figure 3. Case 1: Reconstructed μ_a by (top) Algorithm 1, (middle) Algorithm 2, and (bottom) Algorithm 3. The dashed lines show μ_a in Equation (42).

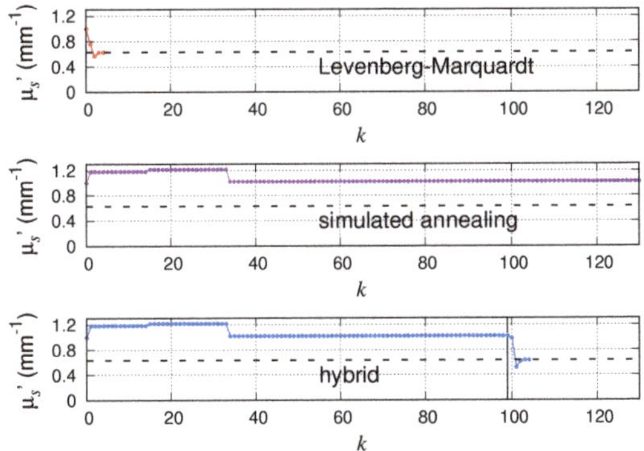

Figure 4. Case 1: Reconstructed μ'_s by (top) Algorithm 1, (middle) Algorithm 2, and (bottom) Algorithm 3. The dashed lines show μ'_s in Equation (42).

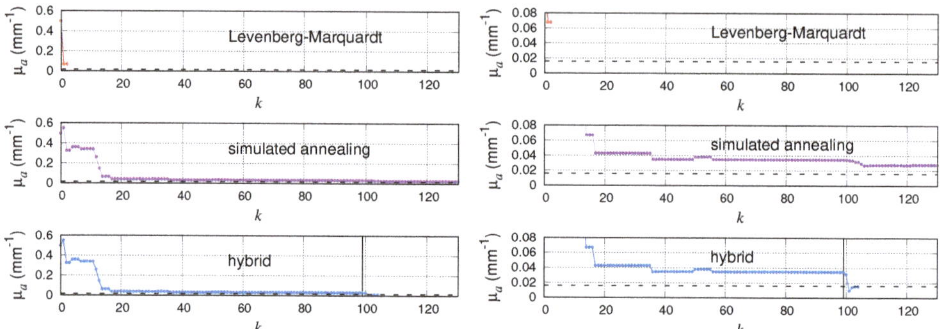

Figure 5. (Left) Case 2: Reconstructed μ_a by (top) Algorithm 1, (middle) Algorithm 2, and (bottom) Algorithm 3. The dashed lines show μ_a in Equation (42). (Right) Same as the left panel but the vertical axes are from 0 to 0.08.

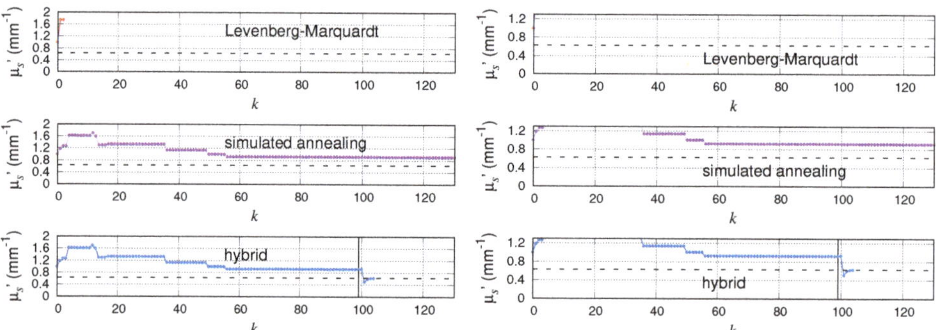

Figure 6. (Left) Case 2: Reconstructed μ_s' by (top) Algorithm 1, (middle) Algorithm 2, and (bottom) Algorithm 3. The dashed lines show μ_s' in Equation (42). (Right) Same as the left panel but the vertical axes are from 0 to 1.3.

The initial guesses for Case 1 are reasonably close to the values found in Equation (42). However, we started with initial guesses which are quite different from the above-mentioned values in Case 2. It is not surprising that Algorithm 1 (LM) does not work for Case 2, but Algorithm 3 (hybrid) can arrive at the correct values. The numerical calculations were performed on Matlab (i7-8700 CPU, 16 GB memory). In the hybrid method, the Metropolis–Hastings MCMC does not reach the steady state at about 100 steps but can be switched to the Levenberg–Marquardt method. For Figures 5 and 6, the computation time was 5 s. The simulated annealing method returns the correct value after about 10,000 steps, but it takes 8 min. The computation for Algorithm 1 (LM) stopped in 0.5 s.

Below we summarize the reconstructed values on Table 1. Although Algorithm 2 (SA) and Algorithm 3 (hybrid) return the same results after a long time, the hybrid scheme is about ten times faster, and moreover there is no need for choosing the low temperature for the latter algorithm. For Algorithm 3, it is not necessary to wait until the burn-in time, but it is enough if the initial rapid change ceases. In Section 3.2, we illustrate that Algorithm 3 works once the Monte Carlo simulation escapes from a local minimum and the algorithm does not require that the calculation reaches the steady state.

Table 1. Reconstructed values of μ_a, μ_s' are shown for Case 1 and Case 2. The units of μ_a, μ_s' are both mm^{-1}. For Algorithm 2, values at 120th Monte Carlo step are shown.

	Case 1 (μ_a, μ_s')	Case 2 (μ_a, μ_s')
initial values	(0.01, 1.0)	(0.5, 1.0)
Algorithm 1 (LM)	(0.016, 0.63)	(0.068, 1.75)
Algorithm 2 (SA)	(0.032, 1.02)	(0.028, 0.92)
Algorithm 3 (hybrid)	(0.016, 0.63)	(0.016, 0.63)

3.2. Determination of Absorption Inhomogeneity

We perform a numerical experiment of diffuse optical tomography using the toy model. Let us consider when the iterative scheme fails. We see $F'(0) = \frac{\partial}{\partial a} u(\rho^j, t^k; \rho^i; a)\big|_{a=0} = 0$. For sufficiently small $\epsilon > 0$, we have $F'(\epsilon) > 0$, $F'(-\epsilon) < 0$, $|F'(\pm\epsilon)| \simeq 0$, and $U - F(\pm\epsilon) < 0$. Therefore, if we start the iteration from $a_0 = \epsilon$, we obtain

$$|a_1| < \epsilon, \quad |a_2| < |a_1|, \quad \dots \tag{44}$$

Thus, the sequence a_k approaches 0 and can never arrive at $\bar{a}\,(= 1.5)$.

Let us consider how the difference $U - F(a)$ depends on a. We introduce

$$h(t) - \frac{1}{t} \exp\left(-\frac{y_0^2}{4Dct}\right)\left[1 - \frac{\sqrt{\pi Dct}}{\ell} e^{\left(\frac{y_0}{2\sqrt{Dct}} + \frac{\sqrt{Dct}}{\ell}\right)^2} \operatorname{erfc}\left(\frac{y_0}{2\sqrt{Dct}} + \frac{\sqrt{Dct}}{\ell}\right)\right]. \tag{45}$$

The following form is obtained using Equation (39). By neglecting noise, we have

$$u(\rho^j, t^k; \rho^i; \bar{a}) - u(\rho^j, t^k; \rho^i; a)$$
$$= \frac{\eta e^{-\mu_{a0} c t^k}}{(2\pi D)^2} \int_0^{t^k} h(t^k - s) h(s) \left[\int_{-\infty}^{\infty} d_{\bar{a}}(a, x') \left(1 - \tanh x'^2\right) e^{-\frac{(x_d^j - x')^2}{4Dc(t^k - s)}} e^{-\frac{(x' - x_s^i)^2}{4Dcs}} dx'\right] ds, \tag{46}$$

where $d_{\bar{a}}(a, x') = \xi(\bar{a}, x') - \xi(a, x')$ with

$$\xi(a, x') = a^2 \left[a + 3\left(1 + \frac{\tanh x'^2}{10}\right)\right]. \tag{47}$$

For a given x', the function $|d_{\bar{a}}(a; x')|^2$ has one global minimum at $a = \bar{a}$, one local minimum at $a = -2\left(1 + \frac{\tanh x'^2}{10}\right)$, and one local maximum at $a = 0$. Therefore, the above expression implies that Algorithm 1 (LM) fails when the initial value a_0 is negative and the correct value $\bar{a} = 1.5$ is reconstructed only for a positive initial guess. Indeed, the computation ends up with -2.05 if we start the iteration from $a_0 = -0.1$ (see below) or -0.01, and the value 1.68 is obtained when $a_0 = 0.01$. Figure 7 shows $|d_{\bar{a}}(a, x)|^2$ for $\bar{a} = 1.5$ and $\tanh(x^2) = 0.5$.

In Figure 8, we plot reconstructed values of a against iteration numbers. The initial value is set to $a_0 = -0.1$. The reconstruction by Algorithm 1 (LM) is shown in the top panel of Figure 8. As we can predict from Figure 7, Algorithm 1 (LM) converges to the local minimum and can never arrive at the global minimum. Monte Carlo simulation can jump over the local maximum and approach the global minimum, but keeps fluctuating as shown in the middle panel of Figure 8. We initially set $(\sigma_e, \varepsilon) = (10^{-6}, 0.5)$ for Algorithms 2 and 3. In Algorithm 2, we set $(\sigma_e, \varepsilon) = (10^{-7}, 0.005)$ after the *temperature* decreases at the k_bth Monte Carlo step. In the bottom panel of Figure 8, Algorithm 3 (hybrid) successfully arrives at the global minimum. Using Matlab, the computation time was 17 min while we needed 3 hr for Algorithm 2 (SA) (1000 steps).

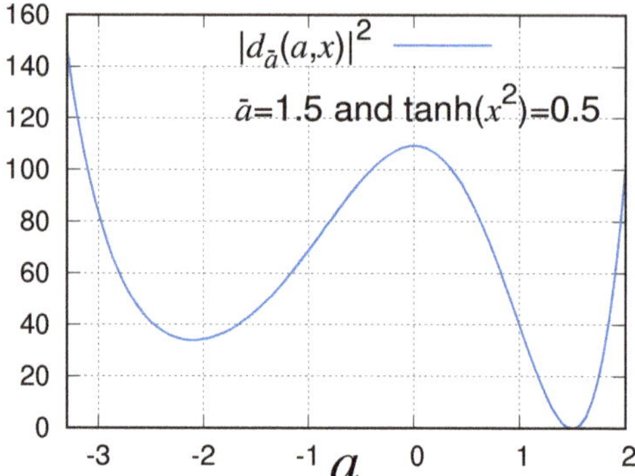

Figure 7. The function $|d_{\bar{a}}(a, x)|^2$ ($\bar{a} = 1.5$) is plotted when $\tanh(x^2) = 0.5$.

After the initial time with large jumps, it is found that we can set $k_b = 99$ in our simulation of the MCMC-iterative hybrid method. The correct value of $\bar{a}$ is reconstructed by Algorithm 3 (hybrid) even when the simulation experiences a local minimum. Starting from $a_{k_b} = 1.8257$, the calculation stops at $a^* = 1.6841$ ($a^* = a_{k_b+3}$). We note that the reconstructed a^* is not exactly 1.5 due to noise.

When the initial guess $a_0 = -0.1$ lies in the valley of the local minimum (see Figure 7), the reconstructed a by Algorithm 3 (hybrid) moves to the valley of the global minimum with the help of Monte Carlo simulation as shown in Figure 8, while the reconstructed a by Algorithm 1 (LM) falls to the local minimum as the nature of Newton-type iterative methods. For Algorithm 3 (hybrid), we obtained $a^* = 1.6841$. There is a possibility that negative reconstructed values are obtained by Algorithm 2 (SA) and Algorithm 3 (hybrid) for the first few iterations if different pseudo-random numbers are used. These negative reconstructed values, however, will turn positive and the behavior of reconstructed a by Algorithm 2 (SA) and Algorithm 3 (hybrid) is always more or less similar to the middle and bottom panels of Figure 8. For the initial guesses of $a_0 = -0.01$ and $a_0 = 0.01$, Algorithm 3 (hybrid) works without any problem and Algorithm 2 (SA) also returns a reasonable result after a sufficiently large number of iterations (results not shown).

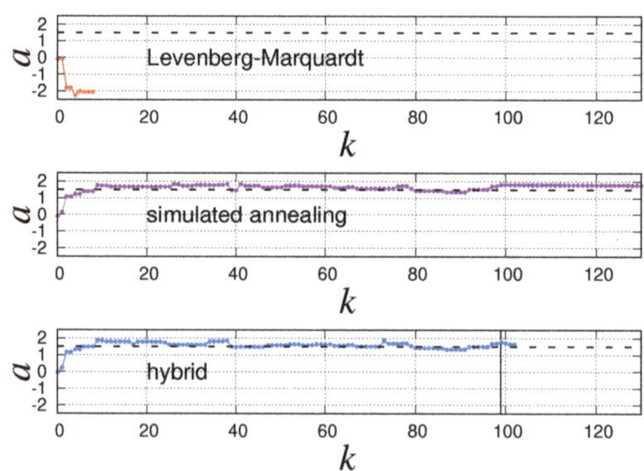

Figure 8. Reconstructed $\bar{a}$ ($= 1.5$) by (top) Algorithm 1 (LM), (middle) Algorithm 2 (SA), and (bottom) Algorithm 3 (hybrid) for the initial value $a_0 = -0.1$.

4. Discussion

In this paper, we have proposed a hybrid numerical scheme which uses Markov chain Monte Carlo in the first step and then uses an iterative method in the second step. We switch from MH-MCMC to LM by observing proposed parameter values. For the typical jump length of parameters in MH-MCMC,

$\varepsilon = 0.1$ was used in Section 3.1 and $\varepsilon = 0.5$ was chosen in Section 3.2. Although these values were set so that the MH-MCMC calculation was efficiently performed, other choices of ε are also possible. The proposed scheme is quite general and can be applied to different inverse problems in NIRS which are solved by iterative methods even when the forward problem must be solved fully numerically with finite difference method or finite element method. It is an interesting future study how the hybrid scheme can be extended to diffuse optical tomography, which has many unknowns.

More sophisticated algorithms than the Metropolis–Hastings Markov chain Monte Carlo used in this paper have been proposed to overcome slow convergence. The delayed rejection scheme reduces the net rejection rate [39]. In the adaptive Metropolis algorithm, parameters in the proposal distribution are adjusted during Monte Carlo steps [40]. The DRAM algorithm, which combines the above-mentioned two schemes, was also proposed [41]. Two-level MCMC algorithms [42,43] and a multi-level MCMC [23] have been investigated to improve the MCMC algorithm. Such Monte Carlo schemes might improve the first step of our hybrid method by finding the valley of the global minimum more easily.

Related to the Metropolis–Hastings Markov chain Monte Carlo algorithm, quantum annealing [44] has been developed in addition to simulated annealing. Aiming at escaping from local minima, brute-force search and genetic algorithm [45] are also well-known optimization algorithms. The introduction of these methods in NIRS may be found useful in the future.

For the clinical use of NIRS, it has been recognized from early days that finding absolute values of the absorption and scattering coefficients is important [2]. In Ref. [15], it is emphasized that the obtained absolute values highly depend on the starting values of the initial estimate for the study of the infant brain with their measurement system. By performing Markov chain Monte Carlo before using standard iterative schemes, we may automatically acquire good initial values. Such clinical application is a natural next step of our research.

Author Contributions: Conceptualization, M.M. and G.N.; methodology, Y.J. and M.M.; software, Y.J. and M.M.; formal analysis, Y.J.; investigation, Y.H.; data curation, Y.J. and Y.H.; writing—original draft preparation, M.M.

Funding: The first author (Y.J.) is supported by National Natural Science Foundation of China (No. 11971121). Y.H. and M.M. acknowledge support from Grant-in-Aid for Scientific Research (17H02081) of the Japan Society for the Promotion of Science. M.M. also acknowledges support from Grant-in-Aid for Scientific Research (17K05572) of the Japan Society for the Promotion of Science and from the JSPS A3 foresight program: Modeling and Computation of Applied Inverse Problems. G.N. was supported by Grant-in-Aid for Scientific Research (15K21766 and 15H05740) of the Japan Society for the Promotion of Science.

Acknowledgments: The authors appreciate Goro Nishimura for fruitful discussion on optical tomography and the use of his computer facilities at Hokkaido University. We thank Tatsunori Emi, Takeshi Iwasaki, and Tomoya Sugiyama for helping related experiments which stimulated the present work.

Conflicts of Interest: The authors declare no conflict of interest. The funders had no role in the design of the study; in the collection, analyses, or interpretation of data; in the writing of the manuscript, or in the decision to publish the results.

References

1. Boas, D.A.; Brooks, D.H.; Miller, E.L.; DiMarzio, C.A.; Kilmer, M.; Gaudette, R.J.; Zhang, Q. Imaging the body with diffuse optical tomography. *IEEE Signal Process. Mag.* **2001**, *18*, 57–75. [CrossRef]

2. Hoshi, Y. Towards the next generation of near-infrared spectroscopy. *Phil. Trans. R. Soc. A* **2011**, *369*, 4425–4439. [CrossRef] [PubMed]

3. Levenberg, K. A method for the solution of certain non-linear problems in least squares. *Q. Appl. Math.* **1944**, *2*, 164–168. [CrossRef]

4. Marquardt, D.W. An algorithm for least-squares estimation of nonlinear parameters. *SIAM J. Appl. Math.* **1963**, *11*, 431–441. [CrossRef]

5. Lange, F.; Tachtsidis, I. Clinical brain monitoring with time domain NIRS: A review and future perspectives. *Appl. Sci.* **2019**, *9*, 1612. [CrossRef]

6.   Chance, B.; Zhuang, Z.; Unah, C.; Alter, C.; Lipton, L. Cognition-activated low frequency modulation of light absorption in human brain. *Proc. Natl. Acad. Sci. USA* **1993**, *90*, 3770–3774. [CrossRef]

7.   Hoshi, Y.; Tamura, M. Detection of dynamic changes in cerebral oxygenation coupled to neuronal function during mental work in man. *Neurosci. Lett.* **1993**, *150*, 5–8. [CrossRef]

8.   Kato, T.; Kamei, A.; Takashima, S.; Ozaki, T. Human visual cortical function during photic stimulate on monitoring by means of near-infrared spectroscopy. *J. Cereb. Blood Flow Metab.* **1993**, *13*, 516–520. [CrossRef]

9.   Villringer, A.; Plank, J.; Hock, C.; Schleikofer, L.; Dirnagl, U. Near-infrared spectroscopy (NIRS): A new tool to study hemodynamic changes during activation of brain function in human adults. *Neurosci. Lett.* **1993**, *154*, 101–104. [CrossRef]

10. Laidevant, A.; da Silva, A.; Berger, M.; Dinten, J.-M. Effects of the surface boundary on the determination of the optical properties of a turbid medium withtime-resolved reflectance. *Appl. Opt.* **2006**, *45*, 4756–4764. [CrossRef]

11. Gibson, A.P.; Hebden, J.C.; Arridge, S.R. Recent advances in diffuse optical imaging. *Phys. Med. Biol.* **2005**, *50*, R1–R43. [CrossRef]

12. Arridge, S.R. Methods in diffuse optical imaging. *Philos. Trans. R. Soc. A* **2011**, *369*, 4558–4576. [CrossRef]

13. Arridge, S.R.; Schweiger, M. A gradient-based optimization scheme for optical tomography. *Opt. Express* **1998**, *2*, 213–226. [CrossRef]

14. Choe, R.; Corlu, A.; Lee, K.; Durduran, T.; Konecky, S.D.; Grosicka-Koptyra, M.; Arridge, S.R.; Czerniecki, B.J.; Fraker, D.L.; DeMichele, A.; et al. Diffuse optical tomography of breast cancer during neoadjuvant chemotherapy: A case study with comparison to MRI. *Med. Phys.* **2005**, *32*, 1128–1139. [CrossRef]

15. Hebden, J.C.; Gibson, A.; Austin, T.; Yusof, R.M.; Everdell, N.; Delpy, D.T.; Arridge, S.R.; Meek, J.H.; Wyatt, J.S. Imaging changes in blood volume and oxygenation in the newborn infant brain using three-dimensional optical tomography. *Phys. Med. Biol.* **2004**, *49*, 1117–1130. [CrossRef]

16. Arridge, S.R.; Schweiger, M. Image reconstruction in optical tomography. *Philos. Trans. R. Soc. Lond. B* **1997**, *352*, 717–726. [CrossRef]

17. Schweiger, M.; Arridge, S. The toast++ software suite for forward and inverse modeling in optical tomography. *J. Biomed. Opt.* **2014**, *19*, 040801. [CrossRef]

18. Zhao, H.; Gao, F.; Tanikawa, Y.; Homma, K.; Yamada, Y. Time-resolved diffuse optical tomographic imaging for the provision of both anatomical and functional information about biological tissue. *Appl. Opt.* **2005**, *44*, 1905–1916. [CrossRef]

19. Arridge, S.R. Optical tomography in medical imaging. *Inverse Probl.* **1999**, *15*, R41–R93. [CrossRef]

20. Arridge, S.R.; Schotland, J.C. Optical tomography: Forward and inverse problems. *Inverse Probl.* **2009**, *25*, 123010. [CrossRef]

21. Arridge, S.R.; Kaipio, J.P.; Kolehmainen, V.; Schweiger, M.; Somersalo, E.; Tarvainen, T.; Vauhkonen, M. Approximation errors and model reduction with an application in optical diffusion tomography. *Inverse Probl.* **2006**, *22*, 175–195. [CrossRef]

22. Bal, G.; Langmore, I.; Marzouk, Y. Bayesian inverse problems with monte carlo forward models. *Inverse Probl. Imaging* **2013**, *7*, 81–105.

23. Langmore, I.; Davis, A.B.; Bal, G. Multipixel retrieval of structural and optical parameters in a 2-d scene with a path-recycling monte carlo forward model and a new bayesian inference engine. *IEEE Trans. Geosci. Remote Sens.* **2013**, *51*, 2903–2919. [CrossRef]

24. Bamett, A.H.; Culver, J.P.; Sorensen, A.G.; Dale, A.; Boas, D.A. Robust inference of baseline optical properties of the human head with three-dimensional segmentation from magnetic resonance imaging. *Appl. Opt.* **2003** *42*, 3095–3108.

25. Groenhuis, R.A.J.; Ferwerda, H.A.; Ten Bosch, J.J. Scattering and absorption of turbid materials determined from reflection measurements. 1: Theory. *Appl. Opt.* **1983**, *22*, 2456–2462. [CrossRef]

26. Egan, W.G.; Hilgeman, T.W. *Optical Properties of Inhomogeneous Materials*; Academic Press: New York, NY, USA, 1979.

27. Carslaw, H.S.; Jaeger, J.C. *Conduction of Heat in Solids*; Oxford University Press: London, UK, 1959.

28. Yosida, K.; Ito, S. *Functional Analysis and Differential Equations*; Iwanami: Tokyo, Japan, 1976. (In Japanese)

29. Hielscher, A.H.; Jacques, S.L.; Wang, L.; Tittel, F.K. The influence of boundary conditions on the accuracy of diffusion theory in time-resolved reflectance spectroscopy of biological tissues. *Phys. Med. Biol.* **1995**, *40*, 1957–1975. [CrossRef]

30. Machida, M.; Nakamura, G. Born series for the photon diffusion equation perturbing the Robin boundary condition. *arXiv* **2017**, arXiv:1706.04500.

31. Ishimaru, A. *Wave Propagation and Scattering in Random Media*; Academic Press: New York, NY, USA, 1978.

32. Fletcher, R. *A Modified Marquardt Subroutine for Nonlinear Least Squares (Report AERE-R 6799)*; The Atomic Energy Research Establishment: Harwell, UK, 1971.

33. Kaipio, J.; Somersalo, E. *Statistical and Computational Inverse Problems*; Springer: New York, NY, USA, 2005.

34. Nakamura, G.; Potthast, R. *Inverse Modeling*; IOP Publishing: Bristol, UK, 2015.

35. Iglesias, M.A.; Lin, K.; Stuart, A.M. Well-posed bayesian geometric inverse problems arising in subsurface flow. *Inverse Probl.* **2014**, *30*, 114001. [CrossRef]

36. Kirkpatrick, S.; Gelatt, C.D., Jr.; Vecchi, M.P. Optimization by simulated annealing. *Science* **1983**, *220*, 671–680. [CrossRef]

37. Gelman, A.; Rubin, D.B. Inference from iterative simulation using multiple sequences. *Stat. Sci.* **1992** 7, 457–472. [CrossRef]

38. Martinez W.L.; Martinez, A.R. *Computational Statistics Handbook with MATLAB*; Chapman and Hall/CRC: London, UK, 2015.

39. Tierney, L. Markov chains for exploring posterior distributions. *Ann. Stat.* **1994**, *22*, 1701–1762. [CrossRef]

40. Haario, H.; Saksman, E.; Tamminen, J. An adaptive metropolis algorithm. *Bernoulli* **2001**, *7*, 223–242. [CrossRef]

41. Haario, H.; Laine, M.; Mira, A.; Saksman, E. Dram: Efficient adaptive mcmc. *Stat. Comput.* **2006**, *16*, 339–354. [CrossRef]

42. Christen, J.A.; Fox, C. Markov chain monte carlo using an approximation. *J. Comput. Graph. Stat.* **2005**, *14*, 795–810. [CrossRef]

43. Efendiev, Y.; Hou, T.; Luo, W. Preconditioning markov chain monte carlo simulations using coarse-scale models. *SIAM J. Sci. Comput.* **2006**, *28*, 776–803. [CrossRef]

44. Kadowaki, T.; Nishimori, H. Quantum annealing in the transverse ising model. *Phys. Rev. E* **1998**, *58*, 5355–5363. [CrossRef]

45. Holland, J.H. *Adaptation in Natural and Artificial Systems: An Introductory Analysis with Applications to Biology, Control and Artificial Intelligence*; University of Michigan Press: Ann Arbor, MI, USA, 1975.

applied sciences

Article

Improving Localization of Deep Inclusions in Time-Resolved Diffuse Optical Tomography

David Orive-Miguel [1,2,*], Lionel Hervé [1,*], Laurent Condat [2] and Jérôme Mars [2]

1   CEA, LETI, MINATEC Campus, F-38054 Grenoble, France
2   Univ. Grenoble Alpes, CNRS, Grenoble INP, GIPSA-Lab, 38000 Grenoble, France;
    Laurent.Condat@gipsa-lab.grenoble-inp.fr (L.C.); jerome.mars@gipsa-lab.grenoble-inp.fr (J.M.)
*   Correspondence: davidorive@gmail.com (D.O.-M.); lionel.herve@cea.fr (L.H.)

Received: 30 September 2019; Accepted: 8 December 2019; Published: 12 December 2019

Abstract: Time-resolved diffuse optical tomography is a technique used to recover the optical properties of an unknown diffusive medium by solving an ill-posed inverse problem. In time-domain, reconstructions based on datatypes are used for their computational efficiency. In practice, most used datatypes are temporal windows and Fourier transform. Nevertheless, neither theoretical nor numerical studies assessing different datatypes have been clearly expressed. In this paper, we propose an overview and a new process to compute efficiently a long set of temporal windows in order to perform diffuse optical tomography. We did a theoretical comparison of these large set of temporal windows. We also did simulations in a reflectance geometry with a spherical inclusion at different depths. The results are presented in terms of inclusion localization and its absorption coefficient recovery. We show that (1) the new windows computed with the developed method improve inclusion localization for inclusions at deep layers, (2) inclusion absorption quantification is improved at all depths and, (3) in some cases these windows can be equivalent to frequency based reconstruction at GHz order.

Keywords: diffuse optical tomography; inverse problem; datatypes; diffusion approximation

---

1. Introduction

Time-resolved diffuse optical technology is an emerging photonic technique to continuously quantify the concentrations of several physiological chromophores such as hemoglobin, lipid or collagen. Successful measurements have been done at different human body locations such as brain [1,2], breast [3] or thyroid [4]. An interesting extension of this technology is to perform diffuse optical tomography [5,6] by computing three-dimensional maps of oxy- and deoxy-hemoglobin; in this approach, photon propagation is modeled in a computer and results are compared with experimental measurements. Then, optical parameters in the model are updated by solving an ill-posed inverse problem until the difference between the model and the data is negligible.

In order to have accurate results, it is important to have a realistic model for photon propagation in tissues. The most accurate approach is to use the integro-differential Radiative Transfer Equation (RTE) [7]. Although RTE has some analytical solutions [8–10] these just hold for simple geometries and cannot be applied to more complex environments, such as an human head or breast models, without making strong assumptions. Apart from classical Monte-Carlo simulations [11,12], new numerical methods have been proposed, some of which are the one-way RTE [13] or hybrid RTE [14]; nevertheless, they are still highly time-consuming for real-time applications.

Due to the previous reasons, usually the first-order time-dependant Diffusion Approximation is used,

$$\frac{1}{c}\frac{\partial \phi(\mathbf{r},t)}{\partial t} - \nabla \cdot (D(\mathbf{r})\nabla\phi(\mathbf{r},t)) + \mu_a(\mathbf{r})\phi(\mathbf{r},t) = S(\mathbf{r},t), \tag{1}$$

where c is the speed of light in the medium, ϕ is the fluence rate, $D = (3\mu_s')^{-1}$[15,16], μ_a and μ_s' are the diffusion, absorption and reduced scattering coefficients respectively, and S is the source function. This approximation, based on spherical harmonics [7], is valid assuming that (1) the reduced scattering coefficient is much greater than absorption ($\mu_s' \gg \mu_a$) and that (2) detectors are sufficiently far from light sources ($>10/\mu_s'$). These assumptions hold in most practical cases and although higher order approximations have also been suggested [17–19] the extra computational cost usually does not compensate the small gain in accuracy.

The time-dependant Equation (1) can be solved using Finite Element Method [20] but it is still a slow process since time has to be discretized in short steps in order to guarantee stability and convergence. For this reason, several authors have proposed to use temporal windows datatypes since they are faster to compute than the fluence rate itself [21,22]. Nevertheless, as indicated in [23], the temporal windows datatypes are faster to compute than the fluence rate if they have the form $w(t) = t^n e^{-pt}$ where $n \in \mathbb{N}$ is in the set of natural numbers and $p \in \mathbb{C}$ is in the set of complex numbers. This formula incorporates the Fourier transform (when $n = 0$ and p is an imaginary number), Laplace transform (when $n = 0$ and p is a complex number), standard moments (when n is an integer and $p = 0$) and Mellin-Laplace moments (when n is an integer and p is a real number).

Some authors prefer to use Fourier Transform (FT) datatypes. In [2] the magnitude and phase of 100 MHz frequency from a time-resolved signal was used for the reconstruction. A similar approach was published in [24] to retrieve the fluorescence lifetime distribution. A related idea is to directly use frequency-domain technology [25–27] to retrieve optical properties or fluorescence lifetime [28,29]. These reconstruction approaches based on solving the inverse problem in frequency domain are a good alternative to temporal windows but are not equivalent since only MHz order frequencies are used.

In this paper, we propose a new technique for computing fluence rate temporal windows datatypes without constraining ourselves to $w(t) = t^n e^{-pt}$ form and preserving low computing times. We analyse different types of windows in terms of temporal selectivity, noise influence and computational efficiency. Described datatypes are compared in simulations with respect to the state-of-the-art reconstruction technique. Finally, we reformulate the new method as a frequency based reconstruction using up to GHz order.

2. Methods

In this section, an introduction to time- and datatypes-based reconstruction will be given and a new method for computing temporal windows will be described. Further on, different temporal windows and their noise influence on reconstruction methods will be studied. After, the setup of the numerical simulations are presented.

2.1. Time- and Datatypes-Based Reconstruction

Time-resolved diffuse optical tomography can be posed in terms of time bins or datatypes; in both cases time-resolved measurements from the detectors are used. Absorption of the medium Ω can be recovered by using iteratively the linearized Born approximation,

$$\delta\phi_{sd}(t) = \int_{\Omega} \left[\phi_s^{(k)} * G_d^{(k)} \right] (\mathbf{r}, t)\, \delta\mu_a^{(k)}(\mathbf{r})\, d\mathbf{r}, \tag{2}$$

where $\delta\phi_{sd} = \phi_{sd} - \phi_{sd}^{(k)}$ is the fluence rate difference at detector d when source at position s was activated; in ϕ_{sd} the experimental factors are included. $G_d^{(k)}(\mathbf{r}, t)$ indicates the value of the Green's function at every point in the domain assuming the source is located at position d. $\phi_s^{(k)}(\mathbf{r}, t)$ is the fluence rate at every point in the domain for a source located at s. $\delta\mu_a^{(k)}(\mathbf{r})$ is the absorption update at each iteration. The superscript indicates that absorption values obtained at iteration k were used. The reconstructed absorption will be the result of adding all $\delta\mu_a$ to the initial absorption estimation, that is, $\mu_a = \delta\mu_a^{(1)} + \delta\mu_a^{(2)} + \ldots$ A simplified sketch of the reconstruction procedure is shown at Figure 1. Equation (2) can also be extended to include changes in scattering.

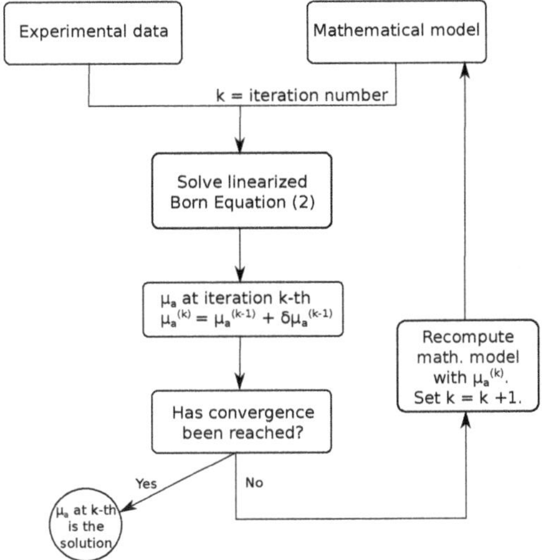

Figure 1. Absorption reconstruction algorithm based on linearized Born equation (Equation (2)).

Equation (2) can be discretized into a full time-resolved linear system,

$$
\begin{bmatrix} L_{sd}(t_1) \\ L_{sd}(t_2) \\ \vdots \\ L_{sd}(t_N) \end{bmatrix} = \begin{bmatrix} J_{sd}(t_1,r_1) & J_{sd}(t_1,r_2) & \cdots & J_{sd}(t_1,r_M) \\ J_{sd}(t_2,r_1) & J_{sd}(t_2,r_2) & \cdots & J_{sd}(t_2,r_M) \\ \vdots & & & \\ J_{sd}(t_N,r_1) & J_{sd}(t_N,r_2) & \cdots & J_{sd}(t_N,r_M) \end{bmatrix} \begin{bmatrix} \delta\mu_a(r_1) \\ \delta\mu_a(r_2) \\ \vdots \\ \delta\mu_a(r_M) \end{bmatrix},
\tag{3}
$$

where $J_{sd}(t,r)$ and $L_{sd}(t)$ are column vectors with one entry for each source-detector pair. J matrix is known as sensitivity matrix and L is the discretization of $\delta\phi_{sd}(t)$. In this case the system matrix size is $(N \cdot S \cdot D) \times M$ where N is the number of time bins considered, S is the number of sources, D is the number of detectors and M is the number of mesh nodes. Expressing the reconstruction in terms of time has two main drawbacks: (1) the number of time bins is high (usually of a few thousands order) which makes the system very large and (2) it is necessary to simulate the distribution of photon time of flight (DTOF) for each source and detector, which is very time-consuming.

Due to previous reasons, it is better to solve the problem of Equation (2) in terms of datatypes. A datatype is a filter that is applied to the DTOF to reduce the size of the problem. For example, when a temporal window filter is applied to the DTOF, the problem is reduced to

$$
\begin{bmatrix} L_{sd}(w_1) \\ L_{sd}(w_2) \\ \vdots \\ L_{sd}(w_W) \end{bmatrix} = \begin{bmatrix} J_{sd}(w_1,r_1) & J_{sd}(w_1,r_2) & \cdots & J_{sd}(w_1,r_M) \\ J_{sd}(w_2,r_1) & J_{sd}(w_2,r_2) & \cdots & J_{sd}(w_2,r_M) \\ \vdots & & & \\ J_{sd}(w_W,r_1) & J_{sd}(w_W,r_2) & \cdots & J_{sd}(w_W,r_M) \end{bmatrix} \begin{bmatrix} \delta\mu_a(r_1) \\ \delta\mu_a(r_2) \\ \vdots \\ \delta\mu_a(r_M) \end{bmatrix},
\tag{4}
$$

whose matrix size is $W \cdot S \cdot D \times M$, where W is the number of windows filters used. In practice, this system will be much smaller than the system of Equation (3) since only a few dozens of windows are usually needed.

The Fourier transform is also a datatype that converts the reconstruction problem to,

$$
\begin{bmatrix} L_{sd}(f_1) \\ L_{sd}(f_2) \\ \vdots \\ L_{sd}(f_F) \end{bmatrix} = \begin{bmatrix} J_{sd}(f_1,r_1) & J_{sd}(f_1,r_2) & \cdots & J_{sd}(f_1,r_M) \\ J_{sd}(f_2,r_1) & J_{sd}(f_2,r_2) & \cdots & J_{sd}(f_2,r_M) \\ \vdots & & & \\ J_{sd}(f_F,r_1) & J_{sd}(f_F,r_2) & \cdots & J_{sd}(f_F,r_M) \end{bmatrix} \begin{bmatrix} \delta\mu_a(r_1) \\ \delta\mu_a(r_2) \\ \vdots \\ \delta\mu_a(r_M) \end{bmatrix},
\tag{5}
$$

where the matrix and the vector on the left hand side are complex. The matrix size is $F \cdot S \cdot D \times M$, where F is the number of frequency bins. It is interesting to notice that although frequency based systems use frequencies in the range of MHz, in time-resolved systems, after applying the FT, the frequencies extend to the range of GHz. As it will be shown later, frequencies of GHz order provide important information for deep inclusions.

For the following discussion, it is important to note the difference between FT- and temporal windows-based reconstruction. In the first case, the reconstruction is done directly in the frequency domain. However, in the second case, the reconstruction is performed by using temporal windows, which are proposed to be computed in the frequency domain, due to computational efficiency reasons.

2.2. A Novel Method to Compute Temporal Windows

Let us define $u(t)$ as the DTOF simulated at a detector and $w(t)$ as an arbitrary shaped temporal window. A datatype Γ is defined as

$$
\int_0^\infty u(t)w(t)\, dt = \Gamma,
\tag{6}
$$

since $u(t) = 0$ for $t < 0$. If $w(t) = t^n e^{-pt}$ then Γ can be calculated directly without computing $u(t)$ explicitly [30]. However, until now, no method has been published that uses a different window form without computing explicitly $u(t)$ for each time step. The approach described in this paper proposes to compute the datatypes in the frequency domain; this can be done by using the Plancherel theorem as follows:

$$
\Gamma = \int_{-\infty}^\infty u(t)\overline{w(t)}\, dt = \int_{-\infty}^\infty U(f)\overline{W(f)}\, df,
\tag{7}
$$

where uppercase character denotes the Fourier transform defined as $U(f) = \int_{-\infty}^\infty u(t)e^{-2\pi i f t}\, dt$ (e.g., $W(f)$ is the Fourier transform of $w(t)$) and $\overline{W(f)}$ denotes the conjugate of $W(f)$. If $\overline{W(f)}$ or $U(f)$ are non-zero at a small interval of the frequency domain, the integral of Equation (7) can be approximated numerically by using few frequencies, see Figure 2. This new approach allows fast computation of a larger set of temporal windows.

To approximate Γ using few frequencies, it is important to ensure that window spectra decay rapidly. Dispersion of the windows after applying Fourier transform can be controlled by using the *uncertainty principle*. This principle states that the narrower a function $g(t)$ is, the more spread its Fourier transform $G(f)$ will be. That is, if it is wanted to have a sharp window in time domain then the integral in frequency domain will be more spread around the zero frequency. Defining the dispersion of a function in time and frequency domain as

$$
D_0(g) = \frac{\int_{-\infty}^\infty t^2 |g(t)|^2\, dt}{\int_{-\infty}^\infty |g(t)|^2\, dt}, \qquad D_0(G) = \frac{\int_{-\infty}^\infty f^2 |G(f)|^2\, df}{\int_{-\infty}^\infty |G(f)|^2\, df}
\tag{8}
$$

the *uncertainty principle* states that if $g(t)$ is absolutely continuous and the functions $tg(t)$ and $g'(t)$ are square integrable (i.e., they are in L^2 space) then [31]

$$
D_0(g)D_0(G) \geq \frac{1}{16\pi^2},
\tag{9}
$$

where G is the Fourier transform of $g(t)$ and the equality holds only for the Gaussian density function centered at $t = 0$.

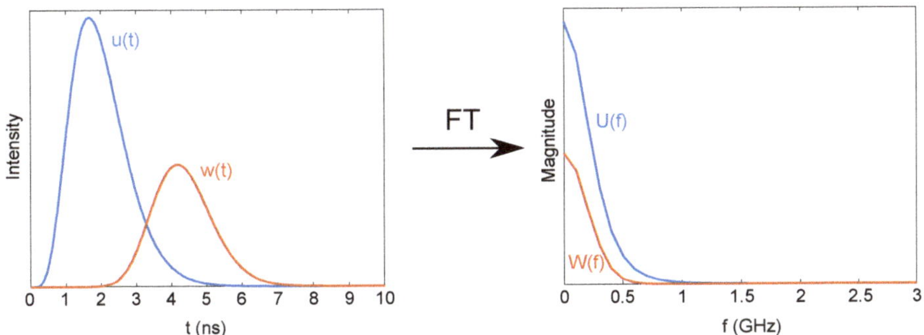

Figure 2. (**Left**) Distribution of photon time of flight (DTOF) $u(t)$ and temporal window $w(t)$ in time domain. (**Right**) Magnitude of the Fourier transforms of $u(t)$ and $w(t)$; it is feasible to aproximate numerically the integral in frequency domain since spectra decay rapidly.

Computational Aspects

The reported technique has to be much faster to compute than the resolution of time-dependant Diffusion Approximation equation in order to be useful in practice. In the following section, some important computational aspects related to the time complexity will be shown.

First, to compute $\phi(\mathbf{r}, t_n)$ in Equation (1), it is needed to know the previous values of $\phi(\mathbf{r}, t_{n-i})$, where t_n is the n-th discrete time bin and $t_{n-i} < t_n$. This implies that to compute $\phi(\mathbf{r}, t_n)$ previously $n - 1$ linear systems have to be solved. However, in the frequency resolved version of Equation (1), there is no such restriction, since the fluence rate at any frequency can be computed independently. This is the main reason why the integral of Equation (7) is faster to compute in the frequency domain. Therefore, the described technique is easily parallelizable and could be implemented in Graphics Processing Units (GPU) hardware; some labs have already implemented state-of-the-art photon propagation models using GPU technology [32,33].

Another important property is that for real functions, Fourier transform coefficients will have symmetric real part and anti-symmetric imaginary part. From this follows that only positive (or negative) frequencies must be computed and only half of the integral needs to be approximated,

$$\int_{-\infty}^{\infty} U(f)\overline{W(f)}\,\mathrm{d}f = 2\int_{0}^{\infty} U(f)\overline{W(f)}\,\mathrm{d}f \approx 2\sum_{i=0}^{N} U(f_i)\,\overline{W(f_i)}\,\Delta f, \tag{10}$$

where N frequencies were used in the approximation.

However, the proposed method also suffers from one limitation. When computing temporal windows with shape $w(t) = t^n e^{-pt}$, the linear system to solve has the form $\mathbf{A}\mathbf{x} = \mathbf{b}^{(n)} + \mathbf{x}^{(n-1)}$, that is, as the order n of the window is increased, only the right part of the systems changes, but the matrix system is constant. This encourages to use, for example, LU factorization of the matrix $\mathbf{A}$ to solve all the systems quickly. However, when solving the Diffusion Approximation equation at frequency domain, the systems have the form $\mathbf{A}(f)\mathbf{x} = \mathbf{b}$ where f is the frequency. In this case, the matrix $\mathbf{A}$ needs to be factorized for each frequency, which implies an increase in the computational cost. Nevertheless, Diffusion Approximation at frequency domain is highly parallelizable because of the independence between frequencies. Therefore, unlike for model of windows $w(t) = t^n e^{-pt}$ the equations at each frequency-domain could be fastly solved in parallel with a GPU.

2.3. Temporal Windows

In the following subsection, most used temporal windows (standard and Mellin-Laplace moments; in this work, the words moments and windows will be used interchangeably) and proposed windows for the new method (Gaussian, Tukey and Poisson windows) are described. An analysis regarding their optimality for numerical integration is also included.

2.3.1. Standard Moments

Standard moments have been used to retrieve the optical properties of homogeneous media [34] and layered models [35]. Standard moments windows are defined as $w(t) = t^n H(t)$, where $n \in \mathbb{N}$ and $H(t)$ is the Heaviside step function. These types of windows can be computed fast with state-of-the-art techniques [22]. The Fourier transform of standard moment window is

$$W(f) = \frac{n!}{(i2\pi f)^{n+1}}, \tag{11}$$

whose magnitude is $n! \cdot (2\pi f)^{-(n+1)}$. Therefore, the standard moments in time domain are equivalent to

$$\int_{-\infty}^{\infty} u(t)\, t^n H(t)\, \mathrm{d}t = \int_{0}^{\infty} u(t)\, t^n\, \mathrm{d}t = \int_{-\infty}^{\infty} U(f)\, \overline{(i2\pi f)^{-n-1} n!}\, \mathrm{d}f, \tag{12}$$

where the term $\overline{(i2\pi f)^{-n-1} n!}$ can be considered as a window in frequency domain. This window magnitude tends to a Dirac delta distribution as n goes to infinity, which is obvious since $t^n H(t)$ will approach a step function with infinity value at $t >= 0$ and the only non-zero magnitude frequency in the spectrum will be the zero frequency. Therefore, the higher the order, the smaller the frequency range covered by the windows.

If the standard moments are centralized by the time of flight $\langle t \rangle = \int_{-\infty}^{\infty} u(t)\, tH(t)\, \mathrm{d}t$ then

$$
\begin{aligned}
\int_{-\infty}^{\infty} u(t)\, (t - \langle t \rangle)^n H(t)\, \mathrm{d}t &= \int_{0}^{\infty} u(t)(t - \langle t \rangle)^n\, \mathrm{d}t \\
&= \int_{0}^{\infty} u(t) \sum_{k=0}^{n} \binom{n}{k} (-1)^{n-k} t^k \langle t \rangle^{n-k}\, \mathrm{d}t \\
&= \sum_{k=0}^{n} \binom{n}{k} (-1)^{n-k} \langle t \rangle^{n-k} \int_{0}^{\infty} u(t) t^k\, \mathrm{d}t,
\end{aligned}
\tag{13}
$$

which is a weighted sum of standard moments. Now, we introduce the state-of-the-art windows known as Mellin-Laplace moments.

2.3.2. Mellin-Laplace Moments

Mellin-Laplace windows are defined as $w(t) = t^n e^{-pt} H(t)$ where $p > 0$ and $n \in \mathbb{N}$. These windows can also be computed fast [21]. Their Fourier transform is given by

$$W(f) = \frac{n!}{(p + i2\pi f)^{n+1}}. \tag{14}$$

In Figures 3 and 4, the windows for different n orders and p values are shown. The higher the Mellin-Laplace order n the narrower the window magnitude will be in frequency domain. As expected from the uncertainty principle, the higher the p value the more spread the window will be in frequency domain.

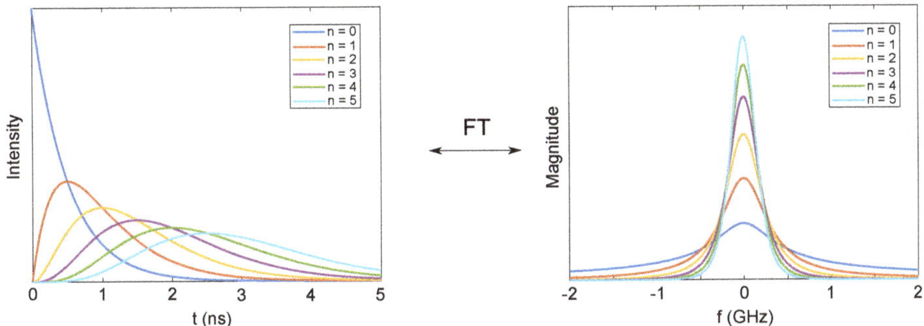

Figure 3. (**Left**) Different n orders of Mellin-Laplace windows for $p = 2$. (**Right**) Fourier transform of those windows; as the order n of the Mellin-Laplace moment is increased, its spectrum magnitude is narrower and has less dispersion.

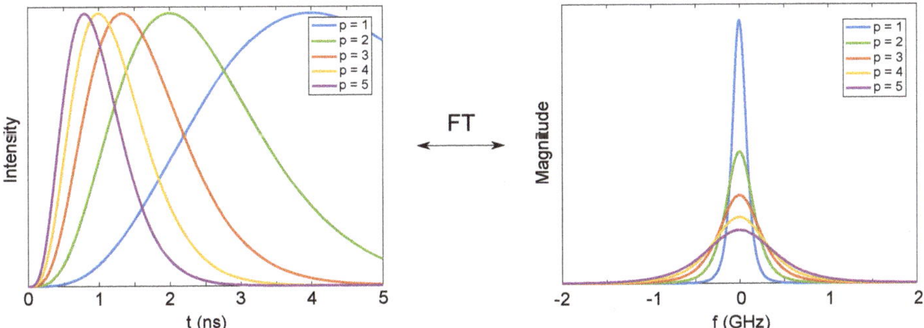

Figure 4. (**Left**) Fifth order ($n = 4$) Mellin-Laplace window with different p values. (**Right**) Fourier transform of those windows; as the p value is increased, the magnitude of the Mellin-Laplace window spectrum is more spread.

Zero order Mellin-Laplace datatypes are Laplace transforms for different p values. Recently, an optical system was proposed [36], which computes Laplace datatypes without the need of measuring the complete DTOF. The system described by Hasnain et al. is quite interesting, since it is much faster than typical time-resolved systems. Nevertheless, Laplace datatypes have poor temporal selectivity (they cover a broad range of time) and are highly correlated, which does not make them the best candidates in terms of tomography capabilities.

2.3.3. Generalized Gaussian Window

The Gaussian function and its Fourier transform are

$$w(t) = e^{-t^2/(2\sigma^2)} \longleftrightarrow W(f) = \sigma\sqrt{2\pi}e^{-2\sigma^2\pi^2 f^2}. \tag{15}$$

As expected from the uncertainty principle, the higher σ, the more spread the window at time domain will be and, therefore, the narrower it will be in the frequency domain, see Figure 5. Note that the centralized Gaussian windows gives the lower bound of the uncertainty principle.

The Laplacian or exponential window (see Figure 6) are defined as

$$w(t) = \exp(-p|t|) \longleftrightarrow W(f) = \frac{2p}{p^2 + (2\pi f)^2}, \tag{16}$$

where p is a parameter.

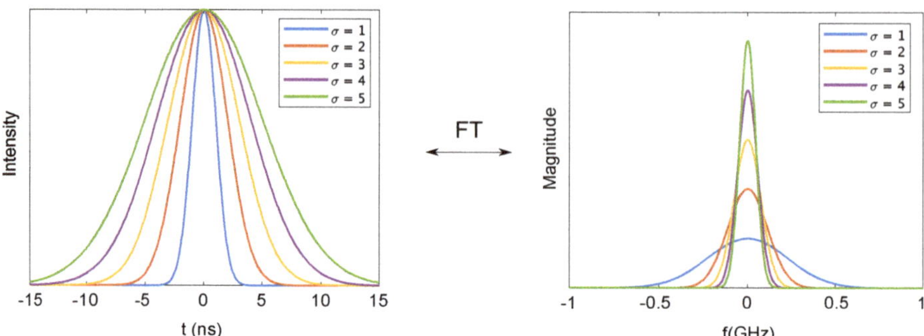

Figure 5. (**Left**) Several Gaussian functions with different σ values. (**Right**) Spectrum magnitude of previous Gaussian functions. The more spread the Gaussian function, the narrower its Fourier transform magnitude.

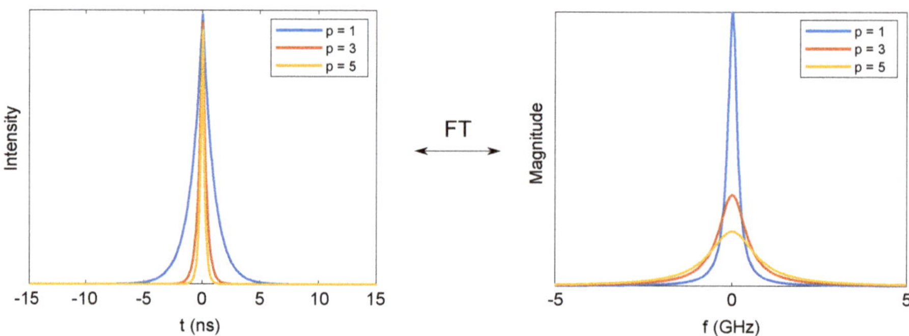

Figure 6. (**Left**) Several exponential windows with different p values. (**Right**) Spectrum magnitude of previous functions.

Both Gaussian and Laplacian windows can be described by the generalized Gaussian window (also known as the generalized normal window),

$$w(t) = \frac{\beta}{2\sigma\Gamma(1/\beta)} e^{-(|t|/\sigma)^\beta} \longleftrightarrow W(f) = \Gamma^{-1}\left(\frac{1}{\beta}\right) \sum_{n=0}^{\infty} \frac{(-1)^n \pi^{2n} f^{2n}}{(2n)!} \sigma^{2n} \Gamma\left(\frac{2n+1}{\beta}\right). \quad (17)$$

where β is a free parameter and Γ is the gamma function. If $\beta = 2$ the function is Gaussian and if $\beta = 1$ is Laplacian. As $\beta \to \infty$ the window converges to a square window.

2.3.4. Tukey Window

The centralized Tukey window [37] is defined as

$$w(t) = \begin{cases} 1, & 0 \le |t| \le \alpha t^* \\ 0.5\left[1 + \cos\left(\pi \frac{t - \text{sign}(t)\,\alpha t^*}{\text{sign}(t)\,(t^* - \alpha t^*)}\right)\right], & \alpha\, t^* \le |t| \le t^*, \\ 0, & \text{elsewhere}, \end{cases} \quad (18)$$

where t^* is the limit of the window and $0 \le \alpha \le 1$ parameter controls the smoothness of the window.

If $\alpha = 1$ the Tukey window is rectangular and its Fourier transform is defined as

$$w(t) = \text{rect}(t/\tau) \longleftrightarrow W(f) = \tau \text{sinc}(\pi f \tau). \tag{19}$$

These windows have a high selectivity of photon arrival time. The spectrum magnitude of rectangular function ($\alpha = 1$) is a sinc function with infinite oscillations, which can make difficult to integrate them numerically. Nevertheless, as the α value decreases, the oscillations at high frequencies disappear without losing much temporal selectivity. Note that before presenting this novel computation method, rectangular windows were only used for functional near-infrared spectroscopy analysis [38] but not for tomography since they could not be calculated without computing the full time-resolved curve.

2.3.5. Trade-Off between Temporal Selectivity and Computation Complexity

An optimal temporal window is the one that is easy to compute (it has few oscillations and small dispersion at frequency domain) and at the same time its temporal selectivity is good enough. For some windows, optimality can be described by $D_0^* = D_0(g)D_0(G)$ (see Section 2.2). The D_0^* of a window quantifies the trade-off between the temporal selectivity and easiness to compute as expressed by uncertainty principle. The lower the value D_0^* is for a given window, the better the window temporal selectivity will be and the less frequencies will be needed to perform the numerical integration.

In Table 1, the main optimality results are summarized.

Table 1. D_0^* for each analyzed window. Standard moments and Tukey windows are not L^2 integrable. See Appendix A for calculations.

Window	D_0^*
Mellin-Laplace	$\frac{1}{16\pi^2} \frac{(2n+2)!}{(2n)!} \left(\frac{2n}{2n-1} - 1 \right), \quad (n \geq 1)$
Gaussian	$1/(16\pi^2)$
Exponential	$1/(8\pi^2)$

The Gaussian is the window that has the lowest D_0^* value; this was expected from the uncertainty principle theorem and this is what makes the Gaussian window one of the most promising windows. The exponential window has a twice bigger value compared to the Gaussian window; this means that the numerical integral could be a bit more difficult to compute but it is still a good candidate. Mellin-Laplace windows will increase its D_0^* value almost linearly for increasing orders although it is independent of p value. Expression D_0^* is not in L^2 space for Tukey windows, nevertheless from Figure 7 it is evident that for some α values is still feasible to perform numerical integration at frequency domain. Standard moments are neither L^2 integrable. Those moments were analyzed in detail in [39,40] and one of the conclusions that was reached is that standard moments high orders are only profitable when more than 10^8 photons are detected. Moreover, since by nature they are highly correlated (see Figure 8a) they will not be considered in this work.

A point that should also be taken into account is the shift theorem of the Fourier transform. It states that given a function $g(t)$, if this function is shifted an amount of t_0, then its Fourier transform is

$$G_{t-t_0}(f) = G(f)e^{-i2\pi f t_0}, \tag{20}$$

where $G_{t-t_0}(f)$ indicates the Fourier transform of shifted function $g(t - t_0) = g(t) * \delta(t - t_0)$ and the complex exponential does not affect the spectral magnitude since $|e^{-i2\pi f t_0}| = 1$ but it affects the phase. This theorem implies that as temporal windows are shifted to later times, more oscillations will appear. However, in practice this phenomenon did not prevent us to perform the numerical integration accurately.

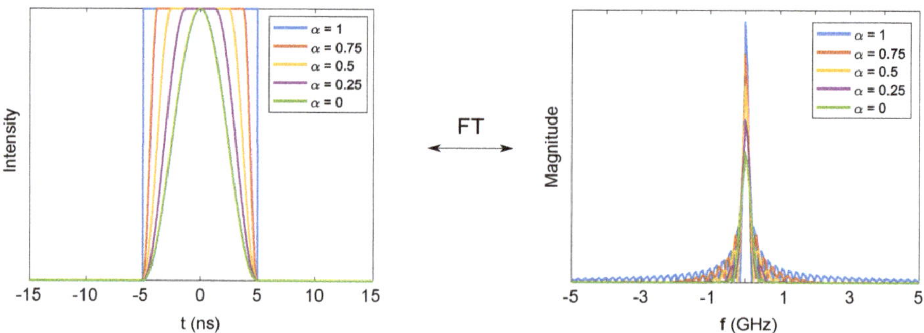

Figure 7. (**Left**) Several Tukey functions with different α values. (**Right**) Spectrum magnitude of previous Tukey functions.

The new method to compute temporal windows is also useful for windows with the form $w(t) = t^n e^{-pt}$, such as Mellin-Laplace or standard moments. Both of these moments suffer from an important handicap if traditional computation methods are used: in order to compute the n-th order window then all the previous orders, $1, ..., n - 1$, must be computed (in the case of Mellin-Laplace if p parameter is modified then all orders must be computed again). These problem can be critical in the case of Mellin-Laplace since as the order of the moments increase, the window temporal selectivity decreases (see Figure 3(left)). If it is wanted to have better sensitivity to later photons (i.e., photons that probabilistically have go deeper into the tissue) then higher p values must be used. There are two ways of doing it: the first option is to use the same p value for all windows; but the main drawback is that many orders will have to be calculated, since a higher p value shifts windows average time-of-flight to earlier photons (see Figure 4). The second option implies to use windows with different p values (e.g., a low p value for early windows and high p values for late windows) which is not computationally efficient since if the p value is changed then all orders up to n will have to be computed again as explained before. Nevertheless, the new approach presented in this paper does not suffer from these bottlenecks since Mellin-Laplace moments can be computed for each p and n value independently (same for n order in the case of standard moments). Therefore, one of the advantages of the presented method is not only that new windows can be computed but also that Mellin-Laplace windows for different p values can be computed more efficiently.

2.4. Noise Correlations

Since DTOF signals are transformed to windows space, the noise will also suffer changes. In this part, noise transformations and correlations are analysed to understand noise influence.

The DTOF signal measured by a detector is theoretically described as

$$u(t) = x(t) + \varepsilon(x(t)), \tag{21}$$

where $x(t)$ is the noiseless signal and ε is the noisy term which depends on $x(t)$ magnitude, uncorrelated (that is, $\mathrm{Cov}(\varepsilon(x(t)), \varepsilon(x(t'))) = 0$ for $t \neq t'$) and follows a Poisson distribution. Applying the Fourier transform to $u(t)$ yields,

$$\begin{aligned} U(f) &= X(f) + Y(f) \\ &= \int_{-\infty}^{\infty} x(t) e^{-i2\pi ft}\, \mathrm{d}t + \int_{-\infty}^{\infty} \varepsilon(x(t)) e^{-i2\pi ft}\, \mathrm{d}t, \end{aligned} \tag{22}$$

where the second term in the right part is the noisy contribution and $U(f)$ can be considered a random variable. As stated before, in time domain, the noise at each time bin is independent, however after applying the Fourier transform this property does not hold anymore. For example, the covariance between two different frequencies is,

$$
\begin{aligned}
E\left[U(f_1) \cdot \overline{U(f_2)}\right] - E\left[U(f_1)\right] \cdot E\left[\overline{U(f_2)}\right] &= E\left[\int_{-\infty}^{\infty} u(t)e^{-i2\pi f_1 t}\,dt \cdot \overline{\int_{-\infty}^{\infty} u(t)e^{-i2\pi f_2 t}\,dt}\right] \\
&\quad - E\left[\int_{-\infty}^{\infty} u(t)e^{-i2\pi f_1 t}\,dt\right] \cdot E\left[\int_{-\infty}^{\infty} u(t)e^{-i2\pi f_2 t}\,dt\right] \\
&= \int_{-\infty}^{\infty}\int_{-\infty}^{\infty} \left(E\left[u(t)\cdot u(t')\right] - E\left[u(t)\right]\cdot E\left[u(t')\right]\right) e^{-i2\pi f_1 t}e^{i2\pi f_2 t'}\,dt\,dt' \\
(\text{Due to TR signal noise independence}) &= \int_{-\infty}^{\infty} E\left[u(t)\right] e^{-i2\pi(f_1-f_2)t}\,dt \\
(E\left[u(t)\right] = x(t),\ \text{due to shot/Poisson noise}) &= \int_{-\infty}^{\infty} x(t)\,e^{-i2\pi(f_1-f_2)t}\,dt \\
&= X(f_1 - f_2),
\end{aligned}
\tag{23}
$$

so the covariance of Fourier transform signal at two different frequencies is equal to the Fourier coefficient of the noiseless signal at frequency $f_1 - f_2$. See Appendix B for an alternative proof.

From this result, the covariance between any temporal window datatypes can be determined. Let us define a window datatype shifted kt_0 in time as,

$$
M_{k,t_0} = \int_0^{\infty} u(t)\,w(t - kt_0)\,dt = \int_{-\infty}^{\infty} U(f)\,\overline{W(f)}e^{i2\pi fkt_0}\,df,
\tag{24}
$$

where t_0 is the shift unit length and k is the number of shifts steps that are taken. Then, the covariance between shifted window datatypes is,

$$
\begin{aligned}
\mathrm{Cov}(M_k, M_l) &= E\left[\int_{-\infty}^{\infty} U(f)\,\overline{W(f)}e^{i2\pi fkt_0}\,df \cdot \overline{\int_{-\infty}^{\infty} U(f)\,\overline{W(f)}e^{i2\pi flt_0}\,df}\right] \\
&\quad - E\left[\int_{-\infty}^{\infty} U(f)\,\overline{W(f)}e^{i2\pi fkt_0}\,df\right] \cdot E\left[\overline{\int_{-\infty}^{\infty} U(f)\,\overline{W(f)}e^{i2\pi flt_0}\,df}\right] \\
&= \int_{-\infty}^{\infty}\int_{-\infty}^{\infty} \left(E\left[U(f)\cdot \overline{U(f')}\right] - E\left[U(f)\right]\cdot E\left[\overline{U(f')}\right]\right)\overline{W(f)}\,e^{i2\pi fkt_0}\,W(f')\,e^{-i2\pi f'lt_0}\,df\,df' \\
&= \int_{-\infty}^{\infty}\int_{-\infty}^{\infty} X(f - f')\,\overline{W(f)}\,W(f')\,e^{i2\pi(kt_0 f - lt_0 f')}\,df\,df'.
\end{aligned}
\tag{25}
$$

From Equation (25), the covariance and correlation matrices of any window can be obtained. Note that for standard (and Mellin-Laplace) moments the shift is already fixed by n (and p) value. In Figure 8, an example of the typical correlation matrices of standard, Mellin-Laplace, Gaussian and Tukey windows are shown. It is evident that, the greater the order of the standard moment, the more correlated it is with the rest of moments, see Figure 8a to get an intuitive idea. In the case of Mellin-Laplace window for a fixed p value, as the order n increases, its get more and more correlated with neighbour windows, see Figure 8b, although not as much as standard moments. However, for Gaussian and Tukey windows, the correlation can be minimized by separating them far enough; in the example given the Gaussian windows are not correlated with any other windows, because the windows did not overlap. The same thing could have been done with Tukey windows.

In time-domain reconstruction, the heteroscedasticity [41] is present in the noise, that is, the noise at each time bin has different variance but they are not correlated with each other (this is due to photon noise physics). After applying overlapping windows, such as Mellin-Laplace or Fourier transform, non-zero covariance values will appear. Therefore, in these cases, Least Squares (LS) will not be a Best Linear Unbiased Estimator (BLUE) anymore. Nevertheless, Generalized Least Squares (GLS) (also known as Aitken's estimator) can be used instead, which is also a BLUE [41]. In theory, LS and GLS will reach the same solution for different noise realizations, however since for GLS the inverse of covariance matrix must be computed, some problems could arise if it is ill-posed. So, noise correlations should be avoided whenever possible.

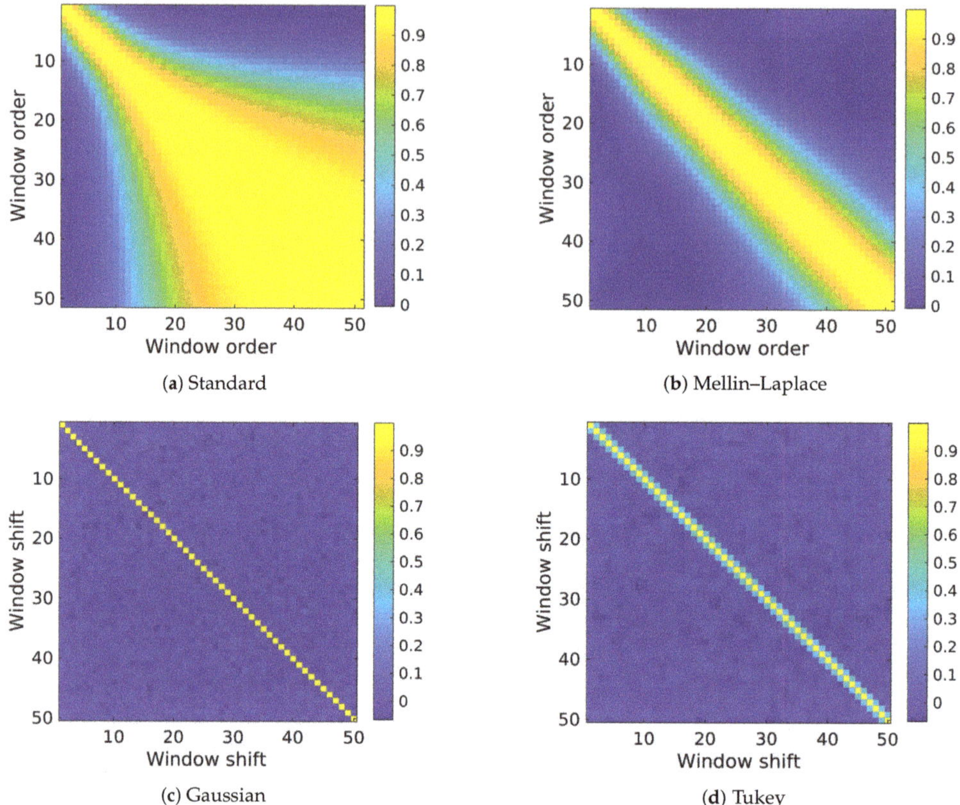

(**a**) Standard

(**b**) Mellin–Laplace

(**c**) Gaussian

(**d**) Tukey

Figure 8. Correlation matrices of standard moments (first 50 orders), Mellin–Laplace (first 50 orders, $p = 7$), Gaussian ($t_0 = 0.16$ ns shifts, $\sigma = 0.05$), and Tukey windows ($t_0 = 0.16$ ns shifts, $\alpha = 0.25$ and $t^* = 0.2$ ns).

2.5. Numerical Simulations

To analyze the reconstruction improvements that could be obtained with the decorrelated windows, some numerical simulations were performed. The used computational phantom was a three-dimensional volume with a spherical inclusion at different depths. This phantom can be consider as an approximation to functional near-infrared spectroscopy experiments where the inclusion represents the activation in the cortex. The computational phantom had a size of $9 \times 9 \times 5$ cm^3. Numerical simulations were performed in a reflectance geometry by solving the time-domain Diffusion Approximation up to 10 ns; the boundary conditions were implemented as described in [42]. To generate the synthetic measurements the domain was discretized using around 32 and 172 thousand nodes and elements respectively, see Figure 9(right). For the reconstruction, a coarser mesh was used (9 and 48 thousand nodes and elements respectively) in order to avoid inverse crime phenomenon. Each simulated curve contained up to 2×10^5 photons, were convoluted with the instrumental response function of a single photon avalanche diode detector (SPAD, FWHM ≈ 160 ps) and corrupted by Poisson noise, see Figure 9(left).

In this work, only absorption inhomogeneities were considered, although it can be easily extended to scattering inhomogeneities. Background absorption was set to $\mu_a = 0.018$ cm^{-1} and the reduced scattering was $\mu_s' = 14.7$ cm^{-1} over all the domain. An inclusion of 0.5 cm radius was included whose optical properties were $\mu_a = 0.337$ cm^{-1} and $\mu_s' = 14.7$ cm^{-1} (these values were taken from the experiment performed at [43] using 550 nm wavelength light and are of similar order as found at several types of human tissue).

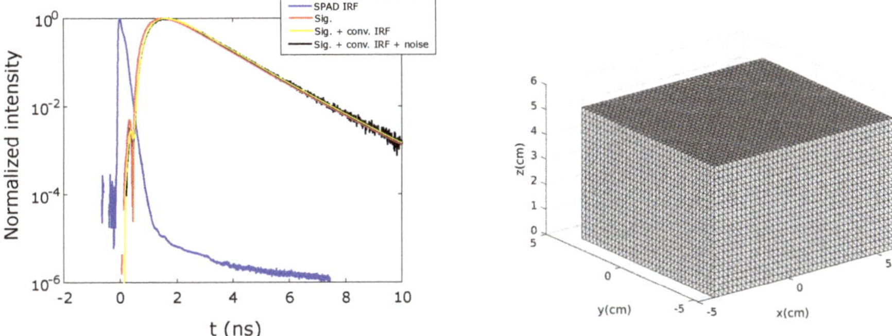

Figure 9. (**Left**) Instrumental response function of a single photon avalanche diode detector (SPAD) detector (blue line). Normalized simulated signal without IRF or noise (red line). Normalized simulated signal convoluted with the IRF (yellow line). Normalized simulated signal convoluted with IRF and corrupted with Poisson noise (black line). (**Right**) Tetrahedral mesh used to generate the synthetic data; it had a size of $9 \times 9 \times 5$ cm^3 and was discretized using around 32 and 172 thousand nodes and elements respectively.

The setup of source and detectors was moved along the top boundary to scan the phantom at multiple positions. The distance between the source and detectors was 3 cm at orthogonal directions, see Figure 10. The scan of the computational phantom with the inclusion was done at 30 different source positions (6 shifts in x-direction separated by 0.75 cm and 5 shifts in y-direction separated by 0.75 cm). The inclusions were set from 1 to 4 cm depth.

During the reconstruction, to avoid large sensitivities in the surface nodes the sensitivity of each node was normalized by a diagonal weight matrix with values proportional to nodes depth. A flowchart of the used algorithm is shown at Figure 11. Simulations and reconstructions were done using a laptop with an Intel Core i7 processor and 4 GB RAM.

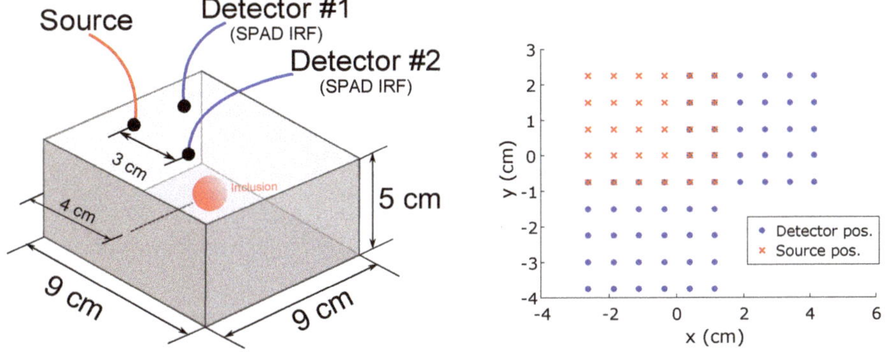

Figure 10. (**Left**) Scheme of the numerical phantom. Simulated signals were convolved with the IRF of a single photon avalanche diode detector SPAD and corrupted by Poisson noise. (**Right**) Source positions; red crosses. Detector positions; blue dots. Note that for each source only the signal at two detectors were considered; 3 cm right and 3 cm down. For example, for source at position $(-2.62, 2.25)$ cm only the signal at detectors $\mathbf{x}_1 = (0.38, 2.25)$ cm and $\mathbf{x}_2 = (-2.62, -0.75)$ cm was used.

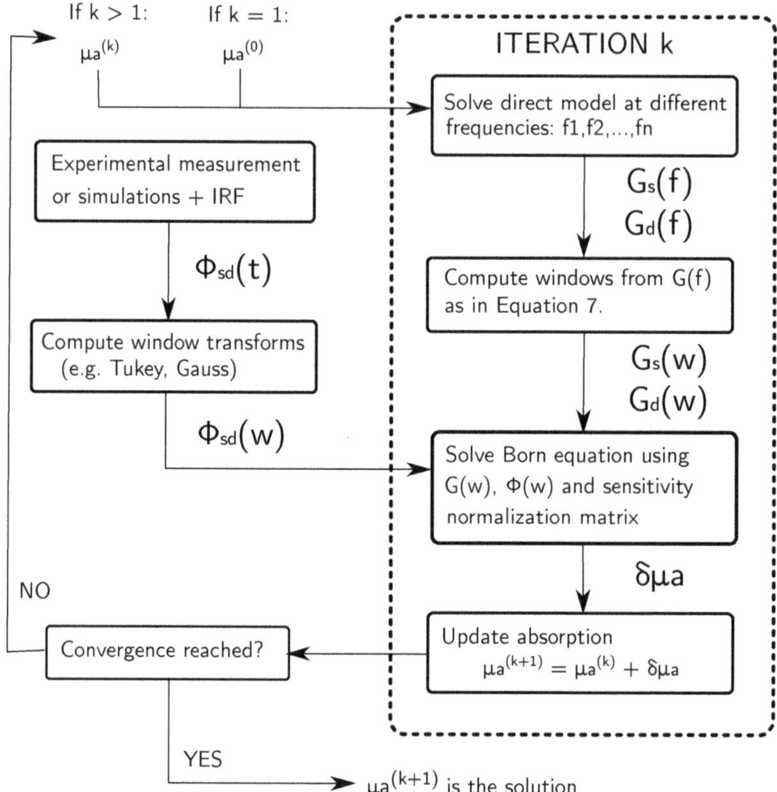

Figure 11. Reconstruction algorithm flowchart.

2.6. Quantitative Evaluation Metrics

Three different evaluation metrics (localization error, average contrast and relative recovered volume) were used to measure the reconstructions quality from different points of view.

Localization error is defined as Euclidean distance between the center of mass of simulated inclusion $\mathbf{x}_s$ and the center of mass of the reconstructed inclusion $\mathbf{x}_r$. If the reconstructed optical properties are identical to the truth, the localization error is zero. The recovered inclusion is determined by selecting the reconstructed absorption changes that are greater than a given threshold defined later.

$$\text{Localization error} = \|\mathbf{x}_s - \mathbf{x}_r\|_2. \tag{26}$$

The average contrast evaluation metric is based on the mean value of the region of interest (i.e., inclusion volume):

$$\text{Average contrast} = \frac{\sum_{i \in N_r} \mu_i}{|N_r|\hat{\mu}} \tag{27}$$

where μ_i denotes the reconstructed absorption at node i, N_r represents the set of nodes at the region of interest, $|N_r|$ denotes the number of nodes within the set and $\hat{\mu}$ is the truth values of absorption at the region of interest. If the reconstructed absorption is identical to the truth, the average contrast value is one.

The last evaluation metric is the relative reconstructed volume (V_{RRV}) which is defined as

$$V_{RRV} = V_r/V_s \times 100\% \tag{28}$$

where V_r and V_s are the volume of the reconstructed inclusion and simulated case respectively. The volume of the reconstructed inclusion, V_r, is computed by thresholding the recovered absorption changes and computing the volume of those elements. If the reconstructed absorption is identical to the truth, the relative reconstructed volume is one.

3. Results and Discussion

In this section, the simulations results are given in order to analyze the performance of tomography algorithms based on temporal windows and Fourier transform datatypes. Subsequently, different perspectives of the same problem are given and results are discussed.

3.1. Comparing State-of-the-Art Windows with Tukey and Gaussian Windows

The purpose of the first simulations is to check whether Tukey or Gaussian windows improve the reconstruction in depth compared to $w(t) = t^n e^{-pt}$ state-of-the-art windows.

For the Mellin–Laplace reconstruction $p = 3$ and the first 35 moments were used (Figure 12a). For Gaussian windows, the value of σ was set to 0.3 and their centers were located from 0.3 to 9.6 ns in steps of 0.3 ns (32 windows in total, Figure 12b). For the Tukey windows, the same centers were used and the parameters were set to $\alpha = 0.25$ and $t^* = 0.3$ ns (Figure 12c). The given parameters were selected so that in all cases the windows totally covered the simulated signal and the number of windows were similar (35 for Mellin–Laplace and 32 for Tukey and Gaussian). The computed frequencies ranged from 0 to 2 GHz in steps of 200 MHz.

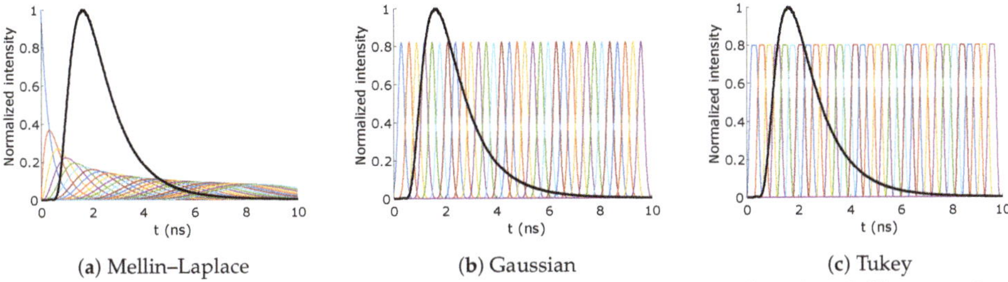

(a) Mellin–Laplace   (b) Gaussian   (c) Tukey

Figure 12. Black curve represents a simulated DTOF. (**a**) Mellin–Laplace windows ($p = 3$, 35 moments), note that they are highly overlapped. (**b**) Gaussian windows ($\sigma = 0.3$). (**c**) Tukey windows ($\alpha = 0.25$, $t^* = 0.3$ ns).

The Figure 13 shows reconstruction using Melin-Laplace, Tukey and Gaussian temporal windows for inclusions that ranged from 1 to 2.5 cm deep (in steps of 0.5 cm). In Figure 14, the reconstructions for inclusions from 3 cm to 4 cm deep are given. In Figure 15, the evaluation metrics results are shown for each method. Significant differences are seen for inclusions shallower than 2.5 cm not just in terms of localization accuracy but also regarding absorption quantification. Localization error for Mellin-Laplace windows is around 0.4 cm for the inclusion located at 2 cm depth. However, for the novel windows the maximum localization error is 0.2 cm. Moreover, the absorption quantification is constantly underestimated by Mellin-Laplace windows, see how average contrast metric decreases exponentially and the contrast of Tukey and Gaussian windows is always greater than Mellin-Laplace windows in the range of 1 to 2.5 cm depth. For inclusions deeper than 3 cm Mellin–Laplace windows reconstruction is significantly worse in terms of absorption quantification. Also note that the localization error for Mellin-Laplace windows is much larger for the inclusion 4 cm deep; these results are in agreement with [44] which shows the same problem beyond 2 cm depth. Moreover, the relative reconstructed volume also increases due to the fact that Mellin-Laplace loses the inclusion localization and absorption quantification, and therefore, it tries to compensate it by reconstructing higher volume inclusions, see for example the inclusion at depth 2.5 cm. This phenomenon occurs due to the diffusion of photons inside the medium. In contrast, the results also show that Tukey and Gaussian windows give very similar reconstructions and their localization error is always less than 0.4 cm.

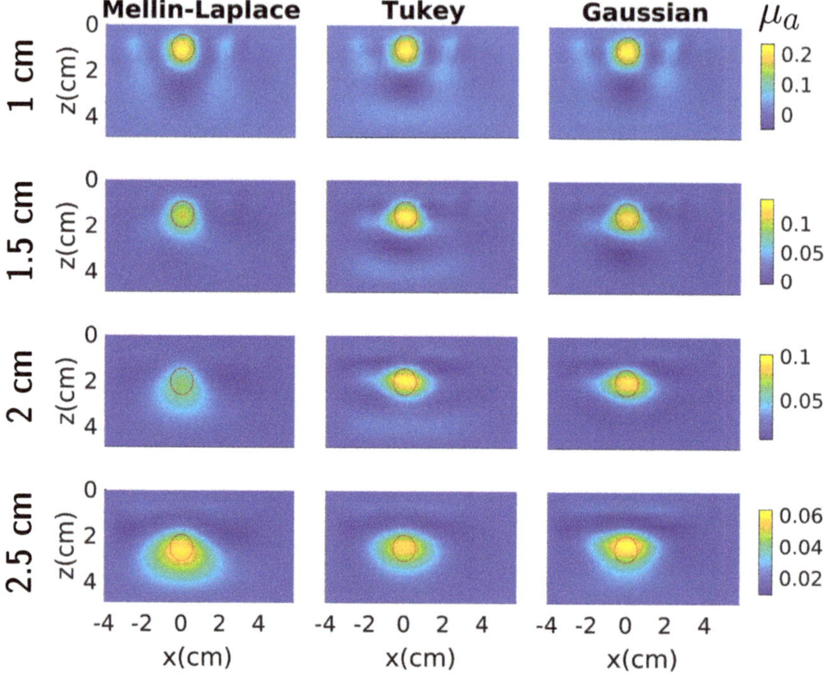

Figure 13. Absorption, μ_a, reconstructions using Mellin–Laplace, Tukey and Gaussian windows for inclusions from 1 cm to 2.5 cm deep. The red circle indicates the correct location of the inclusion.

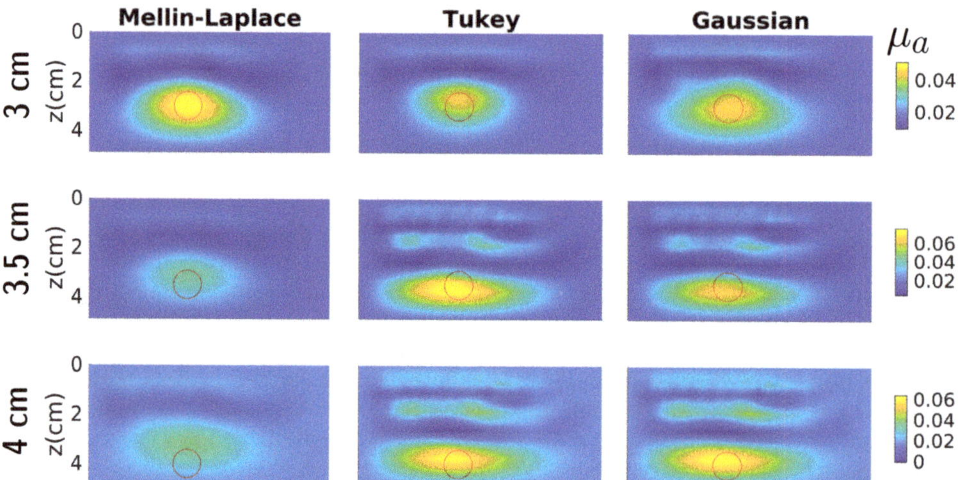

Figure 14. Absorption, μ_a, reconstructions using Mellin–Laplace, Tukey and Gaussian windows for inclusions from 3 cm to 4 cm deep. The red circle indicates the correct location of the inclusion.

The fact that proposed windows outperform Mellin-Laplace state-of-the-art windows for deep inclusions can be explained by the theoretical analysis shown in the previous sections. First, Mellin–Laplace windows have a poor late-arrival photon selectivity and windows are highly correlated. This implies that each Mellin–Laplace window has very little new information by itself. As the order of the Mellin–Laplace window increases the new information that it carries decreases. That implies that a huge amount of window should be used to include all the available information in the DTOFs. This evidence demonstrate the superior performance of Gaussian and Tukey window compared to Mellin–Laplace because of the better photon time arrival selectivity and less noise correlation.

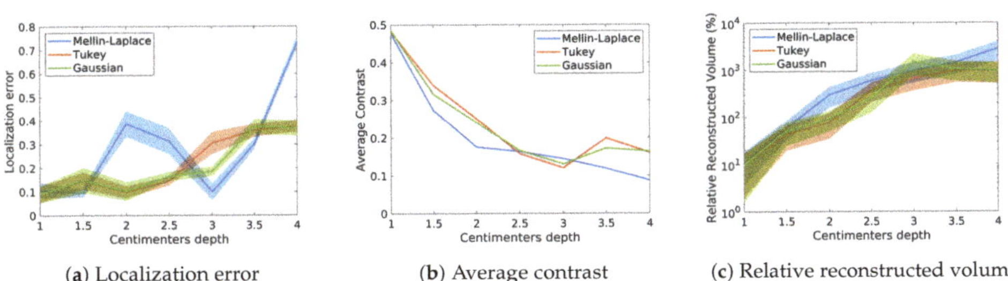

(a) Localization error   (b) Average contrast   (c) Relative reconstructed volume

Figure 15. Mellin-Laplace (blue), Tukey (red) and Gaussian (yellow). Quantitative evaluation metric values for each window and inclusion depth. For localization error and relative reconstructed volume, the standard error shadows indicate the uncertainty given by selecting interest regions whose thresholds range from 60% to 80% of maximum absorption.

Regarding the computation time of the reconstructions, in Figure 16b the number of seconds per iteration is given for each method. To compute windows with form $t^n e^{-pt}$ is much faster than the proposed approach and full-time reconstruction. Specifically, $t^n e^{-pt}$ windows are around six times faster to compute than the proposed approach because *LU* factorization can be applied. However, for the proposed method factorization is not efficient since the matrix changes at each frequency. Full–time resolved approach is 9.5 times slower than $t^n e^{-pt}$ windows and around 1.6 slower than the proposed approach. These results shows the trade-off of the proposed method: more information in depth is obtained but computation time is larger than $t^n e^{-pt}$ windows. However, it should also be taken into account that novel method approach is highly parallelizable, that is, the Diffusion Approximation equation can be solve independently for each frequency. Note that this property does not hold for the direct computation of $t^n e^{-pt}$ windows [22] and full–time approach. Therefore, the computation time of novel approach should considerably improve with GPU usage due to its intrinsic parallelizable nature. In Figure 16a, the number of computed iterations until convergence is given. Convergence was reached when relative difference $\delta\mu_a / \mu_a^{(k)} < 5 \times 10^{-3}$. It can be seen that Mellin–Laplace windows converge faster mainly because it underestimates absorption quantification.

As explained before, the simulated DTOFs were convolved with an SPAD IRF and corrupted with photon noise. These DTOFs were deconvoluted with a Wiener filter before performing the reconstruction. In some papers, such as [21], a different approach is proposed: DTOFs from a reference medium A and objective medium B are used. The optical properties of medium A are already known because an homogeneous medium is taken as reference and it includes implicitly all the information regarding instrumental factors. The medium B is the medium to be reconstructed. Absorption of medium B can be recovered by using iteratively the linearized cross–Born equation,

$$G_{sd}^A(\mathbf{r},t) * M_{sd}^B(\mathbf{r},t) - M_{sd}^A(\mathbf{r},t) * G_{sd}^{B(k)}(\mathbf{r},t) = -M_{sd}^A(\mathbf{r},t) * \int_\Omega \left[G_s^{B(k)} * G_d^{B(k)} \right](\mathbf{r},t)\, \delta\mu_a\, d\mathbf{r}, \quad (29)$$

where $*$ is the convolution operator, M_{sd}^A and M_{sd}^B indicate the measurement obtained at reference and to be reconstructed media by detector d when source s was activated, k indicates the iteration number, G_{sd}^A and $G_{sd}^{B(k)}$ is the simulated Green's function (free of noise and instrumental factors) from source s

at detector d by using the photon propagation mathematical model described in Equation (1) and $G_s^{B(k)}$ indicates the Green's function value at every point in the domain given by the propagation model.

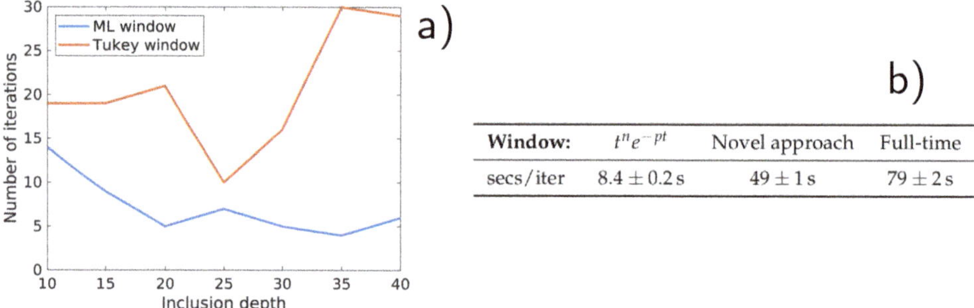

Window:	$t^n e^{-pt}$	Novel approach	Full-time
secs/iter	$8.4 \pm 0.2\,\mathrm{s}$	$49 \pm 1\,\mathrm{s}$	$79 \pm 2\,\mathrm{s}$

Figure 16. (**a**) Number of iterations until convergence for Mellin–Laplace and Gaussian windows. (**b**) Seconds per iteration for Mellin–Laplace windows, novel approach (with Gaussian) and full–time resolved approach.

The cross–Born equation offers the ease of not having to deal with instrumental factors directly (no deconvolution of the measurements is needed). However, it also implies to convolve the data and lose some information in the process. In Figure 17a, reconstructions using Mellin–Laplace and Tukey windows with the cross–Born approach are shown. We chose Tukey windows since they give very similar results to Gaussian windows. In Figure 17b–d, the quantitative metrics for those simulations are given. Results for inclusions as deep as 3 cm are very similar to the deconvolution approach, but for inclusion deeper or equal to 3.5 cm the localisation accuracy is lost, see Figure 17b. Moreover, the contrast is lower than previous reconstructions; although, slight differences are seen between Mellin-Laplace and Tukey windows even using cross-Born equation, see Figure 17c. The relative recovered volume increases considerably for both methods (Figure 17d. These results show that when the cross–Born approach is used, some information in depth is lost for inclusions deeper than 3 cm and to take full advantage of not overlapping datatypes (e.g., Tukey or Gaussian windows), experimental data must be deconvoluted before performing a reconstruction.

From a different point of view, this theoretical analysis can also be viewed as an explanation of full time-resolved tomography limits. That is, Tukey windows-based reconstruction can be seen as an intermediate step between correlated windows-based reconstruction and full time-resolved reconstruction (i.e., reconstruction based on using each time bin as a datatype). In fact, in the last years several labs [32,33], have developed photon propagation models for Graphics Processing Units (GPU). One of the goals of using GPU is to compute the full DTOF curve and to fit it entirely as described in Equation (3). To fit the entire DTOF curve is the limiting case of very narrow Tukey windows; if windows are shrinked to time bin size it will converge to DTOF curve fitting. Therefore, some of the questions that arise from this theoretical work is, how much benefit can be obtained by using full time-resolved reconstructions? Does the computational cost of computing the full curve outweighs the gain of some information? Moreover, since the computation of decorrelated windows, such as Tukey or Gaussian, are easily parallelizable and adaptable to GPU hardware, the authors believe that this computation method is an efficient substitute to full time-resolved reconstructions.

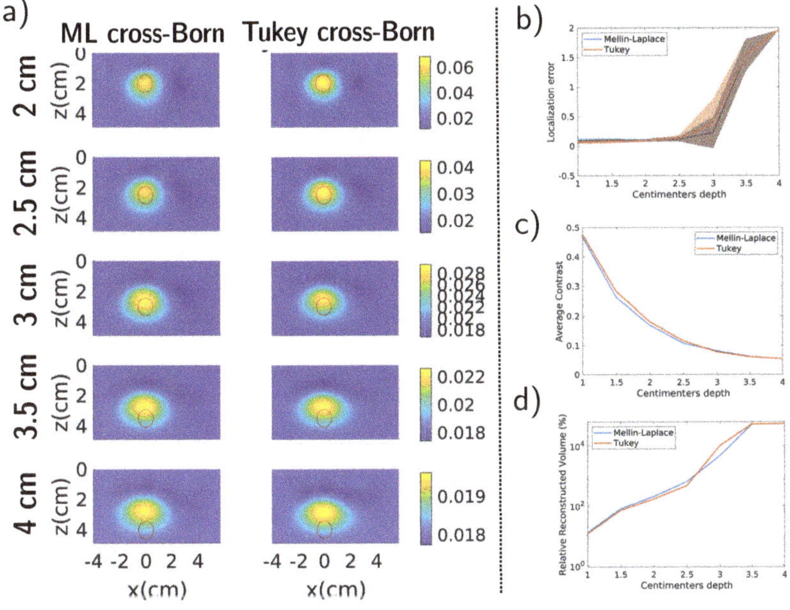

Figure 17. (**a**) Reconstructions for Mellin–Laplace and Tukey windows from same simulated data as Figure 13; cross–Born approach was used to deal with instrumental factors such as IRF or photon noise. The red circle indicates the correct location of the inclusion. (**b–d**) Quantitative metrics for cross-Born based reconstructions.

3.2. Comparing Windows and Frequency-Based Reconstruction

One of the question that could arise now is why not to apply the Fourier transform to DTOFs signals and perform the frequency based reconstruction described at Equation (5). To answer this question, reconstructions were also performed using Fourier transform datatypes for an inclusion 3.5 cm deep with the same optical values in the previous section.

In Figure 18a, the reconstruction obtained by using Fourier transform coefficients from 0 MHz up to 1.5 GHz in steps of 100 MHz is shown. The magnitude of the signal Fourier coefficients are shown at Figure 18b. The results show that using frequencies up to 1.5 GHz yields a similar reconstruction to proposed windows. However, if only frequencies lower than 1.5 GHz are used the reconstructions suffers from instabilities. This agrees with the results of [45] that suggest that useful information can be found at GHz order.

The relative equivalence of frequency and windows based results can be understood by realizing that in both cases, the complex numbers associated to each frequency are fitted. In the case of Fourier transform datatype, this is done directly, but for temporal windows, this is done by windowing the frequency domain. Nevertheless, there are some important differences. First, as was shown in Section 2.4, by applying the Fourier transform to a noisy signal the noise becomes correlated. This does not occur when not overlapping time windows such as Tukey windows are used. Second, when reconstruction is done directly in frequency domain, the linear system to solve is complex while in the temporal windows case it is real. Third, temporal window-based reconstruction is more intuitive to understand, since it is associated to photon arrival times. Therefore, in many cases, a weight matrix could be built in order to weight some windows over other ones, depending on how noisy they are. This fact makes the problem much more easy to stabilize compared to frequency based reconstructions. Last, in the same fashion as the paper published by Hasnain et al. [36], a system that directly measures a gated DTOF (e.g., using rectangular gates) instead of the full curve could be faster than nowadays time-resolved systems. Moreover, those gated DTOFs could be used in the proposed reconstruction process since they are equivalent to Tukey windows. Such a system could be potentially faster than state-of-the-art time-resolved systems without compromising reconstruction quality.

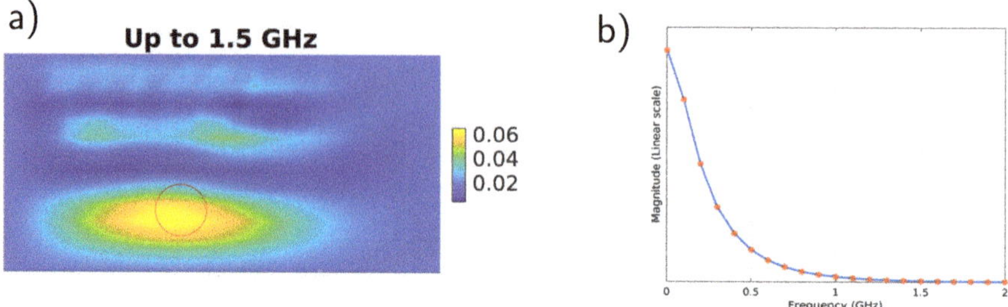

Figure 18. (**a**) Absorption reconstructions for 3.5 cm deep inclusion using up to 2 GHz frequencies. (**b**) Magnitude of the signal Fourier coefficients.

4. Conclusions

In this paper, a new method has been proposed for computing temporal windows. The technique consists of computing the temporal windows in the frequency domain instead of in time domain. The main advantage is that better temporal selectivity windows can be used, without compromising considerably computational efficiency. To illustrate the method, Tukey and Gaussian windows were used. The obtained results in a numerical phantom prove that these new class of windows have better photon time-arrival selectivity and do not correlate with the noise. As a result, the newly developed method improves the localization and absorption quantification of inclusions at the deep layers of a reflectance geometry. These results have been demonstrated to be equivalent, under certain conditions, to frequency-based reconstruction, when frequencies up to GHz order are used.

Author Contributions: Conceptualization, D.O.-M., L.H., L.C. and J.M.; methodology, D.O.-M., L.H., L.C. and J.M.; software, D.O.-M.; validation, D.O.-M.; formal analysis, D.O.-M., L.H., L.C. and J.M.; investigation, D.O.-M.; resources, D.O.-M.; data curation, D.O.-M.; writing–original draft preparation, D.O.-M.; writing–review and editing, D.O.-M., L.H., L.C. and J.M.; visualization, D.O.-M.; supervision, L.H., L.C. and J.M.; project administration, L.H., L.C. and J.M.; funding acquisition, D.O.-M. and L.H.

Funding: This research has received funding from the European Union's Horizon 2020 Marie Skodowska-Curie Innovative Training Networks (ITN-ETN) programme, under Grant Agreement No. 675332 BitMap.

Conflicts of Interest: The authors declare no conflict of interest. The funders had no role in the design of the study; in the collection, analyses, or interpretation of data; in the writing of the manuscript, or in the decision to publish the results.

Abbreviations

The following abbreviations are used in this manuscript:

RTE	Radiative Transfer Equation
FT	Fourier Transform
DTOF	Distribution of Photon Time Of Flight
SPAD	Single Photon Avalanche Diode detector

Appendix A. Windows Selectivity Calculations

Appendix A.1. Gaussian Window

$$
\begin{aligned}
\int_{-\infty}^{\infty} |g(t)|^2 \, \mathrm{d}t &= \int_{-\infty}^{\infty} \left| e^{-t^2/(2\sigma^2)} \right|^2 \, \mathrm{d}t \\
&= \int_{-\infty}^{\infty} e^{-t^2/\sigma^2} \, \mathrm{d}t \\
&= \sigma \int_{-\infty}^{\infty} e^{-x^2} \, \mathrm{d}x \\
&= \sigma\sqrt{\pi}.
\end{aligned}
$$

$$
\begin{aligned}
\int_{-\infty}^{\infty} t^2 |g(t)|^2 \, \mathrm{d}t &= \int_{-\infty}^{\infty} t^2 e^{-t^2/\sigma^2} \, \mathrm{d}t \\
&= \int_{-\infty}^{\infty} t \left(t \cdot e^{-t^2/\sigma^2} \right) \mathrm{d}t \\
&= \left[t \cdot \left(-\frac{\sigma^2}{2} e^{-t^2/\sigma^2} \right) \right]_{-\infty}^{\infty} - \int_{-\infty}^{\infty} -\frac{\sigma^2}{2} e^{-t^2/\sigma^2} \, \mathrm{d}t \\
&= \frac{\sigma^3}{2} \sqrt{\pi}.
\end{aligned}
$$

$$
\begin{aligned}
\int_{-\infty}^{\infty} |G(f)|^2 \, \mathrm{d}f &= 2\pi\sigma^2 \int_{-\infty}^{\infty} e^{-4\sigma^2\pi^2 f^2} \, \mathrm{d}f \\
&= 2\pi\sigma^2 \frac{\sqrt{\pi}}{2\sigma\pi} = \sigma\sqrt{\pi}.
\end{aligned}
$$

$$
\begin{aligned}
\int_{-\infty}^{\infty} f^2 |G(f)|^2 \, \mathrm{d}f &= 2\pi\sigma^2 \int_{-\infty}^{\infty} f^2 e^{-4\sigma^2\pi^2 f^2} \, \mathrm{d}f \\
&= 2\pi\sigma^2 \int_{-\infty}^{\infty} f \left(f \cdot e^{-4\sigma^2\pi^2 f^2} \right) \mathrm{d}f \\
&= 2\pi\sigma^2 \left[f \cdot \left(-\frac{1}{8\sigma^2\pi^2} e^{-4\sigma^2\pi^2 f^2} \right) \right]_{-\infty}^{\infty} - 2\pi\sigma^2 \int_{-\infty}^{\infty} -\frac{1}{8\sigma^2\pi^2} e^{-4\sigma^2\pi^2 f^2} \, \mathrm{d}t \\
&= \frac{1}{8\sigma\pi^{3/2}}.
\end{aligned}
$$

$$
\boxed{D_0(g)D_0(G) = \frac{\sigma^2}{2} \cdot \frac{1}{8\sigma^2\pi^2} = \frac{1}{16\pi^2}.}
$$

Appendix A.2. Laplacian Distribution

$$\int_{-\infty}^{\infty} |g(t)|^2 \, dt = \int_{-\infty}^{\infty} \left| e^{-p|t|} \right|^2 \, dt$$

$$= 2 \int_{0}^{\infty} e^{-2pt} \, dt$$

$$= \frac{2}{-2p} \left[e^{-2pt} \right]_0^{\infty}$$

$$= \frac{1}{p}.$$

$$\int_{-\infty}^{\infty} t^2 |g(t)|^2 \, dt = 2 \int_{0}^{\infty} t^2 e^{-2pt} \, dt$$

$$= 2 \left(\frac{1}{-2p} \left[t^2 e^{-2pt} \right]_0^{\infty} + \frac{2}{2p} \int_{0}^{\infty} t e^{-2pt} \, dt \right)$$

$$= \frac{2}{p} \int_{0}^{\infty} t e^{-2pt} \, dt$$

$$= \frac{2}{p} \left(\frac{1}{-2p} \left[t e^{-2pt} \right]_0^{\infty} + \frac{1}{2p} \int_{0}^{\infty} e^{-2pt} \, dt \right)$$

$$= \frac{1}{2p^3}.$$

$$\int_{-\infty}^{\infty} |G(f)|^2 \, df = \int_{-\infty}^{\infty} \left| \frac{2p}{p^2 + (2\pi f)^2} \right|^2 \, df$$

$$= \frac{4}{(2\pi)^2} \gamma^2 \int_{-\infty}^{\infty} \frac{1}{f^4 + 2\gamma^2 f^2 + \gamma^4} \, df \qquad (\text{where } \gamma = p/(2\pi))$$

$$= \frac{4}{(2\pi)^2} \gamma^2 \frac{\pi}{2|\gamma|^3} = \frac{1}{p}.$$

$$\int_{-\infty}^{\infty} f^2 |G(f)|^2 \, df = \int_{-\infty}^{\infty} f^2 \left| \frac{2p}{p^2 + (2\pi f)^2} \right|^2 \, df$$

$$= \frac{4}{(2\pi)^2} \gamma^2 \int_{-\infty}^{\infty} \frac{f^2}{f^4 + 2\gamma^2 f^2 + \gamma^4} \, df \qquad (\text{where } \gamma = p/(2\pi))$$

$$= \frac{4}{(2\pi)^2} \gamma^2 \int_{-\infty}^{\infty} \frac{f^2}{(f^2 + \gamma^2)^2} \, df$$

$$= \frac{4}{(2\pi)^2} \gamma^2 \int_{-\infty}^{\infty} \frac{1}{f^2 + \gamma^2} - \frac{\gamma^2}{(f^2 + \gamma^2)^2} \, df$$

$$= \frac{4}{(2\pi)^2} \gamma^2 \left[\frac{\pi}{\gamma} - \frac{\pi}{2\gamma} \right]$$

$$= \frac{p}{(2\pi)^2}.$$

$$\boxed{D_0(g) D_0(G) = \frac{1}{2p^2} \cdot \frac{p^2}{(2\pi)^2} = \frac{1}{8\pi^2}.} \tag{A1}$$

Appendix A.3. Rectangle Function

$$\int_{-\infty}^{\infty} |g(t)|^2 \, \mathrm{d}t = \int_{-\infty}^{\infty} |\mathrm{rect}(t/\tau)|^2 \, \mathrm{d}t = \tau.$$

$$\begin{aligned}
\int_{-\infty}^{\infty} t^2 |g(t)|^2 \, \mathrm{d}t &= \int_{-\infty}^{\infty} t^2 \, |\mathrm{rect}(t/\tau)|^2 \, \mathrm{d}t \\
&= 2 \int_{0}^{\infty} t^2 \, |\mathrm{rect}(t/\tau)|^2 \, \mathrm{d}t \\
&= 2 \int_{0}^{\tau/2} t^2 \, \mathrm{d}t = 2 \left[\frac{t^3}{3} \right]_{0}^{\tau/2} = \frac{\tau^3}{12}.
\end{aligned}$$

$$\begin{aligned}
\int_{-\infty}^{\infty} |G(f)|^2 \, \mathrm{d}f &= \int_{-\infty}^{\infty} |\tau \mathrm{sinc}(\pi \tau f)|^2 \, \mathrm{d}f \\
&= \tau^2 \int_{-\infty}^{\infty} |\mathrm{sinc}(\pi \tau f)|^2 \, \mathrm{d}f \\
&= \frac{\tau}{\pi} \int_{-\infty}^{\infty} |\mathrm{sinc}(x)|^2 \, \mathrm{d}x = \tau.
\end{aligned}$$

$$\begin{aligned}
\int_{-\infty}^{\infty} f^2 |G(f)|^2 \, \mathrm{d}f &= \int_{-\infty}^{\infty} f^2 \, |\tau \mathrm{sinc}(\pi \tau f)|^2 \, \mathrm{d}f \\
&= \frac{1}{\pi^3 \tau} \int_{-\infty}^{\infty} x^2 \, |\mathrm{sinc}(x)|^2 \, \mathrm{d}x \\
&= \text{diverges so it is not square integrable.}
\end{aligned}$$

Appendix A.4. Mellin-Laplace Moments

$$\int_{-\infty}^{\infty} |g(t)|^2 \, dt = \int_{-\infty}^{\infty} \left| t^n e^{-pt} H(t) \right|^2 \, dt$$

$$= \int_{0}^{\infty} t^{2n} e^{-2pt} \, dt = \frac{(2n)!}{(2p)^{2n+1}}.$$

$$\int_{-\infty}^{\infty} t^2 |g(t)|^2 \, dt = \int_{-\infty}^{\infty} t^2 \left| t^n e^{-pt} H(t) \right|^2 \, dt$$

$$= \int_{0}^{\infty} t^{2n+2} e^{-2pt} \, dt = \frac{(2n+2)!}{(2p)^{2n+3}}.$$

$$\int_{-\infty}^{\infty} |G(f)|^2 \, df = \int_{-\infty}^{\infty} \left| \frac{n!}{(p+i2\pi f)^{n+1}} \right|^2 \, df$$

$$= \frac{n!^2}{2\pi |p|^{2n+1}} \int_{-\pi/2}^{\pi/2} \frac{\sec^2 x}{|(p+i\tan x)|^{2n+2}} \, dx \qquad (2\pi f = p \tan x)$$

$$= \frac{n!^2}{2\pi |p|^{2n+1}} \int_{-\pi/2}^{\pi/2} \cos^{2n} x \, dx$$

$$= \frac{n!^2}{\pi |p|^{2n+1}} \int_{0}^{\pi/2} \cos^{2n} x \, dx$$

$$= \frac{n!^2}{\pi |p|^{2n+1}} W_{2n}.$$

$$\int_{-\infty}^{\infty} f^2 |G(f)|^2 \, df = \int_{-\infty}^{\infty} f^2 \left| \frac{n!}{(p+i2\pi f)^{n+1}} \right|^2 \, df$$

$$= \frac{(n! \, p)^2}{(2\pi)^3 |p|^{2n+1}} \int_{-\pi/2}^{\pi/2} \frac{\tan^2 x \sec^2 x}{|(p+i\tan x)|^{2n+2}} \, dx \qquad (2\pi f = p \tan x, \quad n \geq 1)$$

$$= \frac{(n! \, p)^2}{(2\pi)^3 |p|^{2n+1}} \int_{-\pi/2}^{\pi/2} \tan^2 x \cos^n x \, dx$$

$$= \frac{(n! \, p)^2}{(2\pi)^3 |p|^{2n+1}} \int_{-\pi/2}^{\pi/2} \sin^2 x \cos^{2n-2} x \, dx$$

$$= \frac{(n! \, p)^2}{(2\pi)^3 |p|^{2n+1}} \left(\int_{-\pi/2}^{\pi/2} \cos^{2n-2} x \, dx - \int_{-\pi/2}^{\pi/2} \cos^{2n} x \, dx \right)$$

$$= \frac{(n! \, p)^2}{(2\pi)^3 |p|^{2n+1}} (2W_{2n-2} - 2W_{2n}) \qquad (n \geq 1).$$

where $W_n = \int_{0}^{\pi/2} \cos^n x \, dx$ is the Wallis integral.

$$D_0(g) = \frac{(2n+2)!}{(2p)^{2n+3}} \cdot \frac{(2p)^{2n+1}}{(2n)!} = \frac{(2n+2)!}{(2n)!} \cdot \frac{1}{(2p)^2}.$$

$$D_0(G) = \frac{(n! \, p)^2}{(2\pi)^3 |p|^{2n+1}} (2W_{2n-2} - 2W_{2n}) \cdot \frac{\pi |p|^{2n+1}}{n!^2 W_{2n}}$$

$$= \frac{p^2}{4\pi^2} \left(\frac{W_{2n-2}}{W_{2n}} - 1 \right)$$

$$= \frac{p^2}{4\pi^2} \left(\frac{2n}{2n-1} - 1 \right).$$

$$\boxed{D_0(g)D_0(G) = \frac{1}{16\pi^2}\frac{(2n+2)!}{(2n)!}\left(\frac{2n}{2n-1}-1\right), \qquad (n \geq 1).}$$

Therefore,

$$\frac{1}{16\pi^2}\frac{(2n+2)!}{(2n)!}\left(\frac{2n}{2n-1}-1\right) > \frac{1}{16\pi^2}, \qquad (n \geq 1).$$

Appendix B. Fourier Transform Influence on Noise

This approach is inspired by the book [46]. Let us call $F \in \mathbb{C}^{M\times M}$ the discrete Fourier transform matrix, which is Hermitian and complex. In the following, we analyze how the noise propagates after applying the Fourier transform.

In the case the data is contaminated by independent Gaussian noise with constant variance, the covariance matrix C is,

$$C = \sigma^2 I.$$

If the Fourier transform is applied then the covariance matrix transforms to

$$C' = \sigma^2 FF^* = m\sigma^2 I.$$

Now, let us assume that the variance of the noise is identical to the signal value s_{exact}; the new covariance matrix is

$$C = \mathrm{diag}(s_{exact}),$$

where $\mathrm{diag}(u)$ indicates a diagonal matrix whose diagonal contains vector u. After applying the Fourier transform, the covariance matrix transforms to

$$C' = F\mathrm{diag}(s_{exact})F^*,$$

and its is known that $F\mathrm{diag}(s_{exact})F^*$ is a circulant matrix with its first column as Fs_{exact}, the Fourier transform of s_{exact}. If the exact signal is dominated by low frequencies, then the largest components of the first column Fs_{exact} will be located at the top and the covariance matrix would be almost diagonal with some nonzero covariance terms near the diagonal. This equation is equivalent to Equation (23).

References

1. Weigl, W.; Milej, D.; Janusek, D.; Wojtkiewicz, S.; Sawosz, P.; Kacprzak, M.; Gerega, A.; Maniewski, R.; Liebert, A. Application of optical methods in the monitoring of traumatic brain injury: A review. *J. Cereb. Blood Flow Metab.* **2016**, *36*, 1825–1843. [CrossRef] [PubMed]
2. Cooper, R.J.; Magee, E.; Everdell, N.; Magazov, S.; Varela, M.; Airantzis, D.; Gibson, A.P.; Hebden, J.C. MONSTIR II: A 32-channel, multispectral, time-resolved optical tomography system for neonatal brain imaging. *Rev. Sci. Instrum.* **2014**, *85*, 053105. [CrossRef] [PubMed]
3. Taroni, P.; Pifferi, A.; Quarto, G.; Spinelli, L.; Torricelli, A.; Abbate, F.; Villa, A.M.; Balestreri, N.; Menna, S.; Cassano, E.; et al. Noninvasive assessment of breast cancer risk using time-resolved diffuse optical spectroscopy. *J. Biomed. Opt.* **2010**, *15*, 060501. [CrossRef] [PubMed]
4. Lindner, C.; Mora, M.; Farzam, P.; Squarcia, M.; Johansson, J.; Weigel, U.M.; Halperin, I.; Hanzu, F.A.; Durduran, T. Diffuse optical characterization of the healthy human thyroid tissue and two pathological case studies. *PLoS ONE* **2016**, *11*, e0147851. [CrossRef]
5. Singer, J.R.; Grünbaum, F.A.; Kohn, P.; Zubelli, J.P. Image Reconstruction of the Interior of Bodies That Diffuse Radiation. *Science* **1990**, *248*, 990–993. [CrossRef]
6. Arridge, S.R.; Schweiger, M.; Delpy, D.T. Iterative reconstruction of near-infrared absorption images. In *Inverse Problems in Scattering and Imaging*; International Society for Optics and Photonics: Bellingham, WA, USA, 1992; Volume 1767.

7. Wang, L.V.; Wu, H. *Biomedical Optics: Principles and Imaging*; Wiley Publisher: Hoboken, NJ, USA, 2007; p. 376.

8. Liemert, A.; Reitzle, D.; Kienle, A. Analytical solutions of the radiative transport equation for turbid and fluorescent layered media. *Sci. Rep.* **2017**, *7*, 3819. [CrossRef]

9. Liemert, A.; Kienle, A. Analytical solution of the radiative transfer equation for infinite-space fluence. *Phys. Rev. A* **2011**, *83*, 015804. [CrossRef]

10. Liemert, A.; Kienle, A. Exact and efficient solution of the radiative transport equation for the semi-infinite medium. *Sci. Rep.* **2013**, *3*, 2018. [CrossRef]

11. Wang, L.; Jacques, S.L.; Zheng, L. MCML-Monte Carlo modeling of light transport in multi-layered tissues. *Comput. Methods Programs Biomed.* **1995**, *47*, 131–146. [CrossRef]

12. Fang, Q. Mesh-based Monte Carlo method using fast ray-tracing in Plücker coordinates. *Biomed. Opt. Express* **2010**, *1*, 165–175. [CrossRef]

13. González-Rodríguez, P.; Kim, A.D. Diffuse optical tomography using the one-way radiative transfer equation. *Biomed. Opt. Express* **2015**, *6*, 2006–2021. [CrossRef] [PubMed]

14. Tarvainen, T.; Vauhkonen, M.; Kolehmainen, V.; Kaipio, J.P. Hybrid radiative-transfer–diffusion model for optical tomography. *Appl. Opt.* **2005**, *44*, 876–886. [CrossRef] [PubMed]

15. Pierrat, R.; Greffet, J.J.; Carminati, R. Photon diffusion coefficient in scattering and absorbing media. *JOSA A* **2006**, *23*, 1106–1110. [CrossRef]

16. Yamada, Y.; Suzuki, H.; Yamashita, Y. Time-domain near-infrared spectroscopy and imaging: A review. *Appl. Sci.* **2019**, *9*, 1127. [CrossRef]

17. Chu, M.; Dehghani, H. Image reconstruction in diffuse optical tomography based on simplified spherical harmonics approximation. *Opt. Express* **2009**, *17*, 24208–24223. [CrossRef] [PubMed]

18. Ma, W.; Zhang, W.; Yi, X.; Li, J.; Wu, L.; Wang, X.; Zhang, L.; Zhou, Z.; Zhao, H.; Gao, F. Time-domain fluorescence-guided diffuse optical tomography based on the third-order simplified harmonics approximation. *Appl. Opt.* **2012**, *51*, 8656–8668. [CrossRef] [PubMed]

19. Xu, Y.; Gu, X.; Fajardo, L.L.; Jiang, H. In vivo breast imaging with diffuse optical tomography based on higher-order diffusion equations. *Appl. Opt.* **2003**, *42*, 3163–3169. [CrossRef]

20. Arridge, S.R.; Schweiger, M.; Hiraoka, M.; Delpy, D.T. A finite element approach for modeling photon transport in tissue. *Med. Phys.* **1993**, *20*, 299–309. [CrossRef]

21. Hervé, L.; Puszka, A.; Planat-Chrétien, A.; Dinten, J.M. Time-domain diffuse optical tomography processing by using the Mellin-Laplace transform. *Appl. Opt.* **2012**, *51*, 5978–5988. [CrossRef]

22. Arridge, S.R.; Schweiger, M. Direct calculation of the moments of the distribution of photon time of flight in tissue with a finite-element method. *Appl. Opt.* **1995**, *34*, 2683–2687. [CrossRef]

23. Arridge, S.R. Optical tomography in medical imaging. *Inverse Probl.* **1999**, *15*, R41. [CrossRef]

24. Soloviev, V.Y.; McGinty, J.; Tahir, K.B.; Neil, M.A.; Sardini, A.; Hajnal, J.V.; Arridge, S.R.; French, P.M.W. Fluorescence lifetime tomography of live cells expressing enhanced green fluorescent protein embedded in a scattering medium exhibiting background autofluorescence. *Opt. Lett.* **2007**, *32*, 2034–2036. [CrossRef] [PubMed]

25. Dehghani, H.; Srinivasan, S.; Pogue, B.W.; Gibson, A. Numerical modelling and image reconstruction in diffuse optical tomography. *Philos. Trans. R. Soc. A* **2009**, *367*, 3073–3093. [CrossRef] [PubMed]

26. Pogue, B.W.; Patterson, M.S.; Jiang, H.; Paulsen, K.D. Initial assessment of a simple system for frequency domain diffuse optical tomography. *Phys. Med. Biol.* **1995**, *40*, 1709. [CrossRef] [PubMed]

27. Gulsen, G.; Xiong, B.; Birgul, O.; Nalcioglu, O. Design and implementation of a multifrequency near-infrared diffuse optical tomography system. *J. Biomed. Opt.* **2006**, *11*, 014020. [CrossRef]

28. Soloviev, V.Y. Mesh adaptation technique for Fourier-domain fluorescence lifetime imaging. *Med. Phys.* **2006**, *33*, 4176–4183. [CrossRef]

29. Milstein, A.B.; Stott, J.J.; Oh, S.; Boas, D.A.; Millane, R.P.; Bouman, C.A.; Webb, K.J. Fluorescence optical diffusion tomography using multiple-frequency data. *J. Opt. Soc. Am. A* **2004**, *21*, 1035–1049. [CrossRef]

30. Arridge, S.R.; Schweiger, M. Photon-measurement density functions. Part 2: Finite-element-method calculations. *Appl. Opt.* **1995**, *34*, 8026–8037. [CrossRef]

31. Pinsky, M.A. *Introduction to Fourier Analysis and Wavelets*; American Mathematical Society: Providence, RI, USA, 2002; Volume 102.

32. Doulgerakis-Kontoudis, M.; Eggebrecht, A.T.; Wojtkiewicz, S.; Culver, J.P.; Dehghani, H. Toward real-time diffuse optical tomography: Accelerating light propagation modeling employing parallel computing on GPU and CPU. *J. Biomed. Opt.* **2017**, *22*, 125001. [CrossRef]
33. Schweiger, M. GPU-accelerated Finite Element Method for Modelling Light Transport in Diffuse Optical Tomography. *J. Biomed. Imaging* **2011**, *2011*, 10. [CrossRef]
34. Liebert, A.; Wabnitz, H.; Grosenick, D.; Möller, M.; Macdonald, R.; Rinneberg, H. Evaluation of optical properties of highly scattering media by moments of distributions of times of flight of photons. *Appl. Opt.* **2003**, *42*, 5785–5792. [CrossRef] [PubMed]
35. Liebert, A.; Wabnitz, H.; Steinbrink, J.; Obrig, H.; Möller, M.; Macdonald, R.; Villringer, A.; Rinneberg, H. Time-resolved multidistance near-infrared spectroscopy of the adult head: Intracerebral and extracerebral absorption changes from moments of distribution of times of flight of photons. *Appl. Opt.* **2004**, *43*, 3037–3047. [CrossRef] [PubMed]
36. Hasnain, A.; Mehta, K.; Zhou, X.; Li, H.; Chen, N. Laplace-domain diffuse optical measurement. *Sci. Rep.* **2018**, *8*, 12134. [CrossRef] [PubMed]
37. Harris, F.J. On the use of windows for harmonic analysis with the discrete Fourier transform. *Proc. IEEE* **1978**, *66*, 51–83. [CrossRef]
38. Wabnitz, H.; Jelzow, A.; Mazurenka, M.; Steinkellner, O.; Macdonald, R.; Milej, D.; Żołek, N.; Kacprzak, M.; Sawosz, P.; Maniewski, R.; et al. Performance assessment of time-domain optical brain imagers, part 2: nEUROPt protocol. *J. Biomed. Opt.* **2014**, *19*, 086012. [CrossRef]
39. Ducros, N.; Hervé, L.; Da Silva, A.; Dinten, J.M.; Peyrin, F. A comprehensive study of the use of temporal moments in time-resolved diffuse optical tomography: Part I. Theoretical material. *Phys. Med. Biol.* **2009**, *54*, 7089. [CrossRef]
40. Ducros, N.; Da Silva, A.; Hervé, L.; Dinten, J.M.; Peyrin, F. A comprehensive study of the use of temporal moments in time-resolved diffuse optical tomography: Part II. Three-dimensional reconstructions. *Phys. Med. Biol.* **2009**, *54*, 7107. [CrossRef]
41. Baltagi, B. *Econometrics*; Springer: Berlin, Germany, 2008.
42. Schweiger, M.; Arridge, S.R.; Hiraoka, M.; Delpy, D.T. The finite element method for the propagation of light in scattering media: Boundary and source conditions. *Med. Phys.* **1995**, *22*, 1779–1792. [CrossRef]
43. Zouaoui, J.; Sieno, L.D.; Hervé, L.; Pifferi, A.; Farina, A.; Dalla Mora, A.; Derouard, J.; Dinten, J.M. Chromophore decomposition in multispectral time-resolved diffuse optical tomography. *Biomed. Opt. Express* **2017**, *8*, 4772–4787. [CrossRef]
44. Puszka, A.; Hervé, L.; Planat-Chrétien, A.; Koenig, A.; Derouard, J.; Dinten, J.M. Time-domain reflectance diffuse optical tomography with Mellin-Laplace transform for experimental detection and depth localization of a single absorbing inclusion. *Biomed. Opt. Express* **2013**, *4*, 569–583. [CrossRef]
45. Wojtkiewicz, S.; Durduran, T.; Dehghani, H. Time-resolved near infrared light propagation using frequency domain superposition. *Biomed. Opt. Express* **2018**, *9*, 41–54. [CrossRef] [PubMed]
46. Hansen, P.C. *Discrete Inverse Problems: Insight and Algorithms*; SIAM: Philadelphia, PA, USA, 2010.

Article

Characteristic Length and Time Scales of the Highly Forward Scattering of Photons in Random Media

Hiroyuki Fujii *,†, **Moegi Ueno, Kazumichi Kobayashi and Masao Watanabe**

Division of Mechanical and Space Engineering, Faculty of Engineering, Hokkaido University, Kita 13 Nishi 8, Kita-ku, Sapporo, Hokkaido 060-8628, Japan; pueinsuki@eis.hokudai.ac.jp (M.U.); kobakazu@eng.hokudai.ac.jp (K.K.); masao.watanabe@eng.hokudai.ac.jp (M.W.)
* Correspondence: fujii-hr@eng.hokudai.ac.jp
† He partially conducted this research while he was a visiting scholar at University of Michigan.

Received: 28 November 2019; Accepted: 17 December 2019; Published: 20 December 2019

Abstract: Background: Elucidation of the highly forward scattering of photons in random media such as biological tissue is crucial for further developments of optical imaging using photon transport models. We evaluated length and time scales of the photon scattering in three-dimensional media. Methods: We employed analytical solutions of the time-dependent radiative transfer, M-th order delta-Eddington, and photon diffusion equations (RTE, dEM, and PDE). We calculated the fluence rates at different source-detector distances and optical properties. Results: We found that the zeroth order dEM and PDE, which approximate the highly forward scattering to the isotropic scattering, are valid in longer length and time scales than approximately $10/\mu_t'$ and $40/\mu_t'v$, respectively, where μ_t' is the reduced transport coefficient and v the speed of light in a medium. The first and second order dEM, which approximate the highly forward-peaked phase function by the first two and three Legendre moments, are valid in the longer scales than approximately $4.0/\mu_t'$ and $6.3/\mu_t'v$; $2.8/\mu_t'$ and $3.5/\mu_t'v$, respectively. The boundary conditions less influence the length scales, while they reduce the times scales from those for bulk at the longer length scale than approximately $4.0/\mu_t'$. Conclusion: Our findings are useful for constructions of accurate and efficient photon transport models. We evaluated length and time scales of the highly forward scattering of photons in various kinds of three-dimensional random media by analytical solutions of the radiative transfer, M-th order delta-Eddington, and photon diffusion equations.

Keywords: radiative transfer equation; highly forward scattering of photons; diffusion and delta-Eddington approximations; characteristic length and time scales of photon transport

---

1. Introduction

Elucidation of the photon scattering and transport in random media such as biological tissue volumes is crucial for biomedical optical imaging using the near-infrared light in the wavelength range from 700 to 1100 nm such as diffuse optical tomography [1–3], because the imaging technique uses light scattered by the media. There are mainly three kinds of systems of the imaging techniques: time-domain, steady-state, and frequency-domain systems. Among them, the time-domain system has several advantages over steady-state and frequency-domain systems thanks to the richest information of the time-domain measurement data [4]. Time-dependent photon transport is basically classified into the two kinds of characteristic regimes (length and time scales): the ballistic and diffusive regimes corresponding to short and long source-detector (SD) distances and times after light incidence, respectively [5]. In the ballistic regime, photon is little scattered and almost goes straight. In the diffusive regime, photon is diffusive due to multiple scattering of photons, and the scattering process can be approximated as isotropic. Between the ballistic and diffusive regimes, photon undergoes a few scattering events in the highly forward direction and travels along the zig-zag paths while changing the direction [6,7]. In this paper, we call this few-scattering-event regime between the ballistic and diffusive regimes as the scattering regime, so that we consider the three kinds of characteristic regimes for photon transport. The radiative transfer equation (RTE), which is the linear Boltzmann equation, can describe photon transport for random media in the three characteristic regimes because the phase function in the RTE can treat the highly forward scattering of photons. However, the numerical calculations of the RTE require high computational loads (times and memories) especially for three-dimensional (3D) random media because the RTE is an integro-differential equation with respect to the light intensity as a function of position vector, unit angular direction vector, and time. To reduce the computational loads, many researches have employed the photon diffusion equation (PDE), which is obtained by the diffusion approximation to the RTE and omits the angular direction. However, the PDE is valid only in the diffusive regime. A coupling model using the RTE and PDE has been constructed [8,9]. For the coupling model, the RTE is calculated in the ballistic and scattering regimes, while the PDE is calculated in the diffusive regime. For constructions of the coupling model or appropriate uses of the PDE, many researches have devoted their work to evaluations of the crossover length and time from the scattering to diffusive regimes [9–15]. For example, a comparison study [10] between actual light measurements and the PDE-results for slab media showed that the crossover length is evaluated as approximately $10/\mu_t'$, where the reduced transport coefficient μ_t' is a sum of reduced scattering coefficient and absorption coefficient. A numerical study [9] using the time-dependent RTE and PDE for 2D media evaluated the crossover length the same as the above study [10], and in addition, evaluated the crossover time as approximately $20/\mu_t'v$, where v is the speed of light in the medium. Although the coupling model can provide accurate and efficient results, a further efficiency improvement of the photon transport model is necessary because the RTE-calculations in the scattering regime still require high computational loads. For modeling the highly forward scattering of photons, the phase function in the RTE is highly forward-peaked and changes exponentially with respect to the scattering angle. Basically, accurate numerical treatments of the highly forward-peaked phase function require a large number of discrete angular directions, resulting in high computational loads of the RTE-calculations. To overcome the difficulty, one needs to investigate the highly forward scattering of photons and influences of the highly forward-peaked phase function on the RTE-results in the scattering regime. For this purpose, in this paper, we employ the M-th order delta-Eddington equation (dEM), which is obtained by the delta-Eddington approximation (dEA) to the RTE.

The dEA decomposes the highly forward-peaked phase function into purely forward-peaked and other components [16,17]. The purely forward-peaked component is given by the delta-function and contributes to a modification of the scattering length scale. The other components are expanded by finite series of Legendre polynomials and used as the phase function in the dEM. The dEM has been widely employed in the field of biomedical optics since Klose and coworkers have introduced from the field of astrophysics [18–22]. It has been shown that the numerical solutions of the dEM agree with those of the RTE while the number of the discrete angular directions is reduced to approximately one-sixth of the required number for the RTE-calculations. Nevertheless, a validity of the dEM is still unclear; Klose and coworkers showed that the second order dEM can provide the accurate results the same as the RTE [21], while Jia and coworkers stated that the zeroth order dEM is sufficient [22], and Boulet and coworkers employed the zeroth and first order dEM [20]. These differences probably come from the fact that the validity of the dEM depends on the SD distances, times after light incidence, and the optical properties, like the validity of the PDE.

Our objective is to examine photon transport especially in the scattering regime for various kinds of random media by using the time-dependent RTE, dEM, and PDE, and to evaluate a regime (length and time scales) for the dEM to be valid as a function of μ_t' and $\mu_t'v$. We employed the analytical solutions instead of the numerical solutions because the analytical solutions are attained much faster than the numerical solutions and allow us to investigate a wide range of the SD distances and optical properties. In addition, we can focus the modification of the scattering coefficient and the approximation of the phase function by the dEA without a discussion of numerical errors induced by the numerical discretization. To the best of our knowledge, no studies have reported to evaluate the regime for the dEM. In this paper, we investigate photon transport for infinite and semi-infinite 3D homogeneous media as a first step of investigations for realistic heterogeneous biological tissue volumes. The investigations for infinite and semi-infinite 3D homogeneous media have been widely discussed for the diffuse optics community [4]. The analytical solutions have been employed for evaluations of the optical properties of various kinds of random media such as biological tissue volumes, colloidal suspensions [23], and agricultural products [24]. The following section describes the three kinds of the photon transport models for 3D random media: the time-dependent RTE, dEM, and PDE; and analytical solutions of the three models for the temporal profiles of the fluence rate. Section 3 investigates the results of the analytical solutions and evaluates the regime for the dEM. Finally, conclusions are described.

2. Materials and Methods

2.1. Time-Domain Photon Transport Models

2.1.1. Radiative Transfer Equation (RTE)

We consider photon transport in the length and time scales of the mean free path and mean time of flight. Then, photon transport can be described by the RTE [25]. For 3D random media, the time-dependent RTE is given by

$$\left[\frac{\partial}{v\partial t} + \mathbf{\Omega} \cdot \nabla + \mu_a(\mathbf{r}) + \mu_s(\mathbf{r})\right] I(\mathbf{r}, \mathbf{\Omega}, t) = \mu_s(\mathbf{r}) \int_{\mathbb{S}^2} d\mathbf{\Omega}' p(\mathbf{\Omega}, \mathbf{\Omega}') I(\mathbf{r}, \mathbf{\Omega}', t) + q(\mathbf{r}, \mathbf{\Omega}, t), \tag{1}$$

where, $I(\mathbf{r}, \mathbf{\Omega}, t)$ in W cm^{-2} sr^{-1} represents the light intensity as a function of spatial position vector $\mathbf{r} = (x, y, z) \in \mathbb{R}^3$ in cm; angular direction (unit direction vector) $\mathbf{\Omega} = (\Omega_x, \Omega_y, \Omega_z) \in \mathbb{S}^2$ in sr; and time t in ps. $\mu_a(\mathbf{r})$ and $\mu_s(\mathbf{r})$ in cm^{-1} are the absorption and scattering coefficients, respectively; v is the speed of light in the medium; $p(\mathbf{\Omega}, \mathbf{\Omega}')$ in sr^{-1} is the phase function with $\mathbf{\Omega}$ and $\mathbf{\Omega}'$ denoting the scattered-out and -in directions, respectively; and $q(\mathbf{r}, \mathbf{\Omega}, t)$ in W cm^{-3} sr^{-1} is a source function.

2.1.2. Henyey-Greenstein Phase Function and Anisotropy Factor

For a formulation of $p(\mathbf{\Omega}, \mathbf{\Omega}')$, the Henyey-Greenstein (HG) phase function [26] is widely employed in biomedical optics to model the forward scattering of photons:

$$p_{HG}(\mathbf{\Omega} \cdot \mathbf{\Omega}') = \frac{1}{4\pi} \frac{1 - g^2}{(1 + g^2 - 2g\mathbf{\Omega} \cdot \mathbf{\Omega}')^{3/2}}, \tag{2}$$

where $g \in [-1, 1]$ is the anisotropy factor, defined by $\int_{\mathbb{S}^2} d\mathbf{\Omega}'(\mathbf{\Omega} \cdot \mathbf{\Omega}') p(\mathbf{\Omega}, \mathbf{\Omega}')$. Mostly, g-values of biological tissue volumes are larger than 0.8 [27], meaning the highly forward scattering. For examples, the g-value of muscle or bone is considered as 0.9; and the g-values of the human liver and of blood vessels are 0.955 [28] and 0.992 [29], respectively. Meanwhile, there are void regions in the human body such as the trachea in the human neck, whose g-value is almost zero, meaning the isotropic scattering. Figure 1 shows the HG phase function (Equation (2)) as a function of $\mathbf{\Omega} \cdot \mathbf{\Omega}' \in [-1, 1]$ in a logarithmic scale. At $g = 0.0$ (isotropic scattering), the phase function is constant over the whole region of $\mathbf{\Omega} \cdot \mathbf{\Omega}'$. Meanwhile, as the g-value approaches to unity from zero (enhancement of the highly forward scattering), the exponential change of p_{HG} with respect to $\mathbf{\Omega} \cdot \mathbf{\Omega}'$ becomes large and the peak of p_{HG} around $\mathbf{\Omega} \cdot \mathbf{\Omega}' = 1.0$ becomes sharp.

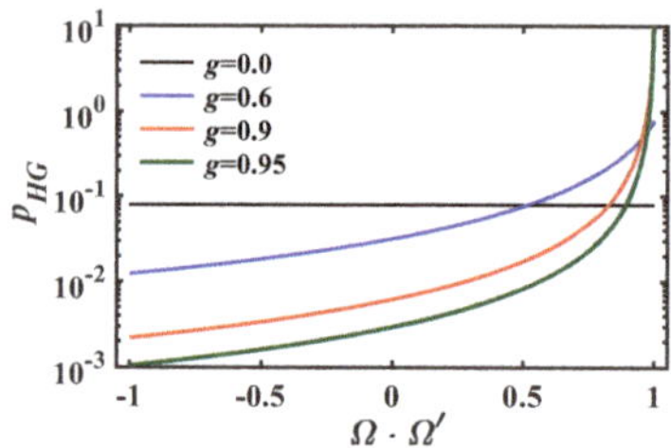

Figure 1. Henyey-Greenstein phase function as a function of $\Omega \cdot \Omega'$ for isotropic scattering ($g = 0.0$), moderately forward scattering ($g = 0.6$), and highly forward scattering ($g = 0.9$ and 0.95), in a logarithmic scale.

The HG phase function is expanded in the infinite series of (unassociated) Legendre polynomials P_l ($l = 0, 1, \cdots, \infty$):

$$p_{HG}(\Omega \cdot \Omega') = \sum_{l=0}^{\infty} \frac{2l+1}{4\pi} g^l P_l(\Omega \cdot \Omega'), \tag{3}$$

where g^l is the l-th order expansion coefficient.

2.1.3. *M*-th Order Delta-Eddington Equation (dEM)

The dEA decomposes the highly forward-peaked phase function into a purely forward-peaked component expressed by the delta function and the other components [16,17]:

$$p_\delta^M(\Omega \cdot \Omega') = \frac{1}{2\pi} h\delta(1 - \Omega \cdot \Omega') + (1 - h)p_{\delta 2}^M(\Omega \cdot \Omega'), \tag{4}$$

where h is a coefficient of the decomposition. $p_{\delta 2}^M$ is a phase function excluding the delta-function component and expanded in the finite series of Legendre polynomials up to the order of M:

$$p_{\delta 2}^M(\Omega \cdot \Omega') = \sum_{l=0}^{M} \frac{2l+1}{4\pi} \sigma_{dEl} P_l(\Omega \cdot \Omega'). \tag{5}$$

In the case of the HG phase function, the expansion coefficient σ_{dEl} and the decomposition coefficient h are determined so as to satisfy the moment conditions up to the order of $M + 1$:

$$\sigma_{dEl} := \int_{\mathbb{S}^2} d\Omega' p_{\delta 2}^M(\Omega \cdot \Omega') P_l(\Omega \cdot \Omega') = \frac{g^l - h}{1 - h}, \quad h = g^{M+1}. \tag{6}$$

The case of $l = 0$ in Equation (6) corresponds to the normalization condition of $p_{\delta 2}^M$: $\int_{\mathbb{S}^2} d\Omega' p_{\delta 2}^M(\Omega \cdot \Omega') = 1$. By using p_δ^M (Equation (4)), the RTE (Equation (1)) is approximated to

$$\left[\frac{\partial}{v\partial t} + \Omega \cdot \nabla + \mu_a(r) + \mu_s^M(r) \right] I(r, \Omega, t) = \mu_s^M(r) \int_{\mathbb{S}^2} d\Omega' p_{\delta 2}^M(\Omega \cdot \Omega') I(r, \Omega', t) + q(r, \Omega, t), \tag{7}$$

where $\mu_s^M(r) = (1 - h)\mu_s(r) = (1 - g^{M+1})\mu_s(r)$. In this paper, Equation (7) is called as the M-th order delta-Eddington equation (dEM). Comparing the RTE (Equation (1)) with the dEM, μ_s is modified to μ_s^M and p is approximated to $p_{\delta 2}^M$. The modification of μ_s to μ_s^M means a change of the scattering length scale. Figure 2a,b plot a scattering length ratio of μ_s^M / μ_s and a relative error E_{phase}^M of the phase functions between the HG function (Equation (2)) and the M-th order dEA (Equation (5)) at different g-values of 0.3, 0.6, 0.9, and 0.95. Here, the error E_{phase}^M is defined by

$$E_{phase}^M = \int_{-1}^{1} d\mu \left| \frac{p_{\delta 2}^M(\mu) - p_{HG}(\mu)}{p_{HG}(\mu)} \right|, \quad \mu = \Omega \cdot \Omega'. \tag{8}$$

As the expansion order M increases, the ratio μ_s^M/μ_s approaches to unity and the error E_{phase}^M decreases, meaning the scattering length scale and the phase function for the dEM converge to those for the RTE. At $g = 0.9$ and 0.95 (highly forward scattering), the convergences are slower than the cases of $g = 0.3$ and 0.6 (moderate forward scattering).

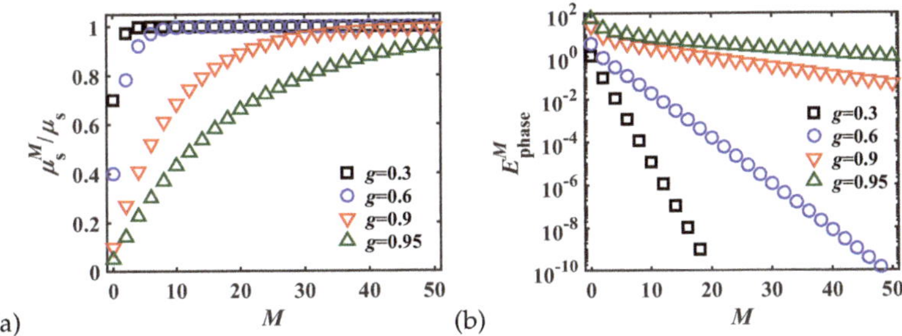

Figure 2. (a) Ratios μ_s^M/μ_s and (b) relative errors E_{phase}^M of the phase functions between the HG function (Equation (2)) and the M-th order dEA (Equation (5)) at different g-values of 0.3, 0.6, 0.9 and 0.95. The error E_{phase}^M is defined by Equation (8).

2.1.4. Photon Diffusion Equation (PDE)

The PDE is obtained by the diffusion approximation to the RTE [30]:

$$\left[\frac{\partial}{v\partial t} - \nabla \cdot D(\boldsymbol{r})\nabla + \mu_a(\boldsymbol{r})\right] \Phi(\boldsymbol{r},t) = q_{DE}(\boldsymbol{r},t), \tag{9}$$

where the fluence rate $\Phi(\boldsymbol{r},t)$ is defined by $\int_{\mathbb{S}^2} d\Omega I(\boldsymbol{r},\boldsymbol{\Omega},t)$; the diffusion coefficient $D(\boldsymbol{r})$ is by $[3(1 - g)\mu_s(\boldsymbol{r}))]^{-1}$; and the isotropic source $q_{DE}(\boldsymbol{r},t)$ is by $\int_{\mathbb{S}^2} d\Omega q(\boldsymbol{r},\boldsymbol{\Omega},t)$.

2.2. Analytical Solutions (AS) of the Time-Domain Photon Transport Models

2.2.1. 3D Infinite Homogeneous Media

In this paper, we mainly employed the analytical solutions of the time-dependent RTE, dEM, and PDE for 3D infinite homogeneous media to calculate the temporal profiles of the fluence rate $\Phi(\boldsymbol{r},t)$ when an isotropic point source is incident at the initial time and at the origin of the xyz-coordinate as shown in Figure 3a. Then, the source-detector (SD) distance r_{sd} is given by $r = |\boldsymbol{r}|$. The analytical form of the fluence rate for the RTE $\Phi_{RTE}(r,t)$ with an arbitrary g-value is derived by Liemert and Kienle [31]:

$$\Phi_{RTE}(r,t) = \Phi_{LK}(r,t;\mu_a,\mu_s,\sigma_{HG},n), \tag{10}$$

where Φ_{LK} represents the analytical form by Liemert and Kienle; σ_{HG} the vector of the expansion coefficients for the HG phase function up to a maximum order of N: $[1,g,g^2,g^3,\cdots,g^N]$; the N-value is determined so as to attain sufficient convergence of the results; and n is the refractive index of the medium. The verifications of $\Phi_{RTE}(r,t)$ were confirmed by several independent works [23,32]. Using Φ_{LK}, we can obtain the analytical form for the dEM $\Phi_{dEM}(r,t)$:

$$\Phi_{dEM}(r,t) = \Phi_{LK}(r,t;\mu_a,\mu_s^M,\sigma_M,n), \tag{11}$$

where σ_M represents the vector of the expansion coefficients for $p_{\delta2}^M$ (Equation (6)) up to $N(> M)$: $[1,\sigma_{dE1},\sigma_{dE2},\cdots\sigma_{dEM},\sigma_{dEM+1} = 0,\cdots,\sigma_{dEN} = 0]$. We confirmed the verification of $\Phi_{dEM}(r,t)$ by comparing with the numerical solutions in Appendix A. The analytical solution for the dE0 ($M = 0$) by Liemert and Kienle is largely oscillated and unstable for several cases such as short SD distances, short times after light incidence, and the low scattering coefficient. In those cases, we employed the analytical solution for the dE0 derived by Paasschens [33] instead. It is noted that Martelli and coworkers employed a heuristic approach to obtain the approximate solution of the RTE for highly forward scattering [34] from the exact solution of the RTE for isotropic scattering by Paasschens [33]. Their approach is identical to the zeroth order dEA, so that the $\Phi(r,t)$-result using their approach coincides with the result of $\Phi_{dE0}(r,t)$. The analytical form for the PDE $\Phi_{PDE}(r,t)$ is given as [35]

$$\Phi_{PDE}(r,t) = \frac{1}{(4\pi Dt)^{3/2}} \exp\left[-\mu_a vt - \frac{r^2}{4Dvt}\right].$$
(12)

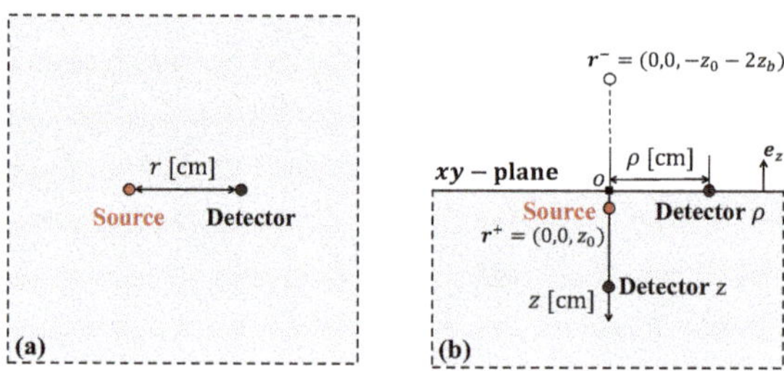

Figure 3. Source and detector positions in 3D (**a**) infinite and (**b**) semi-infinite homogeneous media.

2.2.2. 3D Semi-Infinite Homogeneous Media

To investigate influences of the boundary conditions on the temporal profiles of the fluence rate, we employed the approximate solutions of the RTE, dEM, and PDE for 3D semi-infinite homogeneous media under the diffusive refractive-index mismatched boundary condition:

$$\Phi(r,t) + D\gamma(n)e_z \cdot \boldsymbol{\nabla}\Phi(r,t) = 0,$$
(13)

where e_z represents the outward normal unit vector at the xy-plane in Figure 3b, the coefficient $\gamma(n)$ is give by $2(1 + R_f)/(1 - R_f)$ with $R_f = -1.44n^{-2} + 0.71n^{-1} + 0.69 + 0.064n$ [36]. Using the extrapolation method for the boundary condition (Equation (13)) [37], the analytical solutions of the RTE, dEM, and PDE for semi-infinite homogeneous media are obtained:

$$\Phi_{semi}(r,t;z_0,z_b) = \Phi_{inf}(|r - r^+|,t) - \Phi_{inf}(|r - r^-|,t),$$
(14)

where $r = (\rho,z)$, $\rho = (x,y)$, a real source position of $r^+ = (0,0,z_0)$ with $z_0 = 1/\mu_s'$, an imaginary source position of $r^- = (0,0,-2z_b - z_0)$ with $z_b = \gamma(n)D$, and $\Phi_{inf} = \{\Phi_{RTE}, \Phi_{dEM}, \Phi_{PDE}\}$ (Equations (10)–(12)). Then, the SD distance r_{sd} is given by $|r - r^+|$. As a preliminary study, we compared the $\Phi_{semi}(r,t)$-results for the RTE with those by another diffusion approximation method for the boundary condition [38], and confirmed that both results are almost the same to each other. The verification of the approximate solutions of the RTE has been confirmed by a comparative study with Monte-Carlo simulations [38].

2.3. Optical Properties, Source-Detector (SD) Distances, and Computations of the Analytical Solutions

We investigated photon transport in various kinds of random media at different SD distances. Table 1 lists parameter sets of the optical properties for the media and the SD distances, where the optical properties are in the wavelength range of the near-infrared light. At each parameter set, four parameters are fixed and one parameter is varied, e.g., at the parameter set A, μ_a, μ_s, g, and n are fixed and r_{sd} is varied from 0.15 to 10.0 cm. The optical properties at the parameter set A are often used in the research field of biomedical optics [34,38], and the properties at the parameter set B correspond to those of SiO_2 with epoxy resin [39]. Investigations at the parameter sets C, D, E and F allow us to find the dependence of the analytical results for the fluence rate on μ_a, μ_s, and g, respectively.

Table 1. Optical properties: $\mu_a[\text{cm}^{-1}]$, $\mu_s[\text{cm}^{-1}]$, $g[-]$, and $n[-]$; and SD distance $r_{sd}[\text{cm}]$. Corresponding colors are used in Figures 5, 6 and 8–11.

Parameter Sets: $[\mu_a, \mu_s, g, n, r_{sd}]$	Range	Color
Set A: $[0.10,\ 100,\ 0.90,\ 1.40,\ r_{sd}]$	r_{sd}: 0.15–10.0	Black
Set B: $[0.35,\ 58.0,\ 0.80,\ 1.56,\ r_{sd}]$	r_{sd}: 0.15–10.0	Blue
Set C: $[\mu_a,\ 100,\ 0.90,\ 1.40,\ 0.60]$	μ_a: 0.10–1.00	Red
Set D: $[0.10,\ \mu_s,\ 0.90,\ 1.40,\ 0.60]$	μ_s: 30.0–200	Green
Set E: $[0.10,\ 100,\ g,\ 1.40,\ 0.60]$	g: 0.10–0.95	Yellow
Set F: $[0.10,\ 50.0,\ g,\ 1.40,\ 0.60]$	g: 0.10–0.95	Purple

For the numerical calculations of the analytical solutions of the RTE, dEM, and PDE, we employed MATLAB codes. For the calculation of Φ_{LK}, we modified an open source by Liemert and Kienle [31]. The computational times for analytical solution of the RTE and dEM are less than 30 s when the optical properties and SD distance are given. It is noted that the calculations of the numerical solutions of the RTE and dEM require much higher computational loads than those of the analytical solutions. For example, the computational times to obtain the numerical solutions of the dEM discussed in Appendix A are roughly 26 h, although the parallel computing was carried out.

3. Results

3.1. Classification of the Length and Time Scales of Photon Transport

In this subsection, we investigated the temporal profiles of the fluence rate $\Phi(r, t)$ using the analytical solutions of the RTE (Equation (10)), dE0 (Equation (11) with $M = 0$), and PDE (Equation (12)) for 3D infinite homogeneous media. Also, we investigated the peak time of $\Phi(r, t)$, t_{peak}, because it is one of characteristic times of photon transport, and a difference in $\Phi(r, t)$ is correlated with a difference in t_{peak}. Based on results of t_{peak} at different SD distances $r_{sd} = r$, we classify length and time scales of photon transport into following three kinds of characteristic regimes; (1) ballistic regime: $t_{peak} \sim t_{peak}^{ball}$, (2) scattering regime (few-scattering-event regime): $t_{peak} \propto t_{peak}^{ball}$, t_{peak}^{diff}, and (3) diffusive regime: $t_{peak} \sim t_{peak}^{diff}$, where t_{peak}^{ball} and t_{peak}^{diff} are defined by

$$t_{peak}^{ball} = \frac{r}{v}, \qquad t_{peak}^{diff} = \frac{-3 + 2\sqrt{\frac{9}{4} + \frac{r^2\mu_a}{D}}}{4\mu_a v}. \tag{15}$$

t_{peak}^{ball} represents the arriving time of photons to the detector position without scattering, and t_{peak}^{diff} the peak time for the diffusive photons calculated from the PDE (Equation (12)), respectively. Figure 4 show examples of the temporal profiles of $\Phi(r, t)$ at the three characteristic length regimes. Here, we used the analytical solutions with the optical properties for the parameter set A as listed in Table 1 corresponding to the highly forward scattering of photons. While the RTE can appropriately treat the highly forward scattering, the dE0 and PDE approximate the highly forward scattering to the isotropic scattering. As shown in Figure 4a corresponding to the ballistic regime, the Φ-profiles using the RTE and dE0 are sharp, although both profiles are different from each other. The sharp profiles mean that photons are little scattered and travel almost straight. The peak times t_{peak} for the RTE and dE0 are almost the same as the arriving time t_{peak}^{ball}, whose value is approximately 8.4 ps. Meanwhile, the broad profile of Φ using the PDE is observed, indicating the unphysical multiple scattering of photons. In addition, the t_{peak}-value for the PDE is shorter than the t_{peak}^{ball}-value, meaning the violation of the causality in the PDE. As shown in Figure 4b corresponding to the scattering regime, the profiles using all the equations have smooth peaks. These peaks are formed by scattering of photons, so that these peak times are longer than t_{peak}^{ball}, in other words, the paths of photons are longer than the SD distance. In Figure 4b, the peak times for the dE0 and PDE are different from that for the RTE, indicating that the forward scattering of photons is not approximated to the isotropic scattering in this regime. As shown in Figure 4c corresponding to the diffusive regime, all the profiles are almost the same to each other, meaning the photon scattering becomes diffusive and the isotropic scattering approximation holds.

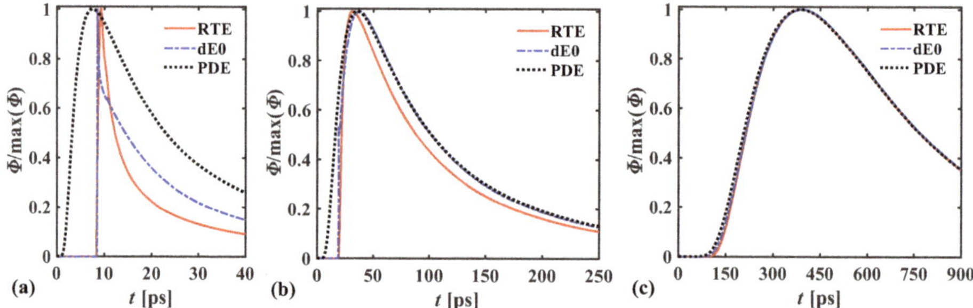

Figure 4. Analytical solutions of the RTE (Equation (10)), dE0 (Equation (11) with $M = 0$), and PDE (Equation (12)) at the three characteristic length regimes of photon transport: (**a**) ballistic regime at the SD distance of $r = 0.18$ cm ($\log_{10} r\mu_t' = 0.26$), (**b**) scattering regime at $r = 0.40$ cm ($\log_{10} r\mu_t' = 0.60$), and (**c**) diffusive regime at $r = 1.60$ cm ($\log_{10} r\mu_t' = 1.20$). The optical properties at the parameter set A are used, which are listed in Table 1.

We evaluated crossover lengths and times from the ballistic to scattering regimes, and from the scattering to diffusive regimes, respectively, by investigating t_{peak} for the RTE, dE0, and PDE. Then, we normalized the SD distance r by μ_t' and the peak time t_{peak} by $\mu_t'v$, respectively, with the reduced transport coefficient of $\mu_t' = \mu_s(1 - g) + \mu_a$. The inverses of μ_t' and $\mu_t'v$ correspond to the characteristic length and time for the dE0 and PDE where photons experience single scattering and absorbing processes. The normalizations allow us to investigate general features of photon transport in various kinds of random media. Figure 5 shows the results of the normalized peak times as a function of the normalized SD distances for the parameter sets A to F listed in Table 1 in a regime of short length and time scales. Here, the colors of the figures correspond to the parameter sets, e.g., the black colored pluses, circles, and squares correspond to the results for the parameter set A. In the regime of $\log_{10} r\mu_t' \lesssim 0.35$ and $\log_{10} t_{peak}\mu_t'v \lesssim 0.35$, the t_{peak}-results for the RTE and dE0 are linearly related to r, while the results for the PDE are not. This result suggests that the regime of the length and time scales correspond to the ballistic regime; and the crossover length and time from the ballistic to scattering regimes are evaluated as approximately $10^{0.35}/\mu_t' \sim 2.2/\mu_t'$ and $10^{0.35}/\mu_t'v \sim 2.2/\mu_t'v$, respectively.

Figure 6a shows the t_{peak}-results in a regime of longer length and time scales than the ballistic regime. In the case of the same parameter set (the same colored figures), the t_{peak}-results for the dE0 and PDE deviate from those for the RTE in the regime of $0.35 \lesssim \log_{10} r\mu_t' \lesssim 1.0$ and $0.35 \lesssim \log_{10} t_{peak}\mu_t'v \lesssim 1.6$, while the results for all the equations are almost the same to each other in the regime of $1.0 \lesssim \log_{10} r\mu_t'$ and $1.6 \lesssim \log_{10} t_{peak}\mu_t'v$. For the further investigation of t_{peak}, we evaluated the relative errors E_{peak} of t_{peak} against the RTE, defined by

$$E_{peak} = \left| \frac{t_{peak} - t_{peak,RTE}}{t_{peak,RTE}} \right| \times 100, \qquad t_{peak} = \{t_{peak,dE0}, t_{peak,PDE}\}. \tag{16}$$

Although we calculated E_{peak} for all the parameter sets, we showed the results only for the parameter set A in Figure 6b because the results for the other parameter sets behave similarly to the results for the parameter set A. As shown in Figure 6b, the E_{peak}-values for the dE0 and PDE are larger than 2% in the regime of $0.35 \lesssim \log_{10} r\mu_t' \lesssim 1.0$. Here the E_{peak}-value of 2% is considered as a thresh to determine whether the dE0 and PDE are valid or not. The large E_{peak}-values indicate the strong influence of modeling the forward scattering on the Φ-results in this regime. Meanwhile, the E_{peak}-values for the dE0 and PDE are less than 2% in the regime of $1.0 \lesssim \log_{10} r\mu'_t$, meaning the isotropic scattering approximation holds. From these results, the crossover length and time from the scattering to diffusive regimes are evaluated as approximately $10^1/\mu'_t = 10/\mu'_t$ and $10^{1.6}/\mu'_t v \sim 40/\mu'_t v$, respectively. The crossover length from the scattering to diffusive regimes has been extensively discussed so far, and our evaluation is consistent with the previous studies such as a comparative study of the measurement data and the PDE-results for 3D slab media [10], and a comparative study of the numerical results using the RTE and PDE for 2D square media [9]. This consistency implies that boundary conditions less influence the crossover length because the previous studies consider the finite media, while this study the infinite media. Meanwhile, the crossover time from the scattering to diffusive regimes is longer than that evaluated by the previous study [9], which is approximately $20/\mu'_t v$. The difference probably comes from the dimension of the medium and boundary effects; the previous study considers 2D finite media, while this study the 3D infinite media.

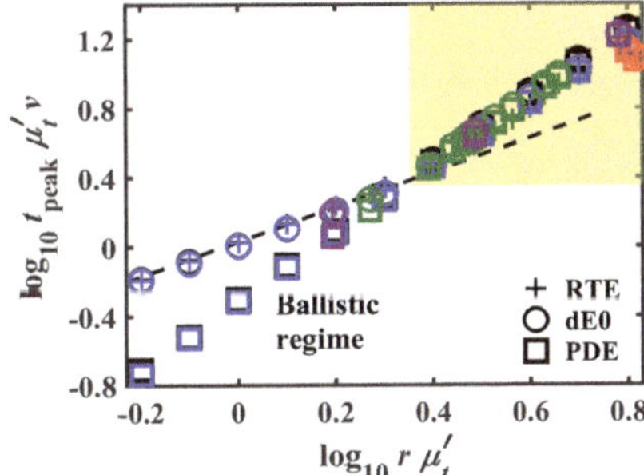

	$[\mu_a,\ \mu_s,\ g,\ n,\ r_{sd}]$
Set A	$[0.10,\ 100,\ 0.90,\ 1.40,\ r_{sd}]$
Set B	$[0.35,\ 58.0,\ 0.80,\ 1.56,\ r_{sd}]$
Set C	$[\mu_a,\ 100.0,\ 0.90,\ 1.40,\ 0.60]$
Set D	$[0.10,\ \mu_s,\ 0.90,\ 1.40,\ 0.60]$
Set E	$[0.10,\ 100,\ g,\ 1.40,\ 0.60]$
Set F	$[0.10,\ 50.0,\ g,\ 1.40,\ 0.60]$

Figure 5. Normalized peak times t_{peak} by $\mu'_t v$ as a function of the normalized SD distances r by μ'_t for the RTE, dE0, and PDE in the ballistic and scattering regimes. $\mu'_t = \mu_s(1-g) + \mu_a$ represents the reduced transport coefficient and v the speed of light in a medium, respectively. Colors of the figures correspond to the results for the parameter sets A to F listed in Table 1 or in the legend at the right-side of the figure. The dashed line represents the linear relation between t_{peak} and r. The yellow colored area corresponds to the scattering regime.

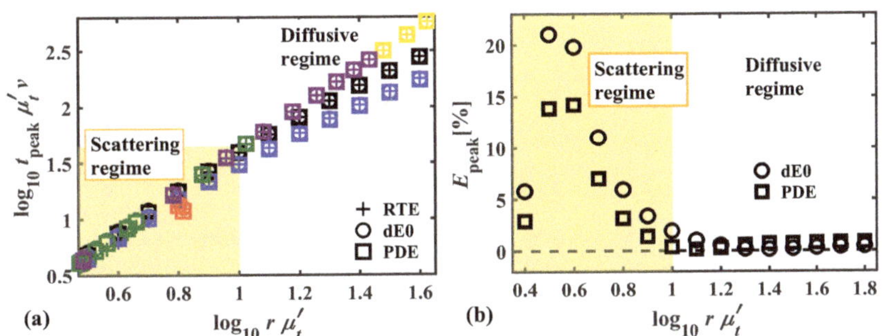

Figure 6. (a) Normalized peak times and (b) relative errors E_{peak} (Equation (16)) for the dE0 and PDE as a function of the normalized SD distances in the scattering and diffusive regimes. The other details are the same as Figure 5.

The dependence of the t_{peak}-results on the parameter sets is observed in the diffusive regime as shown in Figure 6a. This is because that in the diffusive regime, the normalizations of t_{peak} by $\mu_a v$ and of r^2 by μ_a/D are appropriate, which are suggested from the form of t_{peak}^{diff} (Equation (15)). As a preliminary study, we confirmed that when the above normalizations are employed, the t_{peak}-results for all the parameter sets are collapsed on a single curve in the diffusive regime. However, for the evaluation of the crossover length and time from the scattering to diffusive regimes, the above normalizations are not appropriate because the length regime, where the E_{peak}-values are larger than 2%, depends on the parameter sets, so that we can not evaluate the crossover length uniquely for all the parameter sets.

3.2. Length and Time Scales for the dEM to Be Valid in the Scattering Regime

In this subsection, we investigated the temporal profiles of the fluence rate $\Phi(r,t)$ and their peak times t_{peak} for the RTE and dEM in the scattering regime for the purpose of evaluations of the length and time scales for the dEM to be valid. As mentioned in Section 2.1.3, for the dEM, the scattering length scale is modified to the inverse of μ_s^M from the inverse of μ_s, and the highly forward scattering of photons described by p_{HG} is approximated to $p_{\delta2}^M$. These modifications influence the results of $\Phi(r,t)$.

Firstly, we calculated $\Phi(r,t)$ and t_{peak} for the dEM at different g-values of 0.3, 0.6, 0.9, and 0.95 in the scattering length regime, and investigated the dependence of the results of $\Phi(r,t)$ and t_{peak} on the expansion order M and on the g-value. The optical properties and SD distance are set as $\mu_a = 0.10$ cm^{-1}, $\mu_s' = \mu_s(1-g) = 10.0$ cm^{-1}, $n = 1.40$, and $r = 0.32$ cm, where the normalized SD distance is constant as $\sim 10^{0.5}/\mu_t'$. Figure 7a shows the temporal profiles of $\Phi(r,t)$ for the RTE and dEM with $M = 0$, 1, and 2 (dE0, dE1, and dE2) at $g = 0.9$, corresponding to the highly forward scattering. While the Φ-profile for the dE0 largely deviates from that for the RTE, the profiles for the dE1 and dE2 are similar to that for the RTE. Moreover, we investigated the relative error E_{peak}^M of the peak time for the dEM:

$$E_{peak}^M = \left| \frac{t_{peak,dEM} - t_{peak,RTE}}{t_{peak,RTE}} \right| \times 100, \tag{17}$$

where $t_{peak,dEM}$ represents the peak time of $\Phi(r,t)$ for the dEM. As shown in Figure 7b, the E_{peak}^M-values are quite small when the expansion order M is larger than 2 at all the g-values. This result means that the dEM with $M > 2$ can provide the same accuracy as the RTE, almost independently of the g-values. Meanwhile, as shown in Figure 2, the g-dependences are observed in the ratios of μ_s^M/μ_s and the errors of the phase function E_{phase}^M. These results suggest that for the accurate calculations of $\Phi(r,t)$ using the RTE and dEM, it is sufficient to satisfy a few orders moment conditions of the phase function and the exact agreement of the profiles of the phase functions is not required. It is noted that the analytical results of the small E_{peak}^M-values with $M > 2$ are quite different from the numerical calculations. The numerical results for the dEM are usually accurate within a finite range of M, and when the M-value is larger than the maximum value of the finite range, the numerical results for the dEM largely deviate from those for the RTE because of the numerical errors induced by angular discretization.

On the above discussion, we investigated E_{peak}^M for the dEM at the constant normalized SD distance. Next, we investigated E_{peak}^M for the dE0, dE1, and dE2 at the normalized SD distances varied from the scattering to diffusive regimes. Then, we evaluated the crossover lengths for the dE0, dE1, and dE2; in a regime of a smaller length scale than the crossover length, the E_{peak}^M-values are larger than 2%, corresponding to the invalidity of the dEM, while in a regime of a longer length scale, the errors are smaller than 2%, corresponding to the validity of the dEM. Figure 8a–c show the E_{peak}^M-values as a function of the normalized SD distances at the parameter sets A to F listed in Table 1. While the max value of E_{peak}^M for the dE0 is over 20%, the max values for the dE1 and dE2 are around 10% and 4%, respectively. The length regime for the dE2 where the E_{peak}^M-values are larger than 2% is shorter than those for the dE0 and dE1. From the results in the figures (a), (b), and (c), we evaluated the crossover lengths for the dE0, dE1, and dE2 as approximately $10^1/\mu_t' = 10/\mu_t'$, $10^{0.6}/\mu_t' \sim 4.0/\mu_t'$, and $10^{0.45}/\mu_t' \sim 2.8/\mu_t'$, respectively. These results mean that the scattering regime is classified into the three kinds of regimes by the two kinds of the crossover lengths for the dE1 and dE2. Here, we evaluated the crossover lengths for the dE0, dE1, and dE2 as a function of μ'_t for the purpose of the classifications of the scattering regime. As an alternative, we can consider the normalization of the SD distances by $\mu^M_t = \mu^M_s + \mu_a$ because for the dEM, the inverse of μ^M_t is the characteristic length of single scattering and absorption processes. Figure 8d shows E^M_{peak} for the dE0, dE1, and dE2 as a function of the SD distances by μ^M_t. The crossover lengths for the dEM with $M = 0, 1$, and 2 are within the regime of $0.9 \lesssim \log_{10} r\mu^M_t \lesssim 1.0$, less dependently on the expansion order M. It is noted that the crossover length from the scattering and diffusive regimes depends on the expansion order M in the normalization by μ^M_t, so that we cannot evaluate the regime for the dEM to be valid uniquely in the normalization. The rough evaluation of the crossover length for the dEM as $10/\mu^M_t$ means that after ten times scattering and absorption processes for the dEM, the M-th order dEA holds.

Figure 9 shows the normalized peak times for the RTE, dE0, dE1, and dE2 as a function of the normalized SD distances by μ'_t. From the figure and the results for the crossover lengths based on μ'_t, we evaluated the crossover times for the dE0, dE1, and dE2 as approximately $10^{1.6}/\mu'_t v \sim 40/\mu'_t v$, $10^{0.8}/\mu'_t v \sim 6.3/\mu'_t v$, and $10^{0.55}/\mu'_t v \sim 3.5/\mu'_t v$, respectively. Here, the evaluations of the crossover times are based on the t_{peak}-results for the RTE (dashed line in Figure 9 at the parameter set A) because the RTE is the most accurate photon transport model.

3.3. Influence of the Boundary Conditions on the Crossover Lengths and Times

In this subsection, we investigated the boundary effects on the fluence rate $\Phi(r, t)$ and on the crossover lengths and times for the dEM with $M = 0, 1$, and 2 by comparing the analytical solutions for the semi-infinite media with those for the infinite media. We considered the parameter sets A and B listed in Table 1, and varied the SD distances, denoted by $r_{si} = |r - r^+|$, in the two directions: the z-direction with $r = (0, 0, z)$ and ρ-direction with $r = (\rho, 0)$ as shown in Figure 3b. For the case of the detector position in the z-direction, as the z-value increases, the detector position becomes far from the boundary, so that the boundary effects probably become weak. Meanwhile, for the case of the detector position in the ρ-direction, a distance between the detector and the boundary is unchanged, so that the boundary effects are also unchanged. Figure 10a–c show the relative errors E^M_{peak} for the dE0, dE1, and dE2 in the semi-infinite media as a function of the normalized SD distance by μ'_t. The crossover lengths for the dEM in the semi-infinite media are almost the same as those in the infinite media, meaning less influences of the boundary conditions on the crossover lengths.

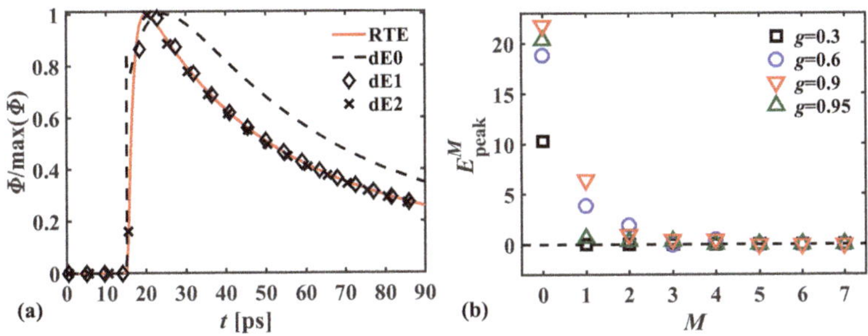

Figure 7. (**a**) Temporal profiles of $\Phi(r, t)$ using the RTE, dE0, dE1, and dE2 at $g = 0.9$. (**b**) Relative errors E^M_{peak} of the peak time of $\Phi(r, t)$ (Equation (17)) as a function of the expansion order M at different g-values of 0.3, 0.6, 0.9, and 0.95. The optical properties and SD distance are set as $\mu_a = 0.10$ cm^{-1}, $\mu'_s = \mu_s(1 - g) = 10.0$ cm^{-1}, $n = 1.40$, and $r = 0.32$ cm.

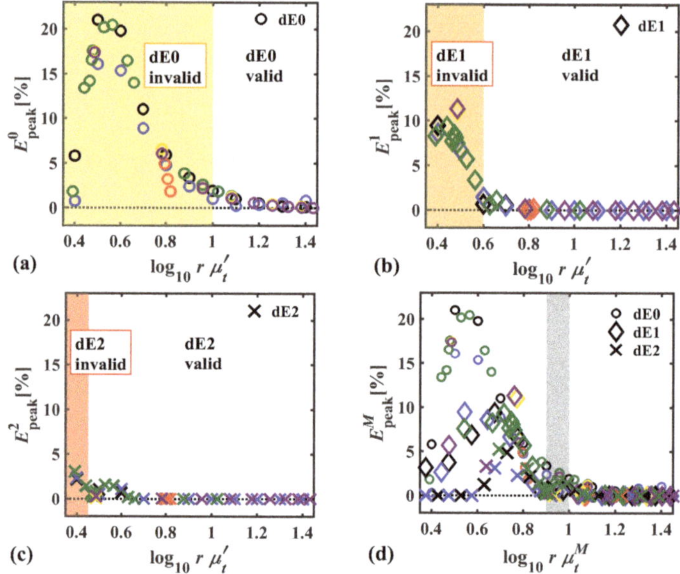

Figure 8. Relative errors E^M_{peak} (Equation (17)) for the (**a**) dE0, (**b**) dE1, and (**c**) dE2 as a function of the normalized SD distances by μ'_t at the parameter sets A to F listed in Table 1. The yellow, orange, and red colored areas correspond to the length regimes where the dE0, dE1, and dE2 are invalid, respectively. (**d**) Normalization of the SD distances by μ^M_t. The gray colored areas correspond to the crossover length for the dEM. The other details are the same as Figure 6.

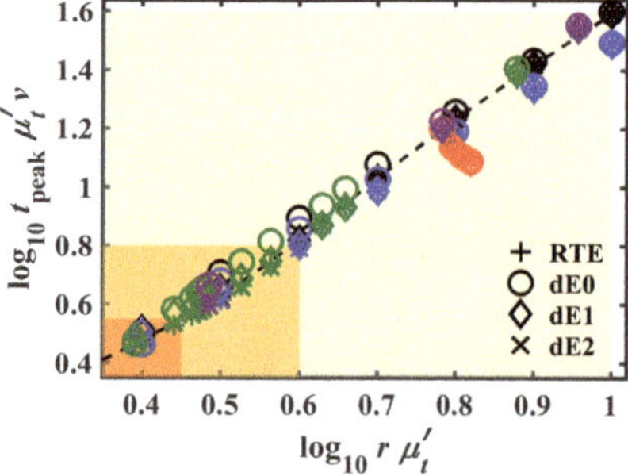

Figure 9. Normalized peak times of $\Phi(r, t)$ for the RTE, dE0, dE1, and dE2 as a function of the normalized SD distances. The dashed line represents the connection line for the RTE-results at the parameter set A. The other details are the same as Figure 8a–c.

Figure 11 shows the normalized peak times for the RTE in the infinite and semi-infinite media as a function of the SD distances r_{sd} for the parameter sets A and B. Here, the SD distances r_{sd} are given by r in infinite media, and by r_{si} in semi-infinite media, respectively. The t_{peak}-results in the infinite media and semi-infinite media at the z-direction behave similarly to each other on the whole regime. Meanwhile, the t_{peak}-results in the semi-infinite media at the ρ-direction (dotted line in Figure 11 for the results at the parameter set A) are smaller than those in the other two cases (dashed line) especially on the regime of the longer length scale than approximately $10^{0.6}/\mu'_t \sim 4.0/\mu'_t$, which is almost the same as the crossover length for the dE1. These results suggest that the boundary conditions influence the crossover time for the dE0, while they have less influence the times for the dE1 and dE2.

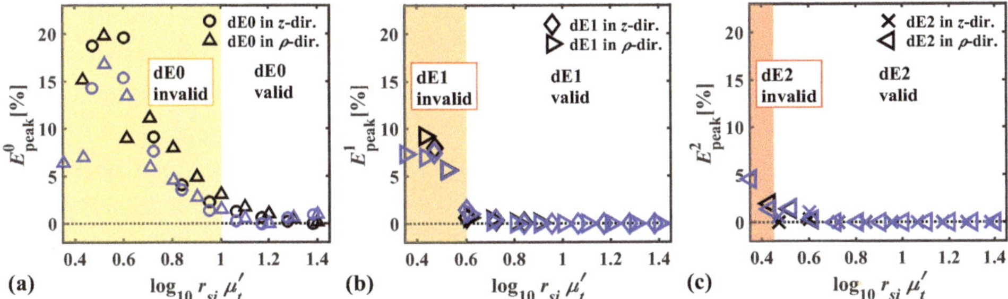

Figure 10. Relative errors E_{peak}^M for the (**a**) dE0, (**b**) dE1, and (**c**) dE2 in the semi-infinite media (Figure 3b) as a function of the normalized SD distances $r_{si} = |r - r^+|$ at the parameter sets A and B listed in Table 1. The SD distances are varied in the z-direction with $r = (0, 0, z)$ and in the ρ-direction with $r = (\rho, 0)$. The other details are the same as Figure 8a–c.

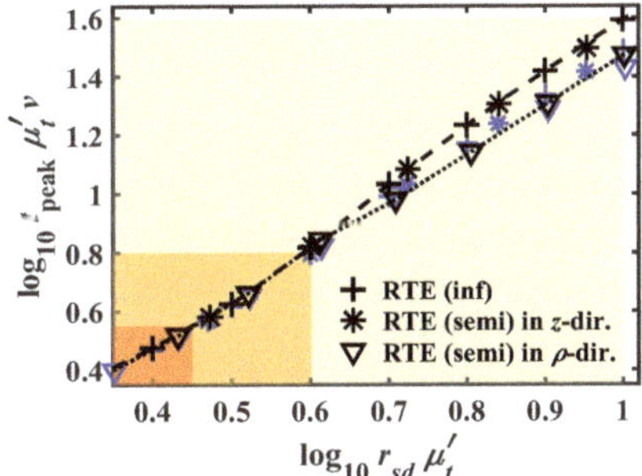

Figure 11. Normalized peak times t_{peak} for the RTE in infinite media, and semi-infinite media at the z- and ρ-directions. The SD distances r_{sd} are given by r in infinite media and by r_{si} in semi-infinite media, respectively. The dashed and dotted lines represent the connection lines for the results in infinite and semi-infinite media at the ρ-direction, respectively, for the parameter set A. The other details are the same as Figure 10.

4. Conclusions

We investigated the temporal profiles of the fluence rate $\Phi(r, t)$ and their peak times t_{peak} for various kinds of random media at different SD distances by analytical solutions of the RTE, dE0, and PDE for 3D infinite homogeneous media. We evaluated the crossover lengths and times from the ballistic to scattering (few-scattering-event) regimes, and from the scattering to diffusive regimes as approximately $2.2/\mu_t'$ and $2.2/\mu_t'v$; $10/\mu_t'$ and $40/\mu_t'v$, respectively.

Next, we investigated $\Phi(r, t)$ and t_{peak} for the dEM mainly in the scattering regime. We found that the results for the dE0 are quite similar to the results for the PDE because the dE0 and PDE approximate the forward scattering of photons as the isotropic scattering. We also found that at the expansion order of M larger than 2, the results for the dEM are almost the same as those for the RTE in the scattering and diffusive regimes, meaning the validity of the dEM with $M > 2$ in the regimes. We evaluated the crossover lengths and times for the dE1 and dE2 as approximately $4.0/\mu_t'$ and $6.3/\mu_t'v$; $2.8/\mu_t'$ and $3.5/\mu_t'v$, respectively.

Finally, we investigated the boundary effects on the characteristic length and time scales of photon transport by comparing the results for the infinite media with those for the semi-infinite media. We found that the boundary conditions less influence on the crossover lengths for the dE0, dE1, and dE2, while the boundary conditions reduce the t_{peak}-values from those for bulk in the regime of the longer length scale than $4.0/\mu_t'$, which is the same as the crossover length for the dE1. Hence, the crossover times for the dE1 and dE2 are less influenced by the boundary conditions. Our findings are useful for constructions of accurate and efficient photon transport models especially in the scattering regime.

Author Contributions: Conceptualization, H.F.; investigation and analysis, H.F. and M.U.; writing—original draft preparation, H.F.; writing—review and editing, H.F., K.K. and M.W.; supervision, K.K. and M.W.; All authors discussed the results and contributed to the final manuscript. All authors have read and agreed to the published version of the manuscript.

Funding: The first author (H.F.) acknowledges support from Grant-in-Aid for Scientific Research (18K13694) of the Japan Society for the Promotion of Science, KAKENHI.

Acknowledgments: The authors would like to thank Go Chiba, Yukio Yamada, and Yoko Hoshi for fruitful discussion. The first author (H.F.) appreciates Leung Tsang, his laboratory members, and staffs at University of Michigan for their kind supports on H.F.'s research work.

Conflicts of Interest: The authors declare no conflict of interest. The funders had no role in the design of the study; in the collection, analyses, or interpretation of data; in the writing of the manuscript, or in the decision to publish the results.

Appendix A. Verification of the Analytical Solutions of the dEM for 3D Infinite Homogeneous Media

We verified the analytical solutions of the dEM with $M = 0$, 1, and 2 for 3D infinite homogeneous media (Equation (11)) by comparing with the numerical solutions for 3D homogeneous cubic media under the refractive-index mismatched boundary condition. We numerically solved the dEM and calculated the temporal profiles of the fluence rate $\Phi(r, t)$ based on the finite-difference and discrete ordinates methods. For spatial and temporal discretization, we employed the 3rd order weighted essentially non oscillatory and the 3rd order total variation diminishing-Runge-Kutta methods, respectively. For accurate treatments of the HG phase function and dE phase functions in a case of the highly forward scattering, we employed the Galerkin quadrature method. Source and detector positions are set inside the medium to suppress the boundary effects because we compare the analytical solutions for infinite media with the numerical solutions for finite media. For the details for the numerical calculations, refer to [32,40]. The optical properties of the medium are set as $\mu_a = 0.20\ \mathrm{cm}^{-1}$, $\mu_s = 100\ \mathrm{cm}^{-1}$, $g = 0.90$, and $n = 1.40$; and the SD distance is set as $r = 0.40$ cm, corresponding to the scattering regime.

Figure A1 compare the analytical and numerical solutions of the dEM with $M = 0$, 1, and 2 for $\Phi(r, t)$. We found good agreements between the analytical and numerical solutions for all the cases, meaning the verification of the analytical solutions of the dEM.

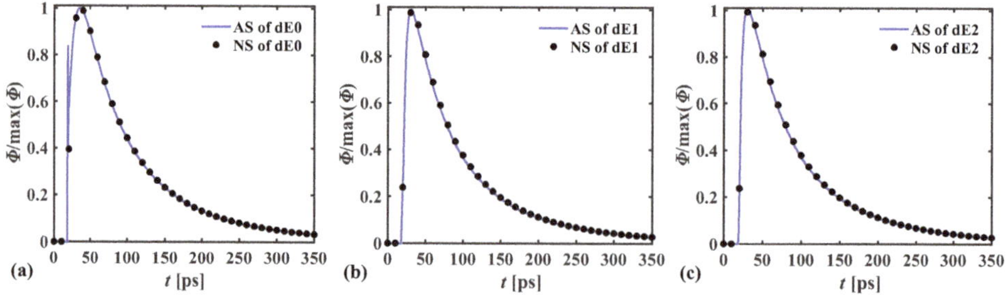

Figure A1. Comparisons of the analytical and numerical solutions of the dEM with (**a**) $M = 0$, (**b**) $M = 1$, and (**c**) $M = 2$ for the fluence rate $\Phi(r, t)$. The optical properties of the medium are set as $\mu_a = 0.20\ \mathrm{cm}^{-1}$, $\mu_s = 100\ \mathrm{cm}^{-1}$, $g = 0.90$, and $n = 1.40$; and the SD distance is set as $r = 0.40$ cm, corresponding to the scattering regime.

References

1. Gibson, A.P.; Hebden, J.C.; Arridge, S.R. Recent advances in diffuse optical imaging. *Phys. Med. Biol.* **2005**, *50*, R1–R43. [CrossRef] [PubMed]
2. Okawa, S.; Hoshi, Y.; Yamada, Y. Improvement of image quality of time-domain diffuse optical tomography with lp sparsity regularization. *Biomed. Opt. Express* **2011**, *2*, 3334–3348. [CrossRef] [PubMed]
3. Yamada, Y.; Okawa, S. Diffuse Optical Tomography: Present Status and Its Future. *Opt. Rev.* **2014**, *21*, 185–205. [CrossRef]
4. Yamada, Y.; Suzuki, H.; Yamashita, Y. Time-Domain Near-Infrared Spectroscopy and Imaging: A Review. *Appl. Sci.* **2019**, *9*, 1127. [CrossRef]
5. Ntziachristos, V. Going deeper than microscopy: The optical imaging frontier in biology. *Nat. Methods* **2010**, *7*, 603–614. [CrossRef] [PubMed]
6. Bizheva, K.K.; Siegel, A.M.; Boas, D.A. Path-length-resolved dynamic light scattering in highly scattering random media: The transition to diffusing wave spectroscopy. *Phys. Rev. E* **1998**, *58*, 7664–7667. [CrossRef]
7. Guerin, W.; Chong, Y.D.; Baudouin, Q.; Liertzer, M.; Rotter, S.; Kaiser, R. Diffusive to quasi-ballistic random laser : Incoherent and coherent models. *J. Opt. Soc. Am. B* **2016**, *33*, 1888–1896. [CrossRef]
8. Tarvainen, T.; Vauhkonen, M.; Kolehmainen, V.; Kaipio, J. A hybrid radiative transfer – diffusion model for optical tomography. *Appl. Opt.* **2005**, *44*, 876–886. [CrossRef]
9. Fujii, H.; Okawa, S.; Yamada, Y.; Hoshi, Y. Hybrid model of light propagation in random media based on the time-dependent radiative transfer and diffusion equations. *J. Quant. Spectrosc. Radiat. Transfer* **2014**, *147*, 145–154. [CrossRef]
10. Yoo, K.M.; Liu, F.; Alfano, R.R. When does the diffusion approximation fail to describe photon transport in random media? *Phys. Rev. Lett.* **1990**, *64*, 2647–2650. [CrossRef]
11. Hielscher, A.H.; Alcouffe, R.E.; Barbour, R.L. Comparison of finite-difference transport and diffusion calculations for photon migration in homogeneous and heterogeneous tissues. *Phys. Med. Biol.* **1998**, *43*, 1285–1302. [CrossRef] [PubMed]
12. Venugopalan, V.; You, J.S.; Tromberg, B.J. Radiative transport in the diffusion approximation: An extension for highly absorbing media and small source-detector separations. *Phys. Rev. E* **1998**, *58*, 2395–2407. [CrossRef]
13. Machida, M.; Panasyuk, G.Y.; Schotland, J.C.; Markel, V.A. The Green's function for the radiative transport equation in the slab geometry. *J. Phys. A Math. Theor.* **2010**, *43*, 065402. [CrossRef]
14. Fujii, H.; Hoshi, Y.; Okawa, S.; Kosuge, T.; Kohno, S. Numerical modeling of photon propagation in biological tissue based on the radiative transfer equation. In Proceedings of the 4th International Symposium on Slow Dynamics in Complex Systems, Sendai, Japan, 2–7 December 2013; Volume 1518, pp. 579–585.
15. Fujii, H.; Okawa, S.; Yamada, Y.; Hoshi, Y.; Watanabe, M. A coupling model of the radiative transport equation for calculating photon migration in biological tissue. In Proceedings of the SPIE for Biophotonics Japan, Tokyo, Japan, 27–28 October 2015; Volume 9792, pp. 1–5.
16. Joseph, J.H.; Wiscombe, W.J.; Weinman, J.A. The delta-Eddington approximation for radioactive flux transfer. *J. Atmos. Sci.* **1976**, *33*, 2452–2459. [CrossRef]
17. Welch, A.; van Gemert, M. *Optical-Thermal Response of Laser-Irradiated Tissue*; Plenum Press: London, UK, 1995.
18. Klose, A.D.; Hielscher, A.H. Modeling Photon Propagation In Anisotropically Scattering Media with the Equation Of Radiative Transfer. *Proc. SPIE* **2003**, *4955*, 624–633.
19. Cong, W.; Shen, H.; Cong, A.; Wang, Y.; Wang, G. Modeling photon propagation in biological tissues using a generalized Delta-Eddington phase function. *Phys. Rev. E* **2007**, *76*, 051913. [CrossRef]
20. Boulet, P.; Collin, A.; Consalvi, J.L. On the finite volume method and the discrete ordinates method regarding radiative heat transfer in acute forward anisotropic scattering media. *J. Quant. Spectrosc. Radiat. Transfer* **2007**, *104*, 460–473. [CrossRef]
21. Klose, A.D.; Hielscher, A.H. Optical tomography with the equation of radiative transfer. *Int. J. Numer. Methods Heat Fluid Flow* **2008**, *18*, 443–464. [CrossRef]
22. Jia, J.; Kim, H.K.; Hielscher, A.H. Fast linear solver for radiative transport equation with multiple right hand sides in diffuse optical tomography. *J. Quant. Spectrosc. Radiat. Transfer* **2015**, *167*, 10–22. [CrossRef]

23. Kamran, F.; Abildgaard, O.H.A.; Subash, A.A.; Andersen, P.E.; Andersson-engels, S.; Khoptyar, D. Computationally effective solution of the inverse problem in time-of-flight spectroscopy. *Opt. Express* **2015**, *23*, 6937–6945. [CrossRef]

24. Qin, J.; Lu, R. Measurement of the optical properties of fruits and vegetables using spatially resolved hyperspectral diffuse reflectance imaging technique. *Postharvest Biol. Technol.* **2008**, *49*, 355–365. [CrossRef]

25. Chandrasekhar, S. *Radiative Transfer*; Dover: New York, NY, USA, 1960.

26. Henyey, L.G.; Greenstein, L.J. Diffuse radiation in the galaxy. *J. Astrophys.* **1941**, *93*, 70–83. [CrossRef]

27. Cheong, W.F.; Prahl, S.A.; Welch, A.J. A review of the optical properties of biological tissue. *IEEE J. Quantum Electron* **1990**, *26*, 2166–2185. [CrossRef]

28. Germer, C.T.; Roggan, A.; Ritz, J.P.; Isbert, C.; Albrecht, D.; Müller, G.; Buhr, H.J. Optical properties of native and coagulated human liver tissue and liver metastases in the near infrared range. *Lasers Surg. Med.* **1998**, *23*, 194–203. [CrossRef]

29. Dehaes, M.; Gagnon, L.; Vignaud, A.; Valabr, R.; Grebe, R.; Wallois, F.; Benali, H. Quantitative investigation of the effect of the extra-cerebral vasculature in diffuse optical imaging : A simulation study. *Biomed. Opt. Express* **2011**, *2*, 680–695. [CrossRef]

30. Furutsu, K.; Yamada, Y. Diffusion approximation for a dissipative random medium and the applications. *Phys. Rev. E* **1994**, *50*, 3634–3640. [CrossRef]

31. Liemert, A.; Kienle, A. Infinite space Green's function of the time-dependent radiative transfer equation. *Biomed. Opt. Express* **2012**, *3*, 543. [CrossRef]

32. Fujii, H.; Yamada, Y.; Chiba, G.; Hoshi, Y.; Kobayashi, K.; Watanabe, M. Accurate and efficient computation of the 3D radiative transfer equation in highly forward-peaked scattering media using a renormalization approach. *J. Comput. Phys.* **2018**, *374*, 591–604. [CrossRef]

33. Paasschens, J.C.J. Solution of the time-dependent Boltzmann equation. *Phys. Rev. E* **1997**, *56*, 1135–1141. [CrossRef]

34. Martelli, F.; Sassaroli, A.; Pifferi, A.; Torricelli, A.; Spinelli, L.; Zaccanti, G. Heuristic Green's function of the time dependent radiative transfer equation for a semi-infinite medium. *Opt. Express* **2007**, *15*, 18168–18175. [CrossRef]

35. Chandrasekhar, S. Stochastic Problems in Physics and Astronomy. *Rev. Mod. Phys.* **1943**, *15*, 1–88. [CrossRef]

36. Egan, W.G.; Hilgeman, T.W. *Optical Properties of Inhomogeneous Materials*; Academic: New York, NY, USA, 1979.

37. Patterson, M.S.; Chance, B.; Wilson, B.C. Time resolved reflectance and transmittance for the non- invasive measurement of tissue optical properties. *Appl. Opt.* **1989**, *28*, 2331–2336. [CrossRef] [PubMed]

38. Simon, E.; Foschum, F.; Kienle, A. Hybrid Green's function of the time- dependent radiative transfer equation for anisotropically scattering semi-infinite media scattering semi-infinite media. *J. Biomed. Opt.* **2013**, *18*, 015001. [CrossRef] [PubMed]

39. Klose, A.D.; Netz, U.; Beuthan, J.; Hielscher, A.H. Optical tomography using the time-independent equation of radiative transfer - Part 1 : forward model. *J. Quant. Spectrosc. Radiat. Transfer* **2002**, *72*, 691–713. [CrossRef]

40. Fujii, H.; Chiba, G.; Yamada, Y.; Hoshi, Y.; Kobayashi, K. Numerical treatment of highly forward scattering on radiative transfer using the delta-M approximation and Galerkin quadrature method. In Proceedings of the 9th International Symposium on Radiative Transfer, RAD-19, Athens, Greece, 3–7 June 2019; pp. 261–268.

MDPI

St. Alban-Anlage 66

4052 Basel

Switzerland

Tel. +41 61 683 77 34

Fax +41 61 302 89 18

www.mdpi.com

Applied Sciences Editorial Office

E-mail: applsci@mdpi.com

www.mdpi.com/journal/applsci